Microgrid

The book discusses principles of optimization techniques for microgrid applications specifically for microgrid system stability, smart charging, and storage units. It also highlights the importance of adaptive learning techniques for controlling autonomous microgrids. It further presents optimization-based computing techniques like fuzzy logic, and neural networks to enhance the computational speed.

Features

- Discusses heuristic techniques and evolutionary algorithms in microgrids optimization problems.
- Covers operation management, distributed control approaches, and conventional control methods for microgrids.
- Presents intelligent control for energy management and battery charging systems.
- Highlights a comprehensive treatment of power sharing in DC microgrids.
- Explains control of low-voltage microgrids with master-slave architecture, where distributed energy resources interface with the grid by means of conventional current-driven inverters.

It is primarily written for senior undergraduates, graduate students, and academic researchers in the fields of electrical engineering, electronics, and communications engineering, computer science and engineering, and environmental engineering.

Smart Engineering Systems: Design and Applications

Series Editor-Suman Lata Tripathi

Internet of Things
Robotic and Drone Technology
Edited by Nitin Goyal, Sharad Sharma, Arun Kumar Rana, Suman Lata Tripathi

Smart Electrical Grid System
Design Principle, Modernization, and Techniques
Edited by Krishan Arora, Suman Lata Tripathi and Sanjeevikumar Padmanaban

Artificial Intelligence, Internet of Things (IoT) and Smart Materials for Energy Applications
Edited by Mohan Lal Kolhe, Kailash J. Karande and Sampat G. Deshmukh

Artificial Intelligence for Internet of Things: Design Principle, Modernization, and Techniques
Edited by N. Thillaiarasu, Suman Lata Tripathi and V. Dhinakaran

Machine Learning and Internet of Things in Solar Power Generation
Edited by Prabha Umapathy, Jude Hemanth, Shelej Khera, Abinaya Inbamani and Suman Lata Tripathi

Microgrid
Design, Optimization, and Applications
Edited by Amit Kumar Pandey, Sanjeevikumar Padmanaban, Suman Lata Tripathi, Vivek Patel, and Vikas Patel

For more information about this series, please visit: https://www.routledge.com/

Microgrid
Design, Optimization, and Applications

Edited by
Amit Kumar Pandey
Sanjeevikumar Padmanaban
Suman Lata Tripathi
Vivek Patel
Vikas Patel

CRC Press
Taylor & Francis Group
Boca Raton London New York

CRC Press is an imprint of the
Taylor & Francis Group, an **informa** business

First edition published 2024
by CRC Press
2385 NW Executive Center Drive, Suite 320, Boca Raton FL 33431

and by CRC Press
4 Park Square, Milton Park, Abingdon, Oxon, OX14 4RN

CRC Press is an imprint of Taylor & Francis Group, LLC

ISBN: 978-1-032-56576-7 (hbk)
ISBN: 978-1-032-77218-9 (pbk)
ISBN: 978-1-003-48183-6 (ebk)

DOI: 10.1201/9781003481836

Typeset in Times
by SPi Technologies India Pvt Ltd (Straive)

Contents

Editors

Amit Kumar Pandey completed his Ph.D in microelectronics and VLSI from MNNIT Allahabad. He earned his M.Tech from NIT Allahabad and B.E in Electronics Engineering from Swami Ramanand Teerth Marathwada University Nanded. He has been an assistant professor at Rajkiya Engineering College, Ambedkar Nagar for the last four years, and has twelve years teaching experience. He has published more than thirty research papers in referred journals such as Springer, Wiley, Emerald, among others. He has also published three patents. His areas of expertise include microelectronics device modeling and characterization, low power VLSI circuit design and VLSI design of testing.

Sanjeevikumar Padmanaban, **PhD**, is a Full Professor of Electrical Power Engineering in the Department of Electrical Engineering, IT and Cybernetics, University of South-Eastern Norway. He has authored 600 plus scientific papers, and is a fellow of the Institution of Engineering and Technology (UK).

Suman Lata Tripathi completed her PhD in microelectronics and VLSI from MNNIT, Allahabad. She completed her post-doc from the Nottingham Trent University, UK in 2023. She is associated with Lovely Professional University as a professor with more than twenty-two years of experience in academics. She has published more than 125 research papers She received the "Research Excellence Award" in 2019 and "Research Appreciation Award" in 2020, 2021, and 2023 at Lovely Professional University, India. Her areas of expertise include microelectronics device modeling and characterization, low power VLSI circuit design, VLSI design of testing, and advanced FET design for IoT, Embedded System Design and biomedical applications.

Vivek Patel is an assistant professor in the Department of Electrical Engineering, faculty of Engineering & Technology, Dr. Shakuntala Mishra National Rehabilitation University Lucknow, India. He graduated from Ideal Institute of Technology, Ghaziabad, India and completed his post- graduation with hons. degree in Power System Engineering specialization in the Department of Electrical Engineering from IIT (ISM) Dhanbad, India. He has authored numerous SCIE, Scopus and pear reviewed research papers in refereed journals such as IEEE, Springer, Elesevier and at conference proceeding. He was also granted two international patents. He received the best paper award at NPSC. His areas of research are power systems and control.

Vikas Patel is pursuing his PhD in Renewable energy from MMMUT, Gorakhpur. He graduated from Ideal Institute of Technology, Ghaziabad, India and completed his post- graduation with hons. degree in Power Electronics and Drives specialization in the Department of Electrical Engineering from Madan Mohan Malaviya University of Technology, Gorakhpur, India. He is currently an assistant professor in the Department of Electrical Engineering, Rajkiya Engineering College Ambedkar Nagar, India. He has more than seven years of experience in academics and research. His areas of expertise include power electronics modeling and characterization, renewable energy, optimization, IoT, and embedded system design.

Contributors

Parvez Ahmad
MNNIT Allahabad
Prayagraj, India

Saqib Ali
NFC Institute of Engineering &
    Technology
Multan, Pakistan

Rakesh Bhadani
Veerayan Polytechnic, Jakhaniya
Mandvi, India

Ashima Bhatnagar Bhatia
Vivekananda Institute of Professional
    Studies-TC
New Delhi, India

M. Kamran Liaqat Bhatti
NFC Institute of Engineering &
    Technology
Multan, Pakistan

Vaibhav Chhabhaiya
Birala Vishvkrama Mahavidhayalay
    Anand
Anand, India

Niraj Kumar Choudhary
MNNIT, Allahabad
Prayagraj, India

Aykut Fatih Güven
Yalova University
Yalova, Turkey

Junho Hong
University of Michigan-Dearborn
Dearborn, USA

Kushagra Jain
MNNIT Allahabad
Prayagraj, India

Y. Lalitha Kameswari
Koneru Lakshmaiah Education
    Foundation
Vaddeswaram, India

Nipon Ketjoy
School of Renewable Energy and Smart
    Grid Technology (SGtech)
Naresuan University
Phitsanulok, Thailand

Irfan Khan
Texas A&M University
Texas, USA

Shama Kouser
Jazan University
Jazan, Saudi Arabia

Prashant Kumar
Rajkiya Engineering College
Ambedkar Nagar, India

Sonu Kumar
Koneru Lakshmaiah Education
    Foundation
Vaddeswaram, India

Tahir Nadeem Malik
HITEC Taxila
Taxila, Pakistan

Shivendu Mishra
Rajkiya Engineering College
Ambedkar Nagar, India

Venugopal Reddy Modhugu
Senior IEEE Member, Independent
  Researcher
USA

Priyanshul Niranjan
MNNIT Allahabad
Prayagraj, India

Shivanshu Pandey
Rajkiya Engineering College
Ambedkar Nagar, India

Kuchan Park
Department of Electrical and Computer
  Engineering
University of Michigan-Dearborn
Dearborn, USA

Vikas Patel
Rajkiya Engineering College
Ambedkar Nagar, India

Vivek Patel
SMNRU Lucknow
Lucknow, India

P. Swati Patro
Parala Maharaja Engineering
  College
Berhampur, India

B. Pragathi
DVR & Dr. HS MIC College of
  Technology
Kanchikacherla, India

Kritika Purohit
Career Point University
Jodhpur, India

Prince Rajpoot
Rajkiya Engineering College
Ambedkar Nagar, India

Challa Krishna Rao
Department of Electrical Engineering
Parala Maharaja Engineering College
Berhampur affiliated to Biju Patnaik
  University of Technology Rourkela
Odisha, India
Department of Electrical and
  Electronics Engineering
Aditya Institute of Technology and
  Management
Tekkali, Andhra Pradesh, India

S. Koteswara Rao
Koneru Lakshmaiah Education
  Foundation
Vaddeswaram, India

Aamer Raza
Islamabad Electric Supply Company
Islamabad, Pakistan

Sarat Kumar Sahoo
Department of Electrical Engineering
Parala Maharaja Engineering College,
  Berhampur
affiliated to Biju Patnaik University of
  Technology
Rourkela, Odisha, India

Nitin Singh
MNNIT Allahabad
Prayagraj, India

Prashant Singh
MNNIT Allahabad
Prayagraj, India

Ravindra Kumar Singh
MNNIT Allahabad
Prayagraj, India

Bevara Srikanth
Parala Maharaja Engineering College
Berhampur, India

Adit Srivastava
Rajkiya Engineering College
Ambedkar Nagar, India

Wencong Su
Department of Electrical and Computer
    Engineering
University of Michigan-Dearborn
Dearborn, USA

Shweta Thumbar
M.S. University Baroda
Baroda, India

Raghvendra Tiwari
G.H. Raisoni Institute of Engineering &
    Technology
Nagpur, India

Suman Lata Tripathi
Lovely Professional University
Phagwara, India

Mayank Velani
DIBER (Defence Research and
    Development Organization)
    Haldwani
Uttrakhand, India

Pawan Whig
Vivekananda Institute of Professional
    Studies-TC
New Delhi, India

Ritika Yaduvanshi
Mahamaya College of Agricultural
    Engineering and Technology
Ambedkar Nagar, India

Franco Fernando Yanine
Faculty of Engineering
Universidad Finis Terrae, Providencia,
    Santiago, Chile

Aydin Zaboli
Department of Electrical and Computer
    Engineering
University of Michigan-Dearborn,
    Dearborn, USA

1 Introduction to Microgrids, Concepts, Definition, and Classifications

Shivanshu Pandey, Prince Rajpoot and Amit Kumar Pandey
Rajkiya Engineering College, Ambedkar Nagar, India

Ritika Yaduvanshi
Mahamaya College of Agricultural Engineering and Technology, Ambedkar Nagar, India

Shivendu Mishra
Rajkiya Engineering College, Ambedkar Nagar, India

1.1 INTRODUCTION

The microgrid concept represents a cutting-edge technological advancement poised to revolutionize our energy infrastructure, enhancing reliability and cost-efficiency. Microgrid systems have the flexibility to operate autonomously or seamlessly integrate with primary grids. This chapter delves into a comprehensive exploration of microgrids and their various types, architectural intricacies, and constituent components. Furthermore, we provide insight into microgrid stations, highlighting their significance within this innovative energy landscape. Upon completing this chapter, you will gain a profound understanding of the manifold benefits offered by microgrid systems and the intricacies of the challenges they pose within our modern energy landscape. We will also provide a glimpse into the present state of microgrids and project their promising future.

1.2 MICROGRID

Microgrids are a modern concept ready to revolutionize our energy landscape, with enormous potential for improving the quality of our energy infrastructure. These adaptable systems can work independently or with the primary grid [1]. A microgrid

DOI: 10.1201/9781003481836-1

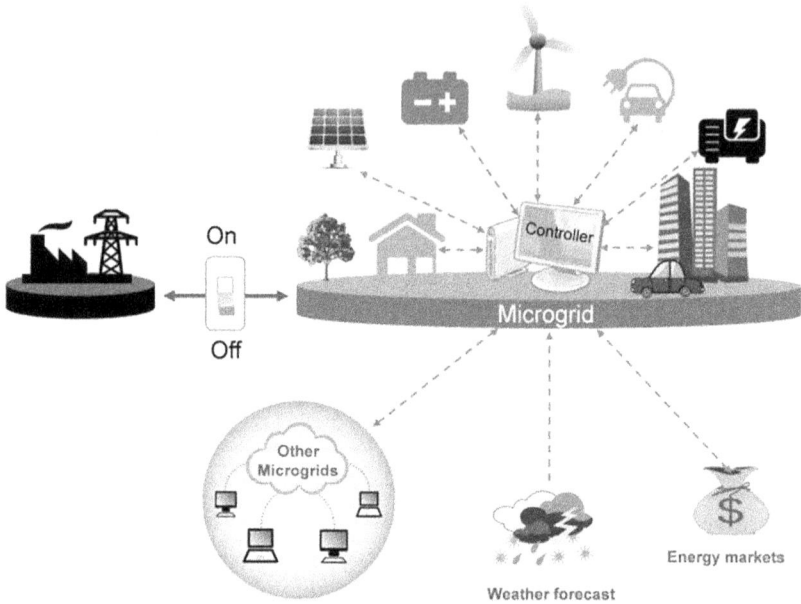

FIGURE 1.1   Microgrid system [4].

is defined as any compact, self-contained power station with its own generation and storage capabilities enclosed by well-defined boundaries [2]. Incorporating such setups into your energy network provides a valuable safety net, ensuring an additional power source is available during power outages caused by various factors. Furthermore, microgrids are essential in lowering energy costs while providing a consistent and dependable energy supply.

Integrating distributed energy resources, including renewable energy sources, energy storage systems, and generators, is the foundation of microgrids. Solar panels, wind turbines, combined heat and power systems, and generators can coexist peacefully within this framework, working in tandem to improve energy resilience and efficiency [3].

The illustration shown in Figure 1.1 depicts a microgrid system scenario.

1.3   CHARACTERISTICS OF MICROGRIDS

A microgrid has the following characteristics.

1.3.1   Local

This indicates that microgrids can generate power for immediate local consumption, highlighting a significant contrast with traditional centralized grids. Conventional grids rely on electricity transport over extensive transmission and distribution lines spanning long distances. Unfortunately, this method is inefficient and incurs substantial power losses, typically between 8% and 15% [3]. However, microgrids offer an elegant solution to mitigate this problem. By incorporating local power generation sources such as generators and solar panels, microgrids effectively eliminate the need for extensive long-distance power transmission, resulting in enhanced efficiency and minimal power loss.

1.3.2 INDEPENDENT

A prominent feature of microgrids is their ability to operate autonomously, even when disconnected from the primary grid. Modern microgrid technology is further enhanced by integrating energy storage devices, enabling them to store excess electricity. This crucial characteristic empowers microgrids to provide a reliable power supply when the primary grid encounters challenges and cannot deliver electricity [3].

1.3.3 INTELLIGENT

Microgrids are inherently intelligent systems driven by a central controller, often called the "brain" of the microgrid. This controller oversees and orchestrates the operations of all devices and systems within the microgrid, including the precise management of electricity supply [3].

1.4 MICROGRID SYSTEM PARTS

A microgrid incorporates various power generation sources, including renewable energy sources like solar and wind and traditional sources like diesel generators. The choice of the optimal power generation source depends on several key factors, such as the microgrid's location, the availability of various energy sources, and the cost of energy production. A microgrid consists of multiple components and technologies designed to provide efficient electricity generation, storage, and distribution. Let's look at the main features of a microgrid system [5, 6].

1.4.1 ENERGY STORAGE

Surplus energy generated by the microgrid is stored, and can then be utilized during periods of low energy production or high demand.

1.4.2 POWER RENOVATION

This includes components like inverters, which play a crucial role in converting the DC power generated by solar panels into AC power suitable for distribution to end-users. This power conversion process is essential for delivering electricity efficiently.

1.4.3 POWER DISTRIBUTION

Within the microgrid, this will include vital components like transformers and switchgear. These elements play a pivotal role in regulating the flow of energy within the microgrid and facilitating power distribution to end-users.

1.4.4 ENERGY MANAGEMENT

Employed to optimize the operation of the microgrid, oversee energy storage and communication, and ensure the safety and reliability of the power delivered to customers. Energy management systems may also incorporate demand response capabilities, enabling efficient energy utilization during peak periods.

1.4.5 CONTROL AND MONITORING

It is instrumental in orchestrating the flow of energy within the microgrid, harmonizing the operation of various components, and guaranteeing the microgrid's safe and dependable performance. The control and monitoring system encompass encompasses sophisticated software and hardware tools, empowering operators to oversee and govern the microgrid from a centralized location remotely. Additionally, it efficiently tends to and manages all activities within the microgrid through a diverse array of software applications [7].

In addition to these core components, microgrids can encompass a variety of additional features, including backup power systems, electric vehicle charging stations, and advanced innovative grid capabilities. The selection of microgrid components is contingent upon factors such as the microgrid's size, its specific application, and the environmental conditions in which it will operate.

1.5 MICROGRID ARCHITECTURE

A microgrid is a complex system composed of several crucial components that are essential for its architecture, including.

1.5.1 DISTRIBUTED SOURCES AND ENERGY STORAGE

Distributed energy sources and storage devices are interconnected to function as a unified and controllable entity within a microgrid system. This unified entity can seamlessly operate in grid-connected and islanding modes, the latter in which power can be supplied even when the grid connection is disrupted. The distributed system resources are a central focus for consumers, and their reliable performance is paramount. Hence, meticulous attention is given to ensuring these distributed resources adhere to electrical boundaries and meet consumer expectations [7].

Sources include:

a. PV panels
b. Diesel generators
c. Wind turbines.
d. Combined heat and power plant

In addition to the mentioned energy sources, microgrids can incorporate heat loads instead of electrical loads when necessary. This adaptability allows for a more versatile and tailored approach to meet specific energy requirements.

Types of storage include:

a. Battery
b. Flywheel
c. Fuel cell
d. Supercapacitor storage

The storage solutions mentioned are essential components used within microgrid systems to manage and store energy as needed efficiently.

1.5.2 Microgrid Central Controller

The Microgrid Central Controller (MGCC) serves as the orchestrator, managing the comprehensive functionalities of the entire microgrid. It issues commands to the local controllers, ensuring coordinated and optimized microgrid operations. Local controllers, aptly named for their role, oversee and regulate the process of individual sources and loads within the microgrid.

1.5.3 Utility Grid

A utility grid, also known as the "grid," is a centralized power generation, transmission, and distribution system that provides electricity to residences, commercial buildings, and industrial facilities over a large area. Energy providers oversee utility grids essential to supplying consumers with electricity. Figure 1.2 shows a radial distribution network, specifically a Medium/Low Voltage Radial Feeder, interconnected with the utility grid. A crucial component known as the Point of Common Coupling

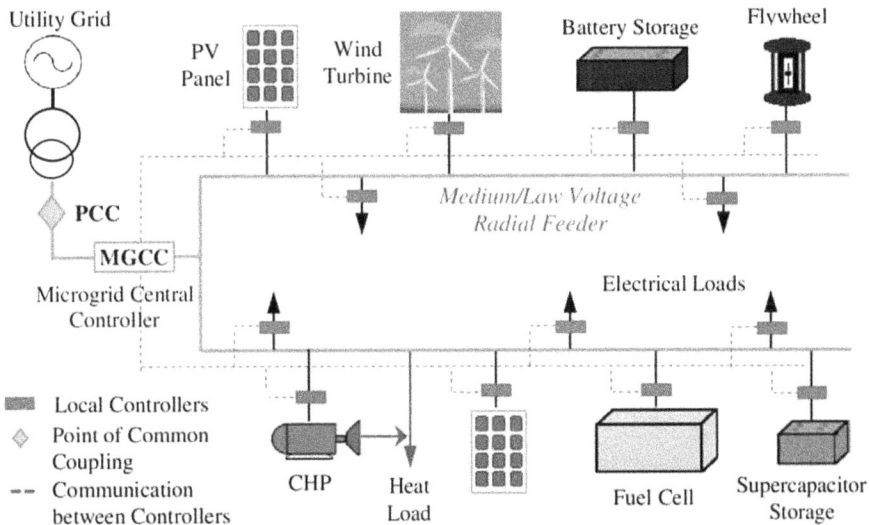

FIGURE 1.2 Microgrid architecture [8].

(PCC) is incorporated in this configuration, typically utilizing a static switch for its implementation.

1.6 MICROGRID CLASSIFICATION

Microgrids can be classified into three types based on their operational frequency [9]:

1. AC Microgrid
2. DC Microgrid
3. Hybrid AC/DC Microgrid

1.6.1 AC MICROGRID

AC microgrids commonly operate at medium or low voltage levels. In these microgrids, distributed sources, energy storage devices, and loads can be interconnected with flexibility, whether using converters or not. The necessity of converters depends on the frequency ratings and the types of generators employed. For instance, diesel generators, microturbines, and wind turbines, which are AC generators, can be directly connected to the AC network without requiring additional converters. However, converters become essential for seamless integration into the microgrid when integrating DC sources like photovoltaic (PV) systems.

1.6.2 DC MICROGRID

In this configuration, energy conservation is achieved by reducing the number of converters, as there is no need for DC-to-AC converters. Instead, converters are crucial interfaces facilitating seamless communication and energy exchange among sources, storage systems, and loads within the microgrid.

1.6.3 HYBRID AC/DC MICROGRID

In this setup, we employ multi-bidirectional converters. This design significantly reduces the reliance on multiple AC-to-DC-to-AC and DC-to-AC-to-DC converters, commonly found in conventional microgrid configurations.

According to the connection, microgrids can be classified into two primary types based on their connection to the main grid:

1. Grid-connected
2. Islanded

1.6.4 GRID-CONNECTED

Grid-connected microgrids are seamlessly integrated with the larger electrical grid, enabling them to function in parallel. These microgrids can import or export energy as required, ensuring a balanced and efficient energy supply.

1.6.5 ISLANDED

Islanded microgrids operate entirely from the larger grid, relying solely on local energy sources to meet their power requirements.

1.7 MICROGRID STATIONS

These stations, often called microgrid substations, play a vital role within the microgrid infrastructure. They are crucial connection points between the microgrid and the broader utility grid, enabling bidirectional power flow. This ensures that the microgrid remains linked to the larger grid, particularly during low energy generation or high demand [10]. Microgrid stations inherently incorporate essential components like transformers and associated equipment. These components serve the crucial function of converting the voltage of the power generated by the microgrid into a form suitable for distribution across the broader utility grid. Additionally, they encompass protective measures, including circuit breakers and fuses, ensuring the microgrid remains isolated from the larger grid in case of a fault. Beyond their role in establishing a connection between the microgrid and the larger grid, these stations also possess the capability to manage power flow within the microgrid itself.

For instance, these stations serve as critical hubs for regulating and optimizing power distribution among various energy resources. They can operate autonomously or in coordination with the primary grid, hosting a variety of distributed energy resources such as solar panels, generators, and inverters. These stations' primary purpose is to ensure efficient and dependable energy delivery to specific locations. As each day passes, their popularity continues to grow, addressing pressing concerns related to energy security, environmental sustainability, and the integration of renewable energy sources into the broader energy landscape. These stations offer a compelling solution for efficiently managing supply and demand dynamics, all while promoting sustainability and locally sourced energy [11].

1.8 CHALLENGES IN MICROGRID

When it comes to the implementation and operation of microgrids, several challenges are bound to arise, such as [12]:

a. **Technical challenges**: Consumers are increasingly drawn to distributed generation systems such as solar panels and small wind turbines. However, the proliferation of these distributed networks presents challenges, particularly in effectively managing complex control and maintenance functions.

b. **Islanding**: Islanding challenges in a microgrid refer to a scenario in which the interconnected grid loses its connection, yet the microgrid continues to operate independently. While this autonomous operation can be advantageous for safety reasons, it can also adversely affect power quality and overall grid stability.

c. **Protection**: When implementing a microgrid, it is crucial to ensure the safety of the isolated island and the stability of the broader power grid. This entails careful control measures to prevent and limit current flow in the two main lines.

d. **In-depth consideration**: The failure of microgrids often stems from an absence or deficiency of thorough review and pre-planning. However, when we engage in meticulous planning, it becomes evident that economic, social, and environmental factors play pivotal roles in microgrid projects' successful implementation and long-term sustainability [13].

e. **E-Sustainable Electricity System**: When embarking on the implementation of a microgrid, we encounter a multitude of challenges that encompass power quality, management, control, stability, efficiency, islanding, protection, and reliability. However, by adeptly applying control principles, sophisticated simulation techniques, meticulous modeling, and in-depth analysis, we can significantly enhance the landscape of modern power distribution [14].

1.9 FUTURE OF MICROGRID

Microgrids are set to become notably more efficient and better tailored to meet consumer needs, incorporating an upgraded and environmentally sustainable power grid. Here are some key areas to note [15].

a. **Electricity Network**: As we look ahead, it is imperative that we prioritize the development of networks that exhibit greater adaptability, accessibility, reliability, and cost-effectiveness. These qualities are essential in our pursuit of advancing the smart grid initiative. The design and architecture of such networks should be regarded as pivotal factors in achieving these objectives.

b. **Greenhouse Gas Emissions**: Enhancing the design and architecture of microgrids can play a significant role in reducing greenhouse gas emissions while simultaneously promoting the utilization of renewable energy sources.

c. **Energy Storage**: This forms a critical component of microgrid implementation and design, emphasizing the imperative utilization of diverse micro sources such as wind, solar, and other renewable alternatives. Energy storage, in particular, stands as a pivotal factor, bolstering microgrid stability and facilitating sustainable operation.

d. **Microgrid as Internet of Energy**: In the contemporary landscape, microgrids are evolving into a dynamic concept known as the Internet of Energy (IoE), enabling the flexible interconnection of numerous generators. This evolution leverages the Embedded Internet of Everything (IoE), which serves as the backbone for managing the smart grid. It is noteworthy that we refrain from employing traditional IoT (Internet of Things) because it does not fully address the demanding challenges inherent to IoE platforms.

1.10 CONCLUSION

Microgrids are poised to play a pivotal role as major energy suppliers in the coming years. As climate change looms and presents a diverse array of threats, businesses and individuals are becoming increasingly aware of the imperative to transition to more sustainable energy systems. In this context, microgrids emerge as a compelling solution. Microgrids possess the transformative potential to revolutionize conventional energy production and distribution, offering a far more efficient approach. Strategic investments in research and development, coupled with the establishment of appropriate regulatory frameworks, will pave the way for a future that is more resilient, sustainable, and conducive to overall wellbeing.

Microgrids represent a significant shift in the generation and distribution of energy, offering several valuable advantages. Microgrids emerge as a desirable and superior option for consumers and businesses seeking to reduce their reliance on the traditional grid. One of their standout features is the ability to seamlessly integrate various energy sources, including renewables like solar and wind. The potential benefits of microgrids are apparent, yet they face several hurdles that must be overcome to unlock their capabilities thoroughly. These challenges encompass cost-related issues, the establishment of conducive regulatory frameworks, and the necessity for advanced control and monitoring systems. Nevertheless, as technology advances in this field, many obstacles will gradually dissipate, paving the way for widespread adoption of microgrids.

REFERENCES

1. "Microgrid Overview" https://www.techtarget.com/whatis/definition/microgrid (Accessed April 15, 2023).
2. "Microgrid Introduction" http://www.teamsustain.in/micro-grid#:~:text=Micro%20Grid%20%E2%80%93%20An%20Outline&text=It%20is%20a%20small%2Dscale, boundaries%20qualifies%20as%20a%20microgrid (Accessed April 7, 2023).
3. "Microgrid Characteristics" https://www.microgridknowledge.com/about-microgrids/article/11429017/what-is-a-microgrid (Accessed April 22, 2023).
4. "Microgrid System" https://nsci.ca/wp-content/uploads/2019/11/Microgrid_en.png (Accessed April 12, 2023)
5. Farhangi, H., 2009. The path of the smart grid. *IEEE Power and Energy Magazine*, 8(1), pp. 18–28.
6. "Microgrid Components" https://www.cummins.com/news/2021/09/23/microgrid-components (Accessed April 18, 2023).
7. "Embracing Microgrid: Applications for Rural and Urban India" (Accessed April 5, 2023).
8. "Microgrid Architecture" https://www.researchgate.net/publication/277022435/figure/fig3/AS:667643115737096@1536189931684/Schematic-diagram-of-a-Microgrid.png (Accessed April 14, 2023).
9. "Classification of Microgrids" https://www.researchgate.net/figure/Classification-of-microgrids_fig1_343518698 (Accessed April 25, 2023).
10. "Microgrid Architecture and Future Trends" https://www.researchgate.net/publication/305828588_Microgrid_Architecture_policy_and_future_trends (Accessed April 9, 2023).

11. "Microgrid Fundamentals, How They Work & the Roles Generators Play" https://www.generatorsource.com/Articles/Other-Information/Microgrid-Fundamentals-Role-Generators-Play.aspx (Accessed April 20, 2023).
12. Akinyele, D., Belikov, J. and Levron, Y., 2018. Challenges of Microgrids in Remote Communities: A STEEP Model Application. *Energies*, *11*(2), p. 432.
13. Sabzehgar, R., 2015, November. A Review of AC/DC Microgrid-Developments, Technologies, and Challenges. In *2015 IEEE Green Energy and Systems Conference (IGESC)* (pp. 11–17). IEEE.
14. Salam, A.A., Mohamed, A. and Hannan, M.A., 2008. Technical Challenges on Microgrids. *ARPN Journal of Engineering and Applied Sciences*, *3*(6), pp. 64–69.
15. Mariam, L., Basu, M. and Conlon, M.F., 2016. Microgrid: Architecture, Policy and Future Trends. *Renewable and Sustainable Energy Reviews*, *64*, pp. 477–489.

2 Modeling AC/DC Microgrid

Sonu Kumar
Koneru Lakshmaiah Education Foundation, Guntur, India
K.S.R.M. College of Engineering, Kadapa, India
Manipur International University, Manipur, India

Y. Lalitha Kameswari
Koneru Lakshmaiah Education Foundation, Guntur, India
MLR Institute of Technology, Hyderabad, India

S. Koteswara Rao
Koneru Lakshmaiah Education Foundation, Guntur, India

B. Pragathi
DVR & Dr. HS MIC College of Technology, Kanchikacherla, India

2.1 INTRODUCTION

When several loads and decentralized energy sources are networked, they form a microgrid that can be managed independently from the larger power grid. It has the capability of both grid-connected and island operation [1]. Microgrids can be more reliable and better able to deal with problems on the grid. As the technology of electric distribution moves into the next century, trends are becoming clear that will change the needs of energy delivery. These changes are occurring because people want more energy, and for it to be used more efficient [2]. Both the demand and supply sides are pushing for these changes. On the demand side, people want more energy to be available and for it to be used more efficiently. The supply side has to be able to accommodate dispersed generation and peak usage [2].

Appliances, heating/cooling systems, and gadgets are powered by the grid [3]. Because everything is connected, fixing one element of the grid impacts everyone. Microgrids can assist. Most microgrids are grid-connected. They can run on local energy generation in emergencies like storms or power outages [4]. Distributed generators, batteries, or solar panels power microgrids. A microgrid system controller is shown in Figure 2.1 and a microgrid power system is shown in Figure 2.2. Depending on electricity and demand, a microgrid might last indefinitely. A microgrid can save

DOI: 10.1201/9781003481836-2

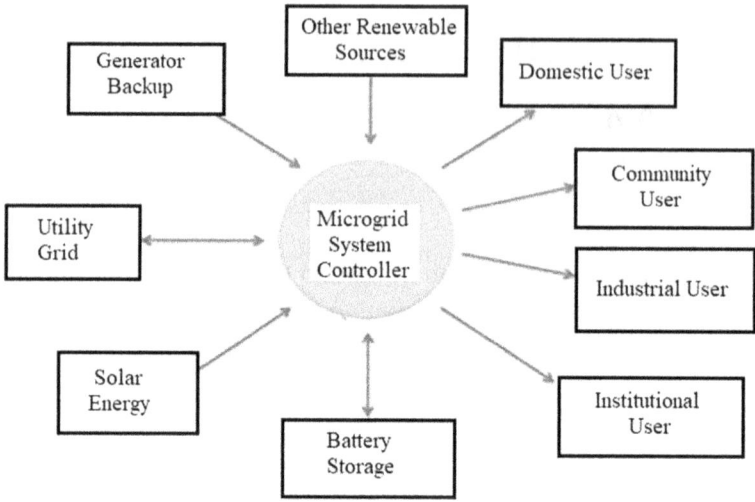

FIGURE 2.1   Microgrid system controller.

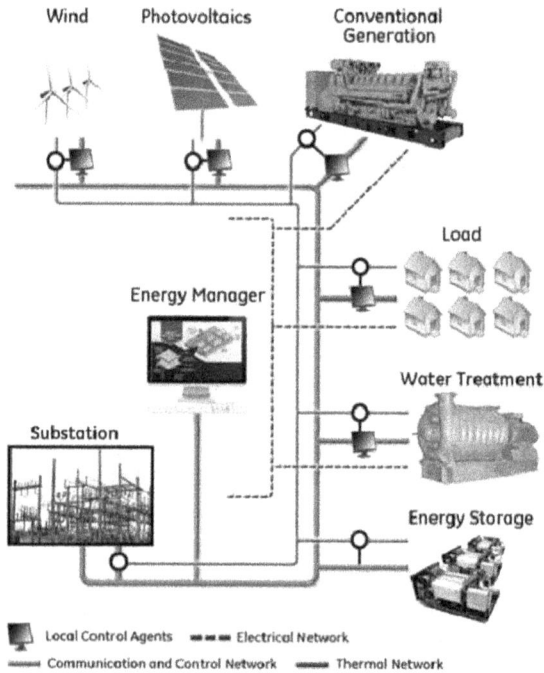

FIGURE 2.2   Microgrid power system [5].

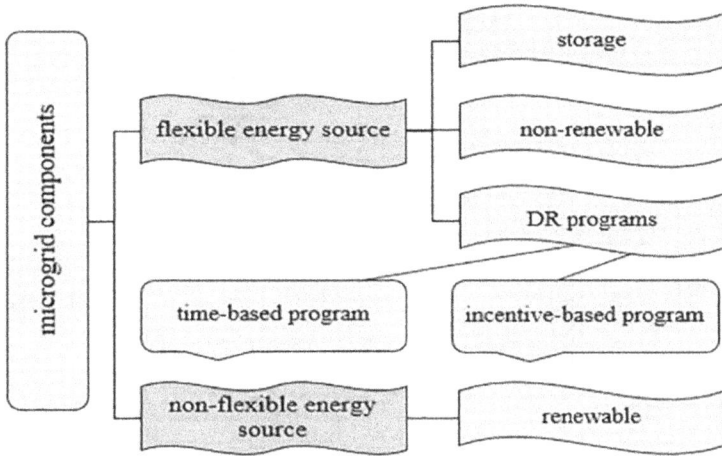

FIGURE 2.3 An energy microgrid schematic showing the many linked power sources [6].

money, connect to a tiny or unstable local supply, and be a backup for the grid in an emergency [4]. Communities can acquire their own electricity and be greener with a microgrid.

In the event of a power outage, microgrids are able to cut themselves off from the grid. If the grid goes down because of something like bad weather or a fallen phone pole, you need to be disconnected from it, or "islanded," so you can keep making and using electricity. As a result, one of the most important things about a microgrid is that it can keep running even when the larger grid goes down. Figure 2.3 shows an energy microgrid with many linked power sources.

Storage units can keep a balance between short-term and long-term reserves. The microgrid is linked to the upstream network, which can get all of its power from the main grid, or just some of it. When connected to a grid, it can both take power from the main grid and put power into it. Therefore, it can increase grid efficiency and provide solutions to specific energy issues [6].

2.2 CLASSIFICATION OF MICROGRIDS

Microgrid classification is represented in Figure 2.4.

2.2.1 DEPENDING ON MODES OF OPERATION

Microgrids can improve the dependability of energy sources by isolating themselves from the grid in the event of a network outage or a decrease in power quality [6]. These modes include grid-connected, transited, island, and reconnection. Since there is no infinite bus, the microgrid has to maintain the reactive power balance when it is operating on its own (islanded). The islanded mode operation is a concern, and the cost function has three parts: (a) the cost of power and energy loss, (b) the cost of installing shunt capacitors, and (c) the cost to the consumer when the

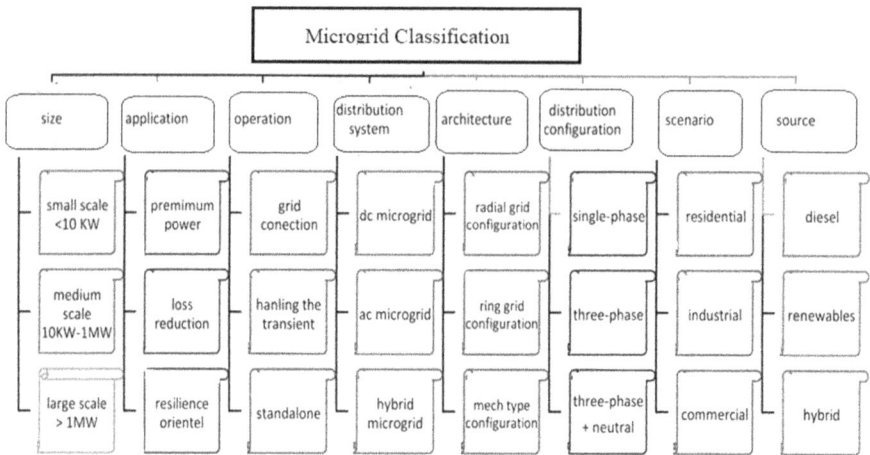

FIGURE 2.4   Microgrids Classification [6].

power goes out. Static transfer switches connect the microgrid to the utility grid in grid-connected mode. In islanded mode, microgrid control is tougher without grid assistance. The host utility grid controls microgrid voltage. Even though electricity can flow both ways in a microgrid, in grid-connected mode, it can swap power with the outside grid to maintain supply. In islanded mode, the microgrid's power supply must fulfill load demand. Interconnected modes are stronger and larger microgrids in isolated mode [6].

2.2.2   DISTRIBUTION-SYSTEM DEPENDENT

A microgrid can be an AC- or DC-based power system, or a hybrid system [6]. Each has pros and cons that become clear in use. Many studies have been done on the pros and cons of both AC and DC microgrids. The DC microgrid can be used with the main power grid or on its own. Figure 2.5 shows how a typical AC microgrid is put together.

A DC microgrid's distribution network can be either monopolar, bipolar, or homopolar. All of the renewable energy sources and loads in an AC microgrid are connected to the same AC bus. The main problem with AC microgrids is that they are hard to control and run. Figure 2.5 represents a typical AC microgrid. Depending on how the power is distributed, there are three types of AC microgrid: single-phase, and three-phase both with and without neutral-point lines) [6]. A DC microgrid is represented in Figure 2.6.

Compared to AC microgrids, DC microgrids are better because they are more reliable, more efficient, and easier to connect to different sources of energy distribution. A microgrid is a network of loads and energy-generating units that are spread out and placed within certain electrical boundaries. They can run their operations on the wide-area grid network or in their island mode, where they run on their own and are not affected by outside forces [7]. In the US, the microgrid business is still fairly new.

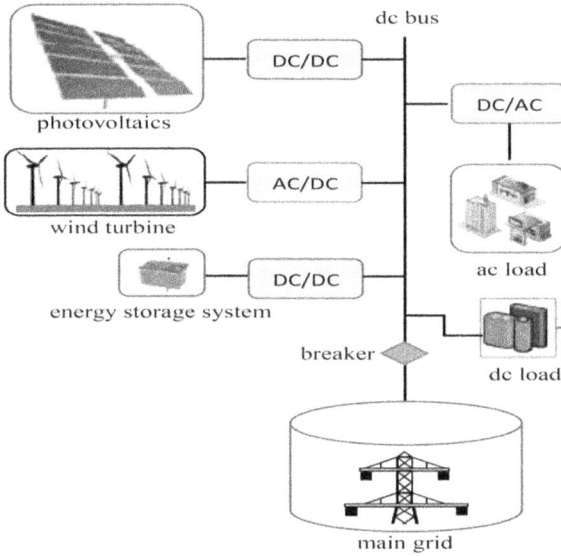

FIGURE 2.5   AC microgrid [6].

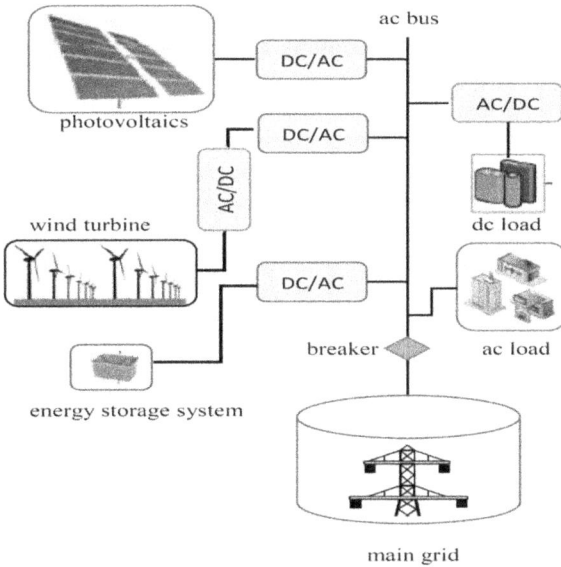

FIGURE 2.6   DC microgrid system [6].

Statistics from the Centre for Climate and Energy Solutions show that there are 160 microgrids up and running in the US. Even though these microgrids currently meet less than 1% of the nation's energy needs, that number is expected to rise in the near future.

AC and DC microgrids operate differently [7]. These varieties and their merits and downsides are summarized below. AC microgrids connect energy sources and users through an AC bus system. Most of the time, AC microgrids are made up of both spread power sources, like renewables, and traditional power sources, like engine-based generators. These many engines are linked by an AC bus system and a way to store energy, such as a battery energy storage system (BESS). DC output is made by renewable sources like solar photovoltaic cells, wind turbines, etc. Power electronic adapters can be used to change this output into AC.

2.2.3 ADVANTAGES OF AC MICROGRIDS

The advantages of microgrids include a) versatility to connect to the regular power grid or work on their own; b) compatibility with AC tools, like motors that use AC loads. The AC from the microgrid makes it easy to run this equipment; c) AC loads don't need an inverter; d) power protection methods that are good value; and e) AC loads can be used more often.

2.2.4 DISADVANTAGES OF AC MICROGRIDS

The disadvantages of microgrids include a) less effective conversion; b) adapters, such as DC–AC adapters, are expensive; c) frequency, voltage regulation, and unbalance adjustment all make it hard to control; d) low power source reliability can impede effectiveness of high-performance equipment that needs a reliable power supply; and e) less efficient at transmitting than DC.

2.2.5 DC MICROGRIDS

DC microgrids work in a similar way to AC microgrids. The main difference between them is that one uses a DC bus network to connect the sources and loads in the network, while the other uses an AC bus. Most of the time, DC buses run at between 350 and 400 Volts. Electronic loads can employ low-voltage buses separate from the main DC bus. However, high-voltage gain DC–DC converters in DC-type microgrids make it simpler to link low-voltage power sources like solar panels (20–45V) to the high-voltage DC bus [7]. They enhance the low-voltage power supply to the high-voltage DC bus. These converters can be put into different groups based on how much voltage gain/power they can handle.

Connecting to the utility grid is managed through the power electronic converter interface [6]. The direction of the power depends on how well load and generation are balanced. Creating hybrid microgrids is an effort to boost network efficiency. They do this by reducing the number of conversion stages, increasing reliability, reducing the number of devices that connect to each other, and lowering the cost of energy. Many studies have been done on how DC, AC, and hybrid microgrids can be used, how they can be protected, and how stable they are. A hybrid microgrid is represented in Figure 2.7.

Renewables help a lot with both saving money and becoming less dependent on traditional energy sources. This method works especially well in industrial areas that are far away, like mines, where gasoline and other traditional power sources were

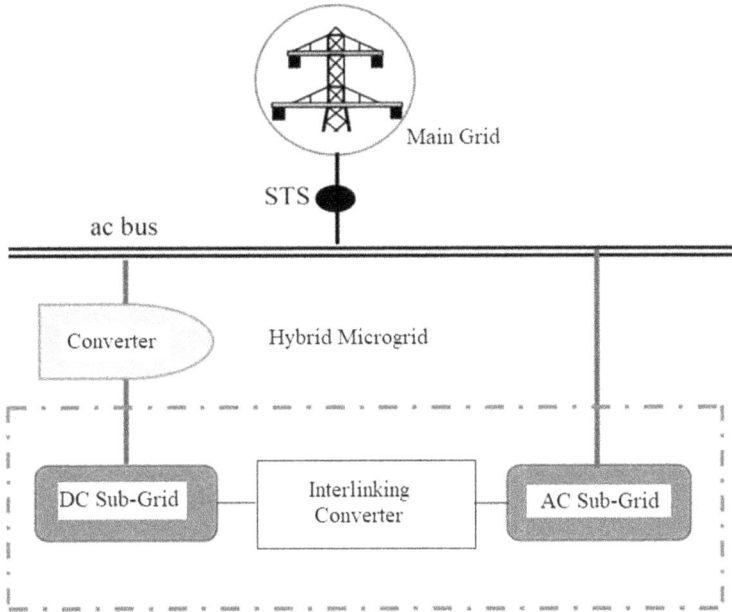

FIGURE 2.7   Hybrid microgrid [7].

once needed. A study by Energy and Mines says that a hybrid power plant with 1 MW of solar power can save 450,000 liters of diesel and US$500,000 per year [7].

According to the United States Energy Information Administration, more than half of the world's energy supply is consumed by industry. They also think that this number will continue to rise. Rising energy costs and the need to reduce carbon emissions from both the sources of energy and the tools that use it put more pressure on both those who make and those who use energy. Microgrids with green energy generators can not only help solve these problems, but they can also be a good place for the industrial sector to make smart investments.

2.2.6   HOW TO CHOOSE THE BEST MICROGRID

Microgrids are becoming more useful for many industries because they help lower our carbon footprint, make us more resilient, and help us save money. Microgrids can be a very important part of making power stable, low-cost, and free of carbon emissions. But no two places are the same because they have different physical, financial, and energy needs.

2.3   ENERGY MANAGEMENT SYSTEM

Through a control and Energy Management System (EMS), microgrid (MG) systems offer users benefits like better power quality, stability, long-term use, and energy that is good for the environment. Integrating these kinds of distributed energy sources into

the power grid makes it possible to set up microgrids. The microgrid is an aggrega-
tion unit that acts as either a generator or a load and needs the right EMS. Figure 2.8
shows the EMS in a microgrid.

Different time scales for microgrid monitoring and power management are
depicted in Figure 2.9. Based on the storage surface capacity and the unique equation
or limitation of each unit, a local controller manages real-time balancing and power
transfer between DGs and ESSs in short-term energy management [7]. On the
medium term, EES includes changing how much renewable energy is made, how
much energy is stored, how much load is expected, and how much energy is stored.

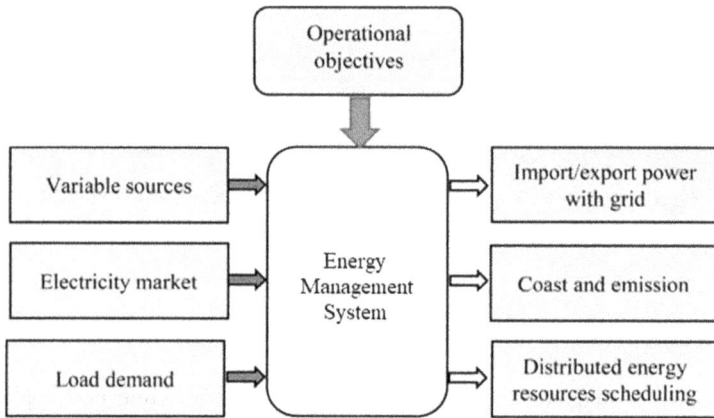

FIGURE 2.8   Energy management system [7].

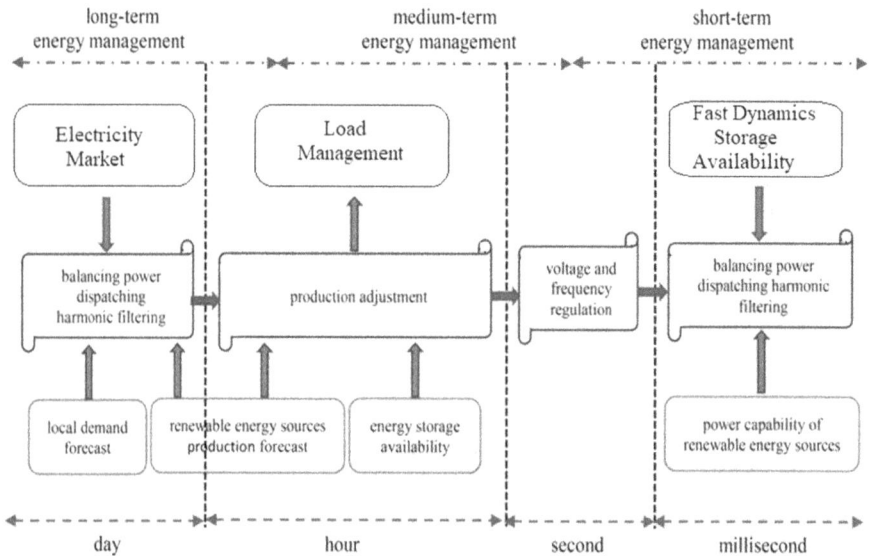

FIGURE 2.9   Functions of microgrid power timing control classification [6].

Long-term, hourly renewable energy output projections are a part of EES, which take into account how the effects of the energy source on the environment change over time. There are two types of microgrid dispatch strategies: those that are optimal and those that are fixed.

When a large number of microgrids connect to the main grid, EMS-controlled energy storage technology can smooth out the random and sporadic power output [8]. Figure 2.9 represents functions of microgrid power timing control classification.

2.4 MICROGRID LOAD FREQUENCY REGULATION

Renewable energy sources are a big part of the modern power system's distributed network, so it makes sense that microgrid load frequency control is a big deal. When it comes to running a power plant and making sure that enough high-quality energy is being supplied, load frequency management is a major concern [9]. For a microgrid to work well, it must be easy to switch between being connected to the grid and running on its own. The most popular types of microgrids fall into the following groups. Figure 2.10 represents microgrid types for specific users.

Most urban microgrid projects can be made sense of by: a) making energy more stable and reliable; b) lowering the cost of energy; and c) maintaining a healthy ecosystem. One possible justification for constructing a microgrid in a city is to improve the security and stability of the power grid [10]. For essential loads like military bases, hospitals, industrial sites, and university laboratories, as well as other users (commercial and residential) that desire better network connectivity when the power grid goes down, this would be an ideal solution [9]. Energy reliability might not be a problem in developed countries like Australia, especially in big cities where distribution grids are more automated and even have self-healing systems. However, the grid is not as strong in small cities and country areas.

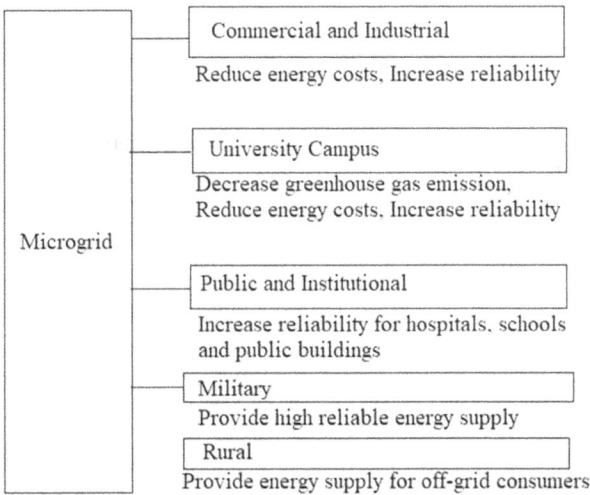

FIGURE 2.10 Types of microgrid.

Distribution networks in developing nations like Brazil are overburdened, resulting in lengthy feeders, voltage control issues, and frequent power outages of varying durations [9]. Natural disasters such as hurricanes, bushfires, and thunderstorms can strike a location at any time. Microgrids can make the energy supply more resilient to minimize power blackouts following these events. So, these situations show that there are chances to set up urban microgrid projects and make the energy supply more reliable and resilient. Another reason to build a microgrid is to save money on energy costs by using local distributed energy resources (DER) inside the microgrid, such as solar energy, biomass, combined heat and power (CHP), and energy storage, which can provide cheaper electricity and thermal energy and keep you from having to pay utilities' high peak-hour tariffs [10].

This is true for both developed and developing countries, since the cost of renewable energy is going down everywhere, but at different rates, and the cost of getting energy to customers is going up for many reasons, like replacing old infrastructure and shutting down cheap coal power plants [10]. The third main driver is environmental sustainability. This has to do with reducing the carbon footprint of buildings, offices, factories, businesses, and government buildings by making appliances more energy efficient, improving cooling and heating systems, and using as much renewable energy as possible instead of fossil fuels [9].

2.5 BENEFITS OF MICROGRID

2.5.1 A Microgrid Makes Electric Power More Reliable

About 8 million people in 15 states and the District of Columbia have been without electricity since 2012, when Superstorm Sandy wreaked havoc, hence electric reliability has been the most talked about microgrid benefit [8]. Some failures lasted more than two weeks. People who were in the dark noticed that lights were still on in nearby houses. There were microgrids in those places. Outages are not only annoying, but they can also be dangerous. During the storm, the backup generators at a Manhattan hospital stopped working. This meant that sick babies had to be carried down long, dark flights of stairs while doctors physically pumped air into their lungs. Microgrids keep the power going by breaking away from the main grid when it starts to fail. This is called "islanding." The microgrid's customers are then served by the generators and, if there are any, batteries, until power is restored on the main grid [8].

2.5.2 A Microgrid Helps with Endurance and Recovery

Energy resilience is a concept that goes hand-in-hand with the idea of electric stability. Reliability is about making sure the power stays on. Resilience is about being able to avoid power outages or get back to normal quickly if they do happen. One of the things that came to light after Superstorm Sandy was how resilient microgrids are. A number of smaller storms since then have shown how important it was. In some cases, a microgrid can restore power to an entire building or business right away, making it hard for the people inside to tell that something went wrong. In other situations, a microgrid is set up to restore only the most important functions in a building. So, a college campus might set up its microgrid to bring power back to research rooms and dining halls, but maybe not to swimming pools or academic offices in the middle of nowhere. When key operations are up and running, the university can keep a minimum level of service that is important. Once the central grid is fixed, the facility can get back to normal activities faster because it did not have to shut down completely [9]. Utility companies are now looking into putting microgrids at their central offices so that they can get power back on faster for their customers during a large power outage. From these wired-up command centers, the utilities would organize their workers and share information [8].

2.5.3 A Microgrid Can Help Consumers and Companies Save Money on Energy Costs

Microgrids can help their users save money and make money at the same time. They keep prices down by managing the supply of energy well. By selling energy and services back to the grid, they bring in money. This gives customers a new way to affect how energy markets work. With their microgrids, they no longer just use energy; they can also make it and control it. This kind of customer is called a prosumer.

Microgrids can make money by helping the core grid in other ways. Ancillary services, like frequency control and spinning reserve, help the grid in other ways. Advanced microgrids are also good at taking advantage of how much energy costs. Prices for electricity change throughout the day based on past and expected demand. An advanced microgrid manager, which is sometimes called the "brain of the system," can use this change to help its customers. The microgrid does this by coordinating how its assets are used with how the price of electricity on the grid goes up and down. When there is a lot of demand for grid energy and prices go up, the manager may tell the microgrid to use more of its own power, so it doesn't have to pay higher prices. If the microgrid has more power than it needs, it could sell some of it back to the grid [8]. The microgrid helps its customers plan both short-term and long-term prices because it uses its assets to get the most value out of them over time. How well it can do this, though, will rest on how smart the microgrid controller is and how the wholesale market in the area works [8]. Participating in state and federal clean energy programs, such as state renewable portfolio standard initiatives or federal production tax credits, is another way microgrids may generate revenue. There are grant programs in some states that are just for microgrids. In places like the Northeast and California, where power costs are high, microgrids have the potential to supply reliable, low-cost power. The customers of a microgrid then get perks like reliability and cleaner energy, but pay less for it [8].

2.5.4 A Microgrid Helps the Climate and Makes Clean Energy More Popular

Many companies and communities set clean energy goals to save energy and lessen the damage their power generation does to the environment. Microgrids can make green electricity in many different ways, including solar power, wind, fuel cells, combined heat and power (CHP) plants, and energy storage [8]. Many CHP plants use natural gas engines, which are cleaner than other fossil fuels. Microgrids use these natural sources of energy in a smart way. They make sure that the variable output of green energy and traditional generation assets are in perfect balance. By doing this, the microgrid mitigates the problem with solar and wind energies, in that they only work when the sun is out, or the wind is blowing [11]. When green energy isn't available, a microgrid can use energy from other sources without any help from a person. Advanced microgrids can also be set up to meet specific environmental goals, such as using the resources with the least amount of carbon, as much as possible [8].

2.5.5   A MICROGRID MAKES THE MAIN GRID STRONGER

When a microgrid is used to strengthen the larger energy grid, it helps not only its own customers, but also its neighbors. Several things enable this. One method is to help the grid work better by taking part in demand response programs or offering additional services. Microgrids can also help the main grid work better during times of high demand. They give grid workers an extra resource they can use during these times. Line loss is also stopped when microgrids are used. As electricity passes across wires, some of it is wasted as heat. Microgrids, in contrast to large central power facilities, are often constructed close to the communities they serve [8]. We use more of the power we make when there is less line loss. We won't have to build more power sources and transmission lines to serve the central grid if we do this.

Compared to new grid substations, transmission lines, or other grid infrastructure, microgrids may be less expensive to construct and maintain. That makes them viable "non-wires alternatives" to costly grid maintenance or upgrades for ensuring consistent power [8].

2.5.6   A MICROGRID IMPROVES PROTECTION

In May 2017, a huge ransomware attack made people all over the world more worried about safety. From hospitals in the United Kingdom to FedEx in the United States, thousands of businesses in 150 nations fell victim to malware [8]. The US power grid did not lose power due to this, but many experts are worried about how vulnerable it is and are taking steps to protect it. One of them is putting in microgrids. Cybersecurity is drawing a lot of interest from military bases and utility companies. Ameren Illinois, for instance, has constructed a 1.47 MW facility that is among the most cutting-edge utility-scale microgrids in the Americas. Ameren is the first utility to install a microgrid controller designed for military use that is also cybersecure [8]. Smart controls and energy sources like wind and solar are all part of the microgrid. A microgrid is less likely to be attacked by hackers because its parts are spread out. If someone tried to attack one generator, the microgrid could still get power from other sources.

2.5.7   A MICROGRID ADDS VALUE TO THE ECONOMY

In many ways, microgrids are good for the economy as a whole. First, they keep goods and workdays from being lost when the power goes out. Second, they get good companies to move to an area. Third, because they are close to homes, they keep jobs in the area. By cutting themselves off from the grid during a blackout and keeping users powered with on-site generators, microgrids keep a lot of money from being lost. Businesses don't have to close, and people can still get to work. Power outages are expensive, especially for places like research labs, data centers, factories, and grocery shops that lose products or services because of the outage. A study by the Congressional Research Service found that outages cost the US economy between $25 billion and $70 billion each year, based on how many storms there are. A medium- or large-sized business or factory could lose more than $15,000 in just 30 minutes of downtime. Microgrids attract companies that need high-quality power since power outages cost money. These places can attract top jobs like data centers and pharma businesses with stable electricity. Building and operating the microgrid provides local jobs, because power isn't generated hundreds of kilometers away [8]. Microgrids boost local economies by using local energy. For all of these reasons, some towns now plan their economies with microgrids in mind.

2.5.8 A MICROGRID IMPROVES THE HEALTH OF A COMMUNITY

If vital services can't function, a power outage can be a nuisance or a health and safety risk. Thus, more communities are installing microgrids to power hospitals, police stations, fire stations, communications centers, and wastewater treatment facilities. Microgrids provide "islands of power." Microgrids service shelters, food stores, and gas stations. The area becomes a safe spot to buy food, water, cell phones, and gas. Public housing residents who can't evacuate immediately receive microgrids in several locations. Others power homeless shelters, elderly homes, and medical satellite centers. Microgrids benefit the world through solar microgrids in Africa and India powering rural communities and boosting the economy. Princeton University's microgrid powered New Jersey during Superstorm Sandy. They warmed up with coffee there. Austin-based Seton Healthcare Family wants stable energy and sustainability. Seton chose a CHP plant microgrid to power its new medical building and heat it using waste heat from combustion [8].

2.6 CONCLUSION

Many people and groups can benefit from a microgrid. When the main grid goes down, it keeps the lights on. It offers a means to improve the distribution grid, make infrastructure more secure, and protect areas that are especially vulnerable. Through smart, automated energy management, a microgrid can increase the use of clean energy and make money for users and the grid as a whole. And finally, using a microgrid with a Battery Management System encourages local control, or what some people call the democratization of energy and the rise of the prosumer. AC/ DC microgrid modeling plays a crucial role in the development and operation of local power systems. Microgrids, whether designed with DC or AC architectures, offer numerous benefits, including reduced environmental impact, improved energy efficiency, enhanced energy security, and the integration of renewable energy sources [12]. However, the implementation of microgrids also presents challenges related to stability, bidirectional power flows, modeling, low inertia, load perturbation, and uncertainty. Extensive research has been conducted on various aspects of microgrids, including classifications, control strategies, protection devices, optimization methods, stability analysis, power sharing, and reactive power compensation techniques. The modeling and analysis of AC/DC microgrids contribute to the reliability, efficiency, and quality of power supply in future energy systems. Further advancements and innovations in this field are essential to address the evolving needs of the power sector and ensure the successful integration of microgrids into the existing electrical infrastructure.

REFERENCES

[1] T. Gangwar, N. P. Padhy and P. Jena, (2022). Planning and dispatch of battery energy storage in microgrid for cycle-life improvement, *IEEE International Conference on Power Electronics, Drives and Energy Systems (PEDES)*, 1–6, doi: 10.1109/PEDES56012.2022.10080085

[2] S. Bose, Y. Liu, K. Bahei-Eldin, J. de Bedout and M. Adamiak, (2007). Tieline controls in microgrid applications, *iREP Symposium – Bulk Power System Dynamics and Control – VII, Revitalizing Operational Reliability*, 1–9. doi: 10.1109/IREP.2007.4410542

[3] S.D. Veeraganti and R. Nittala, (2019). Operation of microgrid and control strategies: Microgrid structure and its control schemes, *Handbook of Research on Smart Power System Operation and Control*, Edited by H.H. Alhelou and G. Hayek, IGI Global, 434–449. doi: 10.4018/978-1-5225-8030-0.ch019

[4] F. B. Setiawan, S. Riyadi, L. H. Pratomo and A. Wibisono, (2022). A 5.4 kWp microgrid laboratory development for higher education and industrial workshop, *14th International Conference on Software, Knowledge, Information Management and Applications (SKIMA)*, 89–94. doi: 10.1109/SKIMA57145.2022.10029640

[5] A. Rasheed and G. Keshava Rao, (2015). Improvement of power quality for microgrid using fuzzy based UPQC controller, *IAES International Journal of Artificial Intelligence (IJ-AI)*, 4(2), 37–44.

[6] G. Shahgholian, (2021). A brief review on microgrids: Operation, applications, modeling, and control, *International Transactions on Electrical Energy Systems*. doi: 10.1002/2050-7038.12885

[7] Veckta, (Accessed 15 June 2023), Diving into the differences between AC microgrids and DC microgrids. https://www.veckta.com/2021/05/27/the-differences-between-ac-microgrids-and-dc-microgrids/

[8] E. Wood, (Accessed 15 June 2023), Microgrid benefits: Eight ways a microgrid will improve your operation. https://www.microgridknowledge.com/about-microgrids/article/11430613/microgrid-benefits-eight-ways-a-microgrid-will-improve-your-operation-and-the-world

[9] A.M. Montero, L.A. Lourdes and C.A. Allende, (2020). The role of social resistance in shaping energy transition policy in Mexico: The case of wind power in Oaxaca, *The Regulation and Policy of Latin American Energy Transitions*, Elsevier. doi: 10.1016/B978-0-12-819521-5.00017-6

[10] M.A.L. Castro, (2020). Chapter 9 - Urban microgrids: Benefits, challenges, and business models, *The Regulation and Policy of Latin American Energy Transitions*, Elsevier, 153–172, doi: 10.1016/B978-0-12-819521-5.00009-7

[11] H.M. Lim, M.H. Nam, Y.M. Kim and Y.K. Seo, (2021). Increasing odontoblast-like differentiation from dental pulp stem cells through increase of β-catenin/p-GSK-3β expression by low-frequency electromagnetic field, *Biomedicines*, 9, 1049. doi: 10.3390/biomedicines9081049

3 Introduction to Optimization Techniques for Microgrid

Mayank Velani
DIBER (Defence Research and Development Organization),
Haldwani, India

Vaibhav Chhabhaiya
Birala Vishvkrama Mahavidhayalay, Anand, India

Rakesh Bhadani
Veerayan Polytechnic, Jakhaniya, Mandvi, India

Shweta Thumbar
M.S. University Baroda, Baroda, India

3.1 INTRODUCTION

The rapid growth of population and technological advancements have increased energy consumption and demand for electricity [1, 2]. Conventional sources of energy are commonly utilized to meet the energy demands of humans. However, these resources are known to have negative environmental effects. Although it is possible to rely on conventional resources to meet the electricity demands, this approach is not an effective resolution. Instead, sustainable energy sources such as PV, marine, geothermal, biofuel, and wind are being evaluated as substitutes [3]. These resources can mitigate the ecological damage, as well as social and financial impacts.

The evolution of electricity transmission and distribution systems can be seen in Figure 3.1. In conventional systems, the energy demand is met mainly by the use of fossil fuels. On the other hand, in modern systems, the energy supply is met by the use of renewable energy sources. The location of power plants and generation units increases the complexity of the transmission system. The traditional electricity grid suffers from various drawbacks, such as high transmission losses and environmental impact. The integration of renewable energy generation has significantly increased over the past couple of decades due to the technological advancements that have occurred in the field. In addition, the idea of distributed energy systems (DES) is

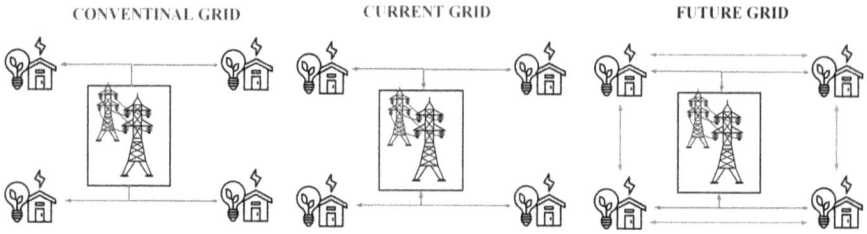

FIGURE 3.1 The evolution of the electricity infrastructure.

gaining popularity. These small-scale electric grids are composed of various distributed energy resources, which can be operated in a coordinated manner to meet the energy demand. These types of electric grids can operate in both islanded and grid-tied systems. In terms of distributed energy resources (DER), this includes generators, storage units, and load management systems [3, 4]. The objective of MGs is to mitigate the economic and social consequences of conventional fossil fuel-based generation by promoting the integration of greater amounts of renewable energy into the electricity grids.

The introduction of the implementation of microgrids in the power system has been seen as a way to envision the increased impact of sustainable energy. With the help of excess renewable generation, these types of electric grids can sell their energy back to the conventional system. It is evident that the expanding use of sustainable energy in the power grid is a contributing factor to the development of sustainable energy systems. Recent technological advancements have shown that it is feasible to achieve the objective of substituting fossil fuels with renewable energy sources. Unfortunately, sustainable energy systems are prone to encountering issues when it comes to modeling their operation due to the unpredictable weather conditions. In conventional systems, the energy sources are wind and solar energy, which are then combined with other factors such as power conditioning units and batteries.

A microgrid is expected to improve the power quality and provide various technical and environmental benefits to both the electric providers and the consumers [5]. It can also manage the imbalance between the generation and consumption of electricity in its local area. In islanded mode, the generation and demand of electricity should be in balance to ensure stability and quality. To address the power imbalance, the microgrid can increase its generation capacity. This can significantly increase the system's environmental and capital costs. The power usage of a residential unit within a microgrid fluctuates throughout the day. If a sustainable energy source such as wind or PV is used, this power plant is likely to operate non-linearly. With the introduction of more affordable solar and wind energy sources, microgrids can now address a portion of their electricity demand. The increasing number of households and the need for more sustainable energy sources have led to the development of energy storage systems & battery technologies. These innovations have created a huge amount of complexity in the management of energy in a microgrid. It is important to adopt a smart and reliable energy management strategy to address this issue.

3.2 MICROGRID STRUCTURE

A microgrid is a type of independent system that consists of a load, a distribution network, and an energy storage facility [6, 7]. In recent years, Mao et al. have presented the main findings of studies on the various aspects of microgrid systems [8]. The integrative power scheme, which includes a wide range of multi-energy carriers, is located near the user's side. These include the micro source, load, and energy storage systems. The goal of a microgrid is to optimize the energy supply and network in a region. Modern systems are becoming more complex, and they are often involved in the design and implementation of multiple power supply systems. In addition to the power supply, various types of integrated energy systems (EIS) are also commonly covered. These include the cooling, gas supply, and heating systems. Figure 3.1 shows the typical structure of a microgrid. Each of its components is described in detail below.

1. A generation system can provide direct energy transmission by operating without converting any energy. This can be done through various means, such as overhead lines and pipelines.
2. A distribution system is an essential component of an energy system, capable of realizing the transformation of energy. Various types of power generation equipment, such as solar and wind turbines, are capable of converting solar and wind energy into electric energy. They can also be used to produce hydrogen. In addition, fuel cells and electric heating pumps can help remove heat from a source of low temperature energy.
3. An energy storage system can help cut down on peak energy consumption and provide a solution to the mismatch between the cooling and heating loads of a gas turbine.
4. Residential, commercial, and industrial users are the energy end users. These entities consume energy, whereupon it is utilized for many applications, such as heating and power.

3.2.1 GENERATION OF ELECTRIC POWER SYSTEM

Different types of electrical power generation systems such as wind, solar, and diesel generators are utilized. The dispatch period is used to manage the load and compensate for the power generated.

3.2.1.1 PV Power Generation System

The outcome of PV power generators is different depending on the conditions of the sky. In clear sky conditions, they exhibit consistent regularity, while in non-clear sky regions, their output characteristics are complex. The external temperature of the PV system and illumination R are the two factors that influence its output power [9].

$$P_{\text{PV}} = P_{\text{PV,STC}} \times \frac{R}{R_{\text{STC}}} \times \left(1 - \lambda \times \left(T_a - T_r\right)\right) \tag{3.1}$$

$$T_a = T_{\text{amb}} + \frac{R}{R_{\text{STC}}} \times \left(T_{\text{NOC}} - 20\right) \tag{3.2}$$

The system outcome of a solar (PV) system is determined by the parameters of Eqs. (3.1) and (3.2), namely, the PPV, STC, and RSTC. The T_a, TNOC, and T_r are the benchmark values of the PV units, respectively, and these are the conditions under which the system operates normally.

3.2.1.2 Wind Turbine Energy Production

Wind energy is a renewable resource that is widely distributed. Due to the continuous technological progress that has occurred in the field, the cost of wind energy turbines has also been reduced [10–12].

$$P_W = \frac{1}{2} \rho \pi R^2 V^3 C_p \tag{3.3}$$

The air density of a wind turbine is determined by ρ the R span of its fan blade and the V for wind velocity. The energy conversion efficiency of a wind turbine is also determined by C_p, θ is the blade pitch angle and λ is blade tip speed ratio.

$$C_p = f(\theta, \lambda) \tag{3.4}$$

The blade tip speed ratio is a calculation that is used in calculating the air density of a wind turbine.

$$\lambda = \frac{W_W R}{V} \tag{3.5}$$

The mechanical velocity of a fan is measured by the W_w.

3.2.1.3 Internal Combustion (IC) Diesel Engine Generator

The fuel consumption of a diesel-powered generator is dependent on the quantity of fuel it utilizes.

$$F = F_0 \times Y_{gen} + F_1 \times P_{gen} \tag{3.6}$$

The consumption rate is expressed as F, the coefficient of intercept is F_0, the slope is F_1, the power is Y_{gen}, and the output is P_{gen}.

There are various constraints that a diesel-powered generator has when it comes to operating power.

$$L_{min} \leq \frac{P_{gen}}{Y_{gen}} \leq 1 \tag{3.7}$$

The minimum load rate that a diesel engine can operate at is L_{min}. The emissions of a diesel-powered generator can be expressed [13].

$$C_O(P_d) = a + b \times P_d + c \times P_d^2 \tag{3.8}$$

Based on practice, the values of the emission coefficients of a, b, and c of a diesel engine are as follows: 28.1444, 1.728, and 0.0017 [9].

3.2.1.4 DC Power by Battery System

A single lead–acid battery can discharge at a depth of about D. Its efficiency determines how much electric energy it can provide.

$$E_b = C_b U_b D \mu_b \times 10^{-3} \qquad (3.9)$$

A lead–acid battery's working voltage can be stable. Its current control is also constant, which indicates that the battery's output is constant.

$$P_b = C_b U_b \times 10^{-4} \qquad (3.10)$$

3.2.2 POWER DISTRIBUTION NETWORK

A power distribution network is composed of a bus bar and a branch that can supply energy to a consumer. It has a variety of cores arranged around a single flat/ribbon cable for data and electric power delivery. The consumer branch is integrated into a housing that has a contact device that's compliant with the principle of penetration technology.

3.2.3 ENERGY STORAGE UNIT

The continuous change in the load caused by renewable energy and the intermittency of the electricity supply can affect the system's stability. The ability to store energy in a microgrid can help maintain its system reliability and improve its power quality. There are two types of energy storage media: energy-type and power-type. The former includes lithium, sodium, and lead-acid batteries. On the other hand, the latter includes flywheel energy storage, supercapacitors, and magnetic energy storage. Due to the complexity of energy storage media, it is not feasible to create a single energy storage solution that can meet all of the applications. Instead, researchers are focused on developing hybrid energy storage systems.

A method for scheduling operation time for micro-electromechanical systems (MEMS) with energy storage was proposed by Tian et al. [14]. The energy flow is composed of a steady-state multi-carrier structure. The heat exchange and thermal capacity effects are comparable to that of energy retention. The energy transfer function and cyclic graph were used to construct the model. The results were then applied to a scheduling problem using the PWL approximation and convex relaxation. The results of the study revealed that the system model and method are generally suitable for the design and operation of a meaningful MEMS system.

An energy storage device can help improve the efficiency of microgrids by converting energy from one source to another. It can be categorized into four different types: gas, cold, electricity, and heat. Its characteristics are determined by the level of energy storage capacity, efficiency, and self-loss.

$$S_t^i = S_{t-1}^i \left(1 - \sigma^i\right) + \left(\eta_{cha}^i P_{cha,t}^i I_{cha,i}^i - \frac{P_{dis,t}^i I_{dis,t}^i}{\eta_{dis}^i} \right) \Delta t \tag{3.11}$$

The energy storage equipment's SOC refers to the state of charge. In addition, it refers to the power of the discharge and storage of energy. D is the length of time that the system is storing and discharging. The superscript t refers to the type of equipment used for different types of energy such as gas, electric, and cold. The status of the equipment is marked with a red and a black subscript. The energy storage loss rate is also denoted by these numbers.

3.2.4 ENERGY CONSUMERS

The operation of customer-side microgrids has been the focus of numerous studies. One of the most important factors that a microgrid needs to consider is the control of its power grid [15]. This is because the control strategy will determine the system's stability and allow it to operate efficiently. The choice of the appropriate control method and the enhancement of the system's operation are the two most important factors that can improve the efficiency and stability of a user-side microgrid. Currently, the most common control methods used in this type of system are master and slave type, peer-to-peer type, and structured. A consumer-side microgrid can operate in island mode with the help of master-slave control. A microgrid usually has multiple generators that provide constant voltage and frequency control. These devices reference the other units in the system for their voltage reference. On the other hand, a subordinate controller is responsible for regulating the operation of the generators. When a system is connected to a power grid, its voltage can be backed by the extensive power grid. Every DG in the system, including the primary control device, utilizes continuous power management throughout the operation [16]. In a peer-to-peer system, all the generators have equal status. A peer-to-peer microgrid can operate without the presence of a single distributed generator. This type of control strategy ensures that the other generators' control strategies are not affected by the access to one of them. Another type of control strategy is hierarchical, which provides a central control device that sends detailed instructions to the various parts of the system.

3.3 OPTIMIZATION FRAMEWORK

3.3.1 GOAL OF OPTIMIZATION

3.3.1.1 Singular Objective

The concept of microgrid management generally focuses on the reduction of operating expenses, energy optimization, and environmental conservation. However, there are conflicts between these goals that can prevent them from being achieved at the same time, which makes it challenging to implement the optimization. The following is a basic list of single objectives.

3.1.1.1.1   Energy Performance

The primary emphasis of the research on microgrid optimization lies in the energy management facets. For analysis of this case conducted by Kong et al. [17], a linear programming model was suggested, that can be used to determine the optimal combination of a gas turbine and a CCHP system. In another case study, El-Sharkh and colleagues proposed a cost-effective optimization solution for the multiple fuel cell power plants that are connected to the grid [18].

3.1.1.1.2   Economic Performance

The optimal economic dispatch of a microgrid can be achieved by establishing a goal that is based on the lowest operation cost. It assumes that the various factors that affect the electricity supply, such as generation power, demand, and exchange rates, are constant [19]. The main cost of running a microgrid is the maintenance cost of its renewable power generation system. This includes the cost of running a microgrid and the replacement of batteries.

$$\min F_1 = \sum_{i=1}^{24} \left[K_{OM} \times P_{pv-i} + K_{OM} \times P_{WT-i} \right] + \sum_{i=1}^{24} \left(F_{buy} - F_{sell} \right) + F_{bat} \qquad (3.12)$$

$$F_{buy} = f P_{buy-i} \qquad (3.13)$$

$$F_{sell} = h P_{sell-i} \qquad (3.14)$$

$$F_{bat} = S \frac{q}{n} \qquad (3.15)$$

The lowest cost of operation is achieved by $\min F_1$. K_{OM} is the maintenance and operation of a renewable power generating unit. P_{pvi} refers to the active power output from a PV cell. P_{WTi}, on the other hand, is the wind turbine's output in the first hour. The price that the large power grid pays for electricity during the first hour is known as the F_{buy}.

3.1.1.1.3   Environmental Protection

A microgrid is an organic component of a smart grid that can provide environmental protection. It can be used to meet the varying needs of its users, such as providing stable and predictable electricity, reducing greenhouse gas emissions, and maximizing the benefits of distributed generation [20]. In a study conducted by Yuan et al., two energy management strategies were identified, that can be utilized in a microgrid. They then analyzed the costs of emission reduction and energy conservation under these strategies [21].

The researchers identified the most cost-effective approach for reducing greenhouse gas emissions in a microgrid [22].

$$F = \frac{C_I + C_O - P_M}{E_O - E_M} \qquad (3.16)$$

The researchers calculated the total costing of system operation and maintenance of a microgrid using the following parameters: total construction cost, total income, and total emissions. The power generation side is responsible for the emission of CO_2 when the grid uses thermal power to meet its load demand.

$$E_{CO_2} = E_{FCO_2} E_E \qquad (3.17)$$

The total emissions of a microgrid are computed by taking into account the E_{CO_2}, E_{FCO_2}, kg/kWh. Researchers then calculated the optimal power generation capacity of the system to meet its emissions reduction goals.

3.3.1.2 Multi-Dimensional Objective

The multi-dimensional objective model is generally higher accurate when it comes to assessing the operation state of microgrids. It can also provide better environmental outcomes and lower the cost of operation [23]. In addition to the environment, factors such as system balance and economic benefits are taken into account to optimize the model.

3.1.1.1.4 The Weighting Factor Method

In a study conducted on the optimal scheduling of a CCHP-MG microgrid, researchers proposed a multi-timescale model that can meet the needs of its users while also suppressing the random fluctuations in the demand and supply sides [24]. This method can help improve the system's stability and economic performance. Yuan et al. [25] presented a method that can optimize the distribution network of energy storage by using a multiple-objective PSO methodology. They also used the top-of-the-system approach to analyze the optimal access scheme. In another study multi-objective scheduling system for MES, which can help increase the efficiency and stability of the electric energy storage system [26]. In a study, the researchers, led by Jiang, utilized a multi-objective model to analyze the controllable load of a microgrid based on the characteristics of its water heaters and air conditioners [27]. They then proposed a strategy that can optimize its effect through a non-discriminatory genetic algorithm.

3.1.1.1.5 Pareto Multi-objective Optimization

One of the most common concepts used in multi-objective optimization is the required condition. This condition determines the optimal solution for a given optimization point. On the other hand, the sufficient condition refers to the condition that the optimization method has to perform well. The optimal approach for implementing Pareto optimization is through a multi-objective optimization method. But a particular point may not be achieved.

The Zhou X. model was designed to address the scheduling problems of multi-microgrid electric systems [28]. It took into account the environment, energy interaction between the different microgrids, and load forecasting uncertainty. The model was transformed using a combination of Latin hypercube sampling and the colony chemotaxis method.

3.3.2 Decision Variable

The decision variables that influence the optimization of microgrids are analyzed in this study. These help the decision-makers identify the optimal strategy and implement it. The researchers utilized a new multi-criteria approach to make decisions. They established an assessment index framework comprising of 18 sub-criteria from different outlooks, such as efficiency, environmental protection, and electric grid reliability. They then calculated the target weights and subjective weights of the best and the worst methods [29]. The researchers presented a variety of evaluation tools and indicators for the design and implementation of CCHP microgrids. These include the energy utilization factor, exergy efficiency, ATE, and fuel saving ratio. Compared to other types of microgrids, a CCHP microgrid offers various environmental and economic benefits [30]. The planning process must take into account the various factors that can affect a facility's efficiency, such as how frequently load changes occur and how much power is generated from renewable sources.

3.3.3 Constraints

Different constraints exist in the optimization of microgrids. For instance, the technical limitations of a distribution system are different from those of a single microgrid. In order to enhance its reliability, a new approach was proposed by Pang et al. [31], which involves designing and operating an adaptable microgrid. This method can be used to expand or narrow the boundary of the system depending on its power generation and demand. In order to improve the efficiency of a microgrid, a model was proposed that can be used to perform various tasks, such as power exchange and balance the system [32]. For instance, the efficient planning of a combined heat and power microgrid can be improved by analyzing the limitations on the thermal and electric power outputs of the power station. the multiple economic constraints of a microgrid based on the power station's output and the balance between thermal and electric power.

The optimization of a microgrid's operation cost is often associated with the economic and environmental protection aspects of its operation. Besides its power balance and generation capacity, other limitations such as the location of its power storage system and transmission capacity are also taken into account to improve its efficiency.

3.3.3.1 Power Balance Limitations

The power balance of a distributed power system is computed based on its minimum operational cost.

$$\sum_{i=1}^{N} P_{\text{gen},j} + P_{\text{buy}}(t) - P_{\text{sell}}(t) = P_{\text{load}} \tag{3.18}$$

In a microgrid, the $P_{\text{gen},j}$ refers to the power that the generating units, such as PV, energy storage provide at any given time.

3.3.3.2 Generation Capacity Constraints

To ensure that its operations are reliable, every generation of microgrid has to meet its output constraints.

$$P_{DGi}^{min} \leq P_{DGi} \leq P_{DGi}^{max} \ (i = 1,2,\ldots,N) \tag{3.19}$$

The maximum and minimum output of a generator are referred to as DGi, in corresponding order.

Transmission capability Limitation with Electric Small-scale grid and Macro grid system.

The enhanced number of distributed power generation facilities, such as microgrids, is driving the need for collaboration between these facilities and large grids.

$$P_{Line}^{min} \leq P_{Line} \leq P_{Line}^{max} \tag{3.20}$$

In terms of transmission capacities, the minimum and maximum are determined by P_{Line}^{min} and P_{Line}^{max}.

3.3.3.3 Microgrid Energy Storage System Location Constraints

When it comes to optimizing the energy storage system's location and capacity, the appropriate range of voltage deviation should also be taken into account.

1. Permissible voltage fluctuation range

$$V_{min} \leq V_{ik} \leq V_{max} \tag{3.21}$$

The maximum and minimum voltage thresholds of a node are determined by V_{max} and V_{min}. The value of the node at the kth moment is known as V_{ik}.

$$P_{store\_minimum} \leq P_{store} \leq P_{store\_maximum} \tag{3.22}$$

In terms of energy storage power, P_{store_min} refers to the minimum power of a facility's energy storage system. On the other hand, P_{store_max} is the system's peak performance.

2. Constraints of the energy storage unit for maintaining energy equilibrium

$$\int_0^T P_{store}(i) \cdot dt = 0 \tag{3.23}$$

3. Power constraint of system

$$P = \sum_{i=1}^{Nbus} P_{load,i} - \sum_{j=1}^{N_{DG}} P_{DG,j} - \sum_{k=1}^{Nstore} P_{store,k} \tag{3.24}$$

The injected power of a grid is known as P. It is followed by the load and PDG, which are the outputs of the first node. The number of linked DGs is called NDG. $P_{store,k}$, is the output of a storage device chosen to provide electricity.

3.4 PERFORMANCE ENHANCEMENT ALGORITHM

One of the most challenging and important research topics in the field of microgrids is the optimization of their configuration. Numerous studies have been conducted on this matter in order to improve their reliability and economy. The literature presents various algorithms that can be utilized to solve this problem. One of the most commonly used algorithms is the genetic algorithm [33–38], which is commonly used to identify the optimal size of components for a microgrid. For instance, in a study conducted by Li et al. [38], who formulated a GA to find the lowest cost of building a microgrid. Bin et al. [33] utilized the NSGA-II algorithm to improve the arrangement of a hybrid AC to DC microgrid. Aiswariya et al. [39] utilized an optimization tool known as the SA algorithm to resolve the battery scheduling issue of a residential microgrid. Zhang et al. developed a search algorithm that combines the PSO and SA algorithms to perform an optimization on a stand-alone microgrid [33–41]. Due to the fuzzy environment in which decisions are made, many such decisions are required. Zhao et al. developed an integrated model that takes into account multiple criteria to perform an optimal evaluation of a battery energy storage solution [29]. Researchers have utilized various methods to enhance the adaptability and flexibility of microgrids, especially considering their robust nature. In a study conducted on the operation of microgrids, Ebrahimi [42] utilized a decision-driven model for optimization. Aside from this, other methods such as the Grey Wolf and moth-flame optimization are also utilized to enhance the efficiency of microgrids. For instance, by using the GWO technique, researchers were able to reduce the energy costs of a microgrid. In a study conducted on the operation of a microgrid, researchers utilized the moth flame optimization technique to improve its efficiency. They then developed a new model that combines the best and worst methods.

Various optimization techniques are commonly used to improve the efficiency of a system. These include the GA, PSO, and fuzzy algorithm.

In the next section, we discuss the various techniques used in optimizing microgrids.

3.4.1 EVOLUTIONARY METHODOLOGY

3.4.1.1 Adaptive Optimization Technique for Microgrids System

The distributed energy system utilizes the power of photovoltaics and wind turbines. The generation and output of these two types of energy are shown in Eqs. (3.25) and (3.26), respectively.

$$P_s(t) = \phi\left(V_m(\beta), I_m(\beta)\right) \tag{3.25}$$

$$P_w = \begin{cases} 0 & v \le v_{in} \text{ or } v \ge v_{out} \\ av + b & v_{in} < v \le v_{rate} \\ P_{rate} & v_{rate} < v < v_{out} \end{cases} \tag{3.26}$$

The variable in Eq. (3.25) refers to the wind turbine's pitch angle. In Equation (3.26), the Parameter is wind velocity. The others are output and v_{rate}.

In order to improve the power supply of a microgrid, the type of battery it uses is very important. The SOC (state of charge) of a battery can be calculated by Eq. (3.27).

$$S_t = S_0 + \frac{\sum_{t=1}^{T} P_{cha,t} X_t \Delta t - \sum_{t=1}^{T} P_{dis,t} Y_t \Delta t}{E_b} \tag{3.27}$$

The SOC of a battery is determined by the following equation: $P_{cha,t}$, and P_{dis}. These three forms denote the power and charging of the battery during certain periods. The capacity and total of a battery are expressed in terms of Y_t and T.

The prices of PV and wind power are shown in terms of C_w and C_v after being sold. The selling price of a microgrid is shown in terms of C_S and C_b.

$$C_w = C_{wn} - C_s + C_b = C_{wn} + (C_b - C_s) \tag{3.28}$$

$$C_v = C_{vn} - C_s + C_b = C_{vn} + (C_b - C_s) \tag{3.29}$$

Economic optimization framework and objective criteria for microgrids.

The objective of the Analysis is to ensure that the interests of various subjects, such as users and power grids, are protected. It performed three total functions: maximizing the grid's benefits, minimizing the cost of renewable energy, and ensuring that the average electricity price of users is kept within a certain range.

$$C_1 = \max \sum_{t=1}^{T} C_b^t \cdot \max \left(P_{G'}^t, 0 \right) \tag{3.30}$$

$$C_2 = \min \sum_{t=1}^{T} C_w \times P_w^t + C_v \times P_v^t \tag{3.31}$$

$$C_3 = \min \frac{1}{\sum_{t=1}^{T} P_L^t} \sum_{t=1}^{T} C^t \tag{3.32}$$

The cost of electricity purchased in a period of time is calculated by C_b^t in Eq. (3.30). The quantity of electricity that the power grid can supply to its microgrid is also represented by P_G^t. Eq. (3.31) illustrates the representation of the cost of renewable energy as C_r^t. The outputs of photovoltaics and wind turbines are shown as P_{vt} and P_{wt}.

In Eq. (3.32), the energy cost is calculated by taking into account the various factors that affect the power supply.

3.4.1.2 Constraints

There are various constraints that affect the power balance of a microgrid. The constraints of renewable energy, the power balance, and the discharge and charge of batteries are also taken into account.

$$P_G^t = P_L^t - P_W^t - P_V^t - P_E^t \qquad (3.33)$$

The output power of a battery is represented by P_E.

$$\begin{cases} P_w \le P_{wt} \\ P_v \le P_{vt} \end{cases} \qquad (3.34)$$

$$S_{min}^t \le S^t \le S_{max}^t \qquad (3.35)$$

The SOC's upper and lower limits are represented by S_{max}^t and S_{min}^t. Limitations of the battery should also be taken into account, such as its discharge and charge power.

3.4.2 PARTICLE SWARM OPTIMIZATION

The PSO algorithm, which is derived from the group cooperation algorithm used by bird foraging, is an iterative optimization method that can be used in optimizing a microgrid. Unlike a general algorithm, it only requires a constant speed update process and doesn't require modifying other parameters. This makes it easier to implement and is often used in studies related to microgrid optimization.

In 2016, a study was conducted by researchers to investigate the feasibility of using ice storage and thermochemical storage to meet the cooling and heating needs of users during severe weather conditions [43]. The study focused on minimizing the cost of these systems by using pollutant emissions constraints.

3.4.2.1 Optimization Model

The study also utilized a microgrid with a main grid and a converter. It was composed of various electric loads, such as cooling and heating loads. Its power system included a variety of equipment, such as wind turbines, a fuel cell, a diesel generator, and battery storage. Wind turbines, PV cells, and fuel cells operate on natural resources. The other equipment, such as the diesel generator and microturbine, require fuel input. The TCS (thermochemical storage system) utilizes the collected and stored ice storage energy to provide air conditioning.

3.4.2.2 Objective Function

The study's objective was to meet the demand for electricity with the lowest emissions and minimize the operation costs of the system. It also considered the constraints of the microgrid to ensure that it can meet its goals.

3.4.2.2.1 Operation Cost

The main grid will supply the additional power if the local consumers' energy demand exceeds its supply. The fee that a microgrid has to pay for electricity is determined by the amount of power that it generates. If the system is capable of supplying enough electricity, the grid will buy it.

$$CF(P) \models \sum_{i=1}^{N} \left(C_i F_i (P_i) + OM_i (P_i) + \text{DCPE}_i - \text{IPSE}_i \right) \tag{3.36}$$

The operating costs of a generator are shown in terms of $CF(P)$. The maintenance and operating costs of an ith generating unit are shown in terms of OM_i. DCPE_i, on the other hand, indicates the cost of electricity purchased from the grid. IPSE_i, on the other hand, shows the income generated by selling the extra power.

3.4.2.2.2 Emission Level

One of the goals of the study was to minimize the pollutants that can be released into the air to avoid ozone depletion.

$$E(P) = \sum_{i=1}^{N} 10^{-2} \left(x_i + y_i P_i + z_i P_i^2 \right) + \alpha_i \exp(\beta_i P_i) \tag{3.37}$$

In Eq. (3.37), x_i, y_i, z_i, α_i, and β_i represent non-negative coefficients associated with the emission characteristic of the ith generator.

3.4.2.2.3 Thermochemical Heat Energy Storage

Thermally reversible reactions are responsible for the thermal energy conversion of solar collectors into electrical power in the form of TCS.

$$p = c_p A v (T_L - T_0) = C_p \rho A v (T_R - T_0) \left[1 - \exp(-hSL / c_p \rho A v) \right]^1 \tag{3.38}$$

The heat transfer coefficient is expressed as h. The circumference, cross section, and length of the tube are shown as A, L, S, and V. The flow rate of the fluid is indicated as V, and the specific heat is defined as c_p. The reactor's temperature is also indicated as T_R.

3.4.2.2.4 Battery Storage

The SOC of a battery is very important in its life cycle. A constraint function can be utilized to increase its life cycle.

$$\begin{cases} \text{SOC}(T+1) = \text{SOC}(T) - \left(\eta_{bc} I_{bc}(T) + \dfrac{1}{\eta_{bd}} I_{bd}(T) \right) P_b(T) \Delta t / C_b \\ I_{bc}(T) P_{c\min} + I_{bd}(T) P_{d\min} \le P_b(T) \le I_{bc}(T) P_{c\max} + I_{bd}(T) P_{d\max} \\ I_{bc}(T) + I_{bd}(T) \in (0,1) \\ \text{SOC}_{\min} \le \text{SOC}(T+1) \le \text{SOC}_{\max} \end{cases} \tag{3.39}$$

The efficiency and charging of a battery are shown in Eq. (3.39). I_{bd} and I_{bc} represent the discharging current and charging respectively.

3.4.2.3 Simulation of Photovoltaic and Wind-type Power Systems

The PV system can be affected by diverse aspects, that module's heat level and weather conditions. It is presumed that the S_{stc} value is 1000 W/m^2 and the T_{stc} value is 25 C. The current I of the PV array is shown as follows.

$$I = I_{SC}\left[1 - C_1\left(e^{\frac{V}{C_2 V_{OC}}} - 1\right)\right]$$

$$C1 = \left(1 - \frac{I_m}{I_{SC}}\right)e^{-\frac{V_m}{C_2 V_{OC}}} \qquad (3.40)$$

$$C2 = \left(\frac{V_m}{V_{OC}} - 1\right)\Big/ \ln\left(1 - \frac{I_m}{I_{SC}}\right)$$

Short-circuit currents in (3.40) are I_{SC}, V_m, and V_{max}. I_m is the current at P_{max}, while V_m, and V correspondingly, the voltage at the specified time.

The power generated by wind turbines is analogous to that discussed in Section 3.2.1.2 The rated power of these turbines, as well as the wind speed, dictate their output power. The speed is also influenced by two factors.

$$f(v) = \frac{k}{c}\left(\frac{v}{c}\right)^{k-1} e^{\left(-\frac{v}{c}\right)} \qquad (3.41)$$

The scale and shape parameters are respectively c and k, while the wind speed is determined by v.

3.4.3 FUZZY DECISION

The concept of fuzzy decision-making is a type of mathematical theory that can be utilized in various applications. It can be applied to solve the configuration problems of hybrid microgrids. an improved version of the TA-MaEA algorithm for optimizing the configuration of a hybrid microgrid system [44]. The researchers conducted the study to find ways to reduce the cost of running a hybrid microgrid system while also reducing its environmental pollutants.

3.4.3.1 Objective Functions

The HMS is same as to the analysis in that it includes a diesel-powered generator, a photovoltaic system, and wind turbines. When the electricity generated by these components is insufficient to meet the demand, a battery is used. The four optimization objectives are discussed below.

3.4.3.1.1 Costs

The operating costs of the system are shown in terms of the expenses of producing electricity from the wind power generators and photovoltaic system, as well as the investment of maintaining the electricity exchange between the main grid and the microgrid.

$$C = \sum_{t=1}^{T} \left(C_g(t) + C_b(t) + P_g(t) \times c_g + P_f(t) \times c_f \right) \tag{3.42}$$

The responsibilities of the main grid and the investment of generating electricity from wind power generator and photovoltaic system are shown in terms of c_g and c_f.

$$C_g = A_w \times P_w(t) \times c_w + A_{pv} \times P_{pv}(t) \times c_{pv} \tag{3.43}$$

$$C_b = P_{b\_dc}(t) \times c_{dc} - P_{b\_ch}(t) \times c_{ch} \tag{3.44}$$

$$P_g(t) = \lambda \times \left(P_{load}(t) - P_{all}(t) - P_b(t) \right) \tag{3.45}$$

$$P_f(t) = 0.246 \times P_{dg}(t) + 0.08415 \times P_{f\_r} \tag{3.46}$$

The equation PV and wind turbine output per unit of time are represented by P_w, P_v, and T. respectively. The cost of electricity generated by these two types of turbines is shown by the c_{pv} and c_w. The output per unit of time of a battery is represented by $P_b(t)$, $P_d(t)$, and $P_b(t)$. The cost of electricity that a diesel generator generates is shown by $c\,ch$, while the profit that a storage system makes from supplying additional power is shown by c_{dc}. The output per load demand split is shown by P_{load}, while the current level in the battery is displayed by P_b. The rating of the diesel generator is shown by P_f.

3.4.3.1.2 Reliability

LPSP refers to a situation wherein the amount of electricity that has been purchased and produced cannot satisfy the demand.

$$\text{LPSP} = \frac{\sum_{t=1}^{T} (P_{load}(t) - P_{all}(t) + P_{b\_min} + P_{dg}(t)}{\sum_{t=1}^{T} P_{load}(t)} \tag{3.47}$$

The minimum amount of electricity that can be stored in a battery is represented by P_{b_min}.

3.4.3.1.3 Pollutant Emissions

The PE function can be used to analyze the emission of pollutants by connecting the generated electricity to the total emissions.

$$\text{PE} = \alpha + \beta \times \sum_{t=1}^{T} P_{dg}(t) + \gamma \times \left(\sum_{t=1}^{T} P_{dg}(t) \right)^2 \tag{3.48}$$

The characteristics of a microgrid are represented by the α, β, and γ symbols.

3.4.3.1.4 Power Balance

The connection of a microgrid to the main grid can improve its self-producing and selling capacity. This can help the system operate more efficiently.

$$PB = \frac{1}{T} \left(\sum_{t=1}^{T} P_{\text{all}}(t) - P_{\text{load}}(t) \right)^2 \tag{3.49}$$

3.4.3.2 Constraints

The initial range of the power supply for wind turbines should be determined according to the situation. Output power can be expressed as follows.

$$P_w(t) = \begin{cases} 0, & V(t) > V_{\text{cut\_out}} \\ \phi \times V^3(t) - \varphi \times P_{w\_r}, & V_{\text{cut\_in}} < V(t) < V_{\text{rated}} \\ P_{w\_r}, & V_{\text{rated}} < V(t) < V_{\text{cut\_out}} \end{cases} \tag{3.50}$$

The output power of a PV panel is expressed as follows: V is the current wind speed, P_w is the standard power, and V_{rated} is the rated power. The initiation and cessation velocities of the wind generator are shown as $V_{\text{cut\_in}}$ and $V_{\text{cut\_out}}$.

$$P_{\text{pv}}(t) = P_{\text{pv}\_} \times \frac{G(t)}{G_{\text{ref}}} \left[1 + K_t \left(T(t) + \left(0.0256 \times G(t) \right) \right) - T_{\text{ref}} \right] \tag{3.51}$$

The horizontal irradiance of the PV power is shown as $G(t)$, while the ambient temperature is indicated by $T(t)$. The battery status is also shown by comparing the G_{ref} and K_t values.

$$P_{b\_ch}(t) = P_{\text{all}}(t) - P_{\text{load}}(t) \tag{3.52}$$

$$P_{b\_dc}(t) = \left(P_{\text{load}}(t) - P_{\text{all}}(t) \right) \vee \left(P_b(t) - P_{b\_min} \right) \tag{3.53}$$

The exchange function of batteries can also be formulated according to the following.

$$P_b(t+1) = \left(P_b(t) + P_{b\_ch}(t) \right) \vee \left(P_b(t) - P_{b\_dc}(t) \right) \tag{3.54}$$

The storage capacity and loading area of batteries for wind generator power systems and solar PV system should be determined within a certain range.

$$A_{w\_min} < A_w < A_{w\_max} \tag{3.55}$$

$$A_{pv\_min} < A_{pv} < A_{pv\_max} \tag{3.56}$$

$$p_{b\_min} < P_b(t) < P_{b\_cap} P_{b\_cap} < P_{b\_max} \tag{3.57}$$

3.4.4   Robust Method

The optimization of microgrid systems can help enhance their independence and adaptability, as well as reduce the likelihood of natural disasters. This topic has been studied in detail and is a vital part of any microgrid project. In order to minimize the variability in the costs associated with implementing a robust microgrid project, Yu and his team utilized a multi-objective approach. The goal of the study is to reduce the project's cost and improve its environmental performance [45].

3.4.4.1   Resilient Economic Efficiency for Framework for Microgrid Development

The study proposes a protopty]pw that utilizes an IC engine, a combustion turbine, a fuel cell, and a power generator. These innovations can provide local heat and power to users through their distributed energy resources. Most of these resources are renewable, and they can be stored in different energy storage facilities. Other components of the study include wind turbines and photovoltaic panels.

The following is a multi-objective function that takes into account the robust optimization.

$$f = u_E \times L + u_{WC} \times (1 - L) 0 \le L \le 1 \tag{3.58}$$

where $u_E = \sum_s p_s \cdot c_s$, and $u_{WC} \ge c_s$

3.4.4.2   Boundaries or Limitations

3.4.4.2.1   Financial Equilibrium

The total expenditure on electric power is computed by taking into account the various factors that affect the cost of power generation, such as the investment cost, maintenance and operation cost, and fuel cost.

$$c_s = \mathrm{cinv}_s + \mathrm{com}_s + c\,\mathrm{cfull}_s + \mathrm{cctax}_s + \mathrm{cbuyn}_s - \mathrm{rsal}_s \tag{3.59}$$

The revenue that a microgrid generates when it sells electricity to the utility grid is also taken into account.

The terms of the polynomial Eq. (3.59) can be given as Eqs. (3.60)–(3.65):

$$\mathrm{cinv}_s = \sum_i \mathrm{cap}_i \cdot \mathrm{FC}_i \cdot \left(\mathrm{Ir} / \left(1 - \left(1 / (1 + \mathrm{Ir})^{LT_i} \right) \right) \right)$$
$$+ \sum_q \mathrm{cap}_q \cdot \mathrm{FCS}_q \cdot \left(\mathrm{Ir} / \left(1 - \left(1 / (1 + \mathrm{Ir})^{QT_q} \right) \right) \right) \tag{3.60}$$

$$\mathrm{com}_s = \sum_i \mathrm{cap}_i \cdot \mathrm{OMF}_i + \sum_{i,m} \mathrm{allot}_{i,m,s} \cdot \mathrm{OMV}_i \tag{3.61}$$

$$\text{cfuel}_s = \sum_{i,m,f} \left(\text{fder}_{i,f,m,s} + \text{fboi}_{f,m,s} \right) \cdot \text{FP}_{f,s} \tag{3.62}$$

$$\text{cbuyn}_s = \sum_{m} \text{ebuyn}_{m,s} \cdot \text{EP}_{m,s} \tag{3.63}$$

$$\text{rsal}_s = \sum_{i,m} \text{esal}_{i,m,s} \cdot \text{SP}_s \tag{3.64}$$

$$\text{cctak}_s = \begin{cases} \left(dm_s - \text{CLIM} \right) \cdot \text{CTAXs} & dm_s \geq \text{CLIM} \\ 0 & dm_s \leq \text{CLIM} \end{cases} \tag{3.65}$$

3.4.4.2.2 Load Balance

The power load of a microgrid should be determined by taking into account the various components of its operation, such as batteries and DERs.

$$\text{EL}_{p,q} \cdot T \leq \sum_{i} \text{eder}_{i,p,q} + \text{disc}_{\text{batt},p,q} - \sum_{i} \text{save}_{i,\text{batt},p,q} + \text{ebuyn}_{p,q} - \sum_{i} \text{esal}_{i,p,q} \tag{3.66}$$

$$\text{EL}_{m,s} \cdot T \cdot \varphi_s \leq \sum_{i} \text{eder}_{i,p,q} + \text{disc}_{\text{batt},p,q} - \sum_{i} \text{save}_{i,\text{batt},p,q} \tag{3.67}$$

$$\text{HL}_{m,s} \cdot T \leq \sum_{i} h\text{der}_{i,p,q} + \text{disc}_{\text{thermo},p,q} - \sum_{i} \text{save}_{i,\text{thermo},p,q} + h\text{boi}_{p,q} \tag{3.68}$$

$$\text{HL}_{m,s} \cdot T \cdot \varphi_s \leq \sum_{i} h\text{der}_{i,p,q} + \text{disc}_{\text{thermo},p,q} - \sum_{i} \text{save}\, e_{i,\text{thermo},p,q} \tag{3.69}$$

The energy generation of a microgrid should fulfill a specific proportion of its energy requirements. The supply and demand balance of heat are also indicated by Eqs. (3.68) and (3.69).

3.4.4:2.3 Energy Allocation for DER

The distributed energy resources (DER) of a microgrid are composed of three components. These include the power that is used to satisfy the load, the excess power that is stored in the batteries, and the power that is sold to the utility grid.

$$\text{allot}_{i,m,s} = \text{eder}_{i,m,s} + \text{save}_{i,\text{batt},m,s} + \text{easl}_{i,m,s} \tag{3.70}$$

The constraint is shown by plotting the various components of a microgrid's distributed energy resources.

$$\eta \cdot \text{EL}_{m,s} \cdot T \cdot y_{i,m,s} \leq \text{allot}_{i,m,s} \leq y_{i,m,s} \cdot A \tag{3.71}$$

In terms of the constraints that are related to the wind turbine electricity generation and solar PV system, they bear resemblance to those observed in prior research.

The capacity of each DER can be calculated by taking into account the various regulations and budget constraints that affect its installation.

$$allot_{i,m,s} / T \leq cap_i \leq MAXE_i \tag{3.72}$$

Energy storage limitation

When a DES is chosen to store heat or power, the amount of energy discharged can be computed as Eq. (3.73).

$$disc_{q,m,s} = \varepsilon_q \cdot lose_{q,m,s} \tag{3.73}$$

The monthly inventory balance of DES can be calculated by taking into account the various items in its warehouse.

$$store_{q,m,s} = \begin{cases} 1/2 \cdot caps_q \cdot T & m \in \{Jan\} \\ \zeta \cdot store_{q,m-1,s} + \sum_i save_{i,q,m}m-1,s - \sum_i disc_{q,m-1,s} & m \notin \{Jan\} \end{cases} \tag{3.74}$$

It is assumed that January marks the start of the new year, and that there will be half of the storage capacity needed. The monthly capacity of DES is calculated based on the rough specification.

The total energy that has been discharged is shown below.

$$lose_{q,m,s} \leq store_{q,m,s} \leq caps_q \cdot T \leq MAXS_q \tag{3.75}$$

The store q, m, s, and $caps_q$ represent the Stock of DES and the monthly production capacity respectively. The upper bound is represented by the MAXSq.

$$ysaves_{q,m,s} + y disc_{q,m,s} \leq 1 \tag{3.76}$$

$$ysave_{q,m,s} \cdot MINS_q \leq \sum_i save_{i,q,m,s} \leq ysve_{q,m,s} \cdot MAXS_q \tag{3.77}$$

$$ydisc_{q,m,s} \cdot MINS_q \leq disc_{i,q,m,s} \leq ydisc_{q,m,s} \cdot MAXS_q \tag{3.78}$$

The bounded limits of discharged and stored energy are given as Eqs. (3.77) and (3.78) respectively.

3.5 OTHER ALGORITHMS

Besides the usual algorithms, other suitable ones have also been utilized in studies related to the optimization of microgrids. Some of these include the moth flame algorithm and the GWO algorithm.

Mirjaili's [46, 47] meta-heuristic algorithm known as GWO is useful in optimizing the energy management of a microgrid. It can be used to detect and prevent premature convergence, and it has the advantage of having small computations and solution accuracy. In addition, it can be used to improve the efficiency of BESSs. According to a study conducted by Kutaiba et al. [48] Intermittency is one of the main factors that has made BESSs an integral feature of microgrids. The authors of this study proposed an approach that uses the GWO algorithm to address the requirements of microgrids. It ensures that the utilization of renewable resources is maximized, and conventional fuel consumption is minimized.

3.5.1 Optimization Criterion

The analysis sought to minimize overall costs while meeting all the requirements. The overall expense of electricity production from various sources can be ascertained.

$$\min F(X) = \sum_{t=1}^{T} f_t + OM_{DG} + TCPD_{BES}b \tag{3.79}$$

Where:

$$ft = \sum_{t=1}^{T} Cost_{grid,t} + Cost_{DG,t} + Cost_{BES,t}$$
$$+ SUC_{MT,t} + SUC_{FC,t} + SDC_{MT,t} + SDC_{FC,t} \tag{3.80}$$

Each of the elements of the polynomial can be specified as follows:

$$Cost_{grid,t} = \begin{cases} B_{grid,t}P_{grid,t} & P_{grid,t} > 0 \\ (1-tax)B_{grid,t}P_{grid,t} & P_{grid,t} < 0 \\ 0 & P_{grid,t} = 0 \end{cases}$$

$$Cost_{DG,t} = B_{MT,t}P_{MT,t}\mu_{MT,t} + B_{FC,t}P_{FC,t}\mu_{FC,t}$$
$$Cost_{BES,t} = B_{BES,t}P_{BES,t}\mu_{BES,t}$$
$$SUC_{MT,t} = SU_{MT} \times \max(0, \mu_{MT,t} - \mu_{MT,t-1})$$
$$SUC_{FC,t} = SU_{FC} \times \max(0, \mu_{FC,t} - \mu_{FC,t-1})$$
$$SDC_{MT,t} = SD_{MT} \times \max(0, \mu_{MT,t-1} - \mu_{MT,t})$$
$$SDC_{FC,t} = SD_{FC} \times \max(0, \mu_{FC,t-1} - \mu_{FC,t})$$
$$OM_{DG} = (OM_{MT} + OM_{FC} + OM_{PV} + OM_{WT}) \times T \tag{3.81}$$

The total operation cost of a microgrid is computed by taking into account the costs of running the electricity distribution grid and the costs of running fuel cells and micro-turbines. It also takes into account the maintenance and start-up costs of the generators and the storage batteries.

$$Cost_{BES} = (FC_{BES} + MC_{BES}) \times C_{BES,max} \tag{3.82}$$

3.5.2 CONSTRAINTS

The operational cost of electricity is computed by taking into account the various constraints of the electric load demand, such as the DG, grid, operation reserve, and BES, and their respective equations.

$$P_{D,t} = P_{MT,t}u_{MT,t} + P_{FC,t}u_{FC,t} + P_{PV,t} + P_{WT,t} + P_{BES,t}u_{BES,t} + P_{grid,t} \tag{3.83}$$

$$\begin{cases} P_{MT,min} \leq P_{MT,t} \leq P_{MT,max} \\ \overline{P}_{FC,min} \leq P_{FC,t} \leq P_{FC,max} \\ P_{PV,min} \leq P_{PV,t} \leq P_{PV,max} \\ P_{WT,min} \leq P_{WT,t} \leq P_{WT,max} \end{cases} t = 1,\cdots,T \tag{3.84}$$

$$P_{grid,min} \leq P_{grid,t} \leq P_{grid,max} \, t = 1,\ldots,T \tag{3.85}$$

$$P_{MT,t}u_{MT,t} + P_{FC,t}u_{FC,t} + P_{PV,t} + P_{WT,t} + P_{BES,t}u_{BES,t} + P_{grid,t} \geq P_{D,t} + OR_t \tag{3.86}$$

There are two types of battery constraints: discharging and charging. The limits of the power that can be delivered by the batteries in these modes are shown.

$$P_{BES,min} \leq P_{BES,t} \leq P_{BES,max} t = 1,\ldots,T \tag{3.87}$$

The minimum and maximum power of a BESS are also shown.

$$P_{BES,t\,max} = \max \left\{ P_{BES,max}, \frac{(C_{BES,t} - C_{BES,min})\Delta d}{\Delta t} \right\} t = 1,\ldots,T \tag{3.88}$$

$$P_{BES,t\,min} = \min \left\{ P_{BES,max}, \frac{(C_{BES,t} - C_{BES,min})\Delta d}{\Delta t} \right\} t = 1,\ldots,T \tag{3.89}$$

3.6 CONCLUSION

The target of this research is to summarize the different advancements in the functioning of microgrids. After reviewing the system configuration, we first introduced a standard microgrid system. It comprises an energy storage device, a power generation system, a distribution system, and an end-user energy system. We then introduced the microgrid optimization framework, which contains the necessary decisions and constraints to achieve the best possible performance. After conducting a comprehensive review of the various optimization algorithms used in the operation of microgrid systems, we discovered that the most commonly used are the SA and GA algorithms. In the future, the development of optimization algorithms for microgrid systems will be significantly affected by two factors. First, due to the lack of a comprehensive understanding of the various aspects of the system, it will be difficult to implement effective methods and models. Second, due to the complexity of the operating environment, it will be even more difficult to optimize the system. Artificial intelligence and machine learning are likely to be utilized in this area to address this issue. Although the development of microgrid systems' optimization algorithms has been rapid, it is still not yet feasible to fully optimize them.

REFERENCES

[1] S. R. Madeti and S. N. Singh, "Monitoring system for photovoltaic plants: A review," *Renewable and Sustainable Energy Reviews*, vol. 67, pp. 1180–1207, Jan. 2017, doi: 10.1016/J.RSER.2016.09.088

[2] H. Rezk, I. Tyukhov, M. Al-Dhaifallah, and A. Tikhonov, "Performance of data acquisition system for monitoring PV system parameters," *Measurement*, vol. 104, pp. 204–211, Jul. 2017, doi: 10.1016/J.MEASUREMENT.2017.02.050

[3] N. L. Panwar, S. C. Kaushik, and S. Kothari, "Role of renewable energy sources in environmental protection: A review," *Renewable and Sustainable Energy Reviews*, vol. 15, no. 3, pp. 1513–1524, Apr. 2011, doi: 10.1016/J.RSER.2010.11.037

[4] C. Marnay et al., "Microgrid evolution roadmap engineering, economics, and experience.

[5] F. Katiraei, R. Iravani, N. Hatziargyriou, and A. Dimeas, "Microgrids management," *IEEE Power and Energy Magazine*, vol. 6, no. 3, pp. 54–65, 2008, doi: 10.1109/MPE.2008.918702

[6] R. H. Lasseter and P. Paigi, "Microgrid: A conceptual solution," *2004 IEEE 35th Annual Power Electronics Specialists Conference (IEEE Cat. No.04CH37551)*, vol. 6, pp. 4285–4290, 2004.

[7] B. Hong, W. Miao, Z. Liu, and L. Wang, "Architecture and functions of micro-grid energy management system for the smart distribution network application," *Energy Procedia*, vol. 145, pp. 478–483, Jul. 2018, doi: 10.1016/J.EGYPRO.2018.04.095

[8] B. Lei et al., "A review of optimization for system reliability of microgrid," *Mathematics*, vol. 11, no. 4, 2023, doi: 10.3390/math11040822

[9] Y.-Y. Hong, Y.-M. Lai, Y.-R. Chang, Y.-D. Lee, and P.-W. Liu, "Optimizing capacities of distributed generation and energy storage in a small autonomous power system considering uncertainty in renewables," *Energies (Basel)*, vol. 8, no. 4, pp. 2473–2492, 2015, doi: 10.3390/en8042473

[10] A. Arcos Jiménez, C. Q. Gómez Muñoz, and F. P. García Márquez, "Machine learning for wind turbine blades maintenance management," *Energies (Basel)*, vol. 11, no. 1, 2018, doi: 10.3390/en11010013

[11] F. P. García Márquez, I. Segovia Ramírez, and A. Pliego Marugán, "Decision making using logical decision tree and binary decision diagrams: A real case study of wind turbine manufacturing," *Energies (Basel)*, vol. 12, no. 9, 2019, doi: 10.3390/en12091753

[12] F. P. García Márquez, A. Pliego Marugán, J. M. Pinar Pérez, S. Hillmansen, and M. Papaelias, "Optimal dynamic analysis of electrical/electronic components in wind turbines," *Energies (Basel)*, vol. 10, no. 8, 2017, doi: 10.3390/en10081111

[13] H. Liu, D. Li, Y. Liu, M. Dong, X. Liu, and H. Zhang, "Sizing hybrid energy storage systems for distributed power systems under multi-time scales," *Applied Sciences*, vol. 8, no. 9, 2018, doi: 10.3390/app8091453

[14] L. Tian, L. Cheng, J. Guo, and K. Wu, "System modeling and optimal dispatching of multi-energy microgrid with energy storage," *Journal of Modern Power Systems and Clean Energy*, vol. 8, no. 5, pp. 809–819, 2020, doi: 10.35833/MPCE.2020.000118

[15] R. K. Pandey, "Intelligent Agent Based Micro Grid Control," MURALI KRISHNA KOULURI (MURALI KRISHNA KOULURI (M.Tech)) Under the Supervision of."

[16] M. Saleem, B.-S. Ko, S.-H. Kim, S. Kim, B. S. Chowdhry, and R.-Y. Kim, "Active disturbance rejection control scheme for reducing mutual current and harmonics in multi-parallel grid-connected inverters," *Energies (Basel)*, vol. 12, no. 22, 2019, doi: 10.3390/en12224363

[17] X. Q. Kong, R. Z. Wang, and X. H. Huang, "Energy optimization model for a CCHP system with available gas turbines," *Appl Therm Eng*, vol. 25, no. 2–3, pp. 377–391, Feb. 2005, doi: 10.1016/J.APPLTHERMALENG.2004.06.014

[18] M. Y. El-Sharkh, A. Rahman, and M. S. Alam, "Short term scheduling of multiple grid-parallel PEM fuel cells for microgrid applications," *Int J Hydrogen Energy*, vol. 35, no. 20, pp. 11099–11106, Oct. 2010, doi: 10.1016/J.IJHYDENE.2010.07.033

[19] O. Sadeghian, A. Moradzadeh, B. Mohammadi-Ivatloo, M. Abapour, and F. P. Garcia Marquez, "Generation units maintenance in combined heat and power integrated systems using the mixed integer quadratic programming approach," *Energies (Basel)*, vol. 13, no. 11, 2020, doi: 10.3390/en13112840

[20] J. Sahebkar Farkhani, M. Zareein, A. Najafi, R. Melicio, and E. M. G. Rodrigues, "The power system and microgrid protection—A review," *Applied Sciences*, vol. 10, no. 22, 2020, doi: 10.3390/app10228271

[21] Y. Yuan, Y. Cao, Z. Fu, X. Xie, and S. Guo, "Assessment on energy-saving and emission reduction benefit of microgrid and its operation optimization," *Dianwang Jishu/Power System Technology*, vol. 36, pp. 12–18, May 2012.

[22] M. Shahab, S. Wang, and A. K. Junejo, "Improved control strategy for three-phase microgrid management with electric vehicles using multi objective optimization algorithm," *Energies (Basel)*, vol. 14, no. 4, 2021, doi: 10.3390/en14041146

[23] Y. Mei, B. Li, H. Wang, X. Wang, and M. Negnevitsky, "Multi-objective optimal scheduling of microgrid with electric vehicles," *Energy Reports*, vol. 8, pp. 4512–4524, Nov. 2022, doi: 10.1016/J.EGYR.2022.03.131

[24] Z. Cheng, D. Jia, Z. Li, J. Si, and S. Xu, "Multi-time scale dynamic robust optimal scheduling of CCHP microgrid based on rolling optimization," *International Journal of Electrical Power & Energy Systems*, vol. 139, p. 107957, Jul. 2022, doi: 10.1016/J.IJEPES.2022.107957

[25] H. Yuan, H. Ye, Y. Chen, and W. Deng, "Research on the optimal configuration of photovoltaic and energy storage in rural microgrid," *Energy Reports*, vol. 8, pp. 1285–1293, Nov. 2022, doi: 10.1016/J.EGYR.2022.08.115

[26] W. Zheng, H. Xiao, Z. Liu, W. Pei, and M. Beshir, "Multi-scale coordinated optimal dispatch method of electricity-thermal-hydrogen integrated energy systems," *IET Energy Systems Integration*, 2003, doi: 10.1049/esi2.12100

[27] S. Sun, C. Wang, Y. Wang, X. Zhu, and H. Lu, "Multi-objective optimization dispatching of a micro-grid considering uncertainty in wind power forecasting," *Energy Reports*, vol. 8, pp. 2859–2874, Nov. 2022, doi: 10.1016/J.EGYR.2022.01.175

[28] X. Zhou, Q. Ai, and M. Yousif, "Two kinds of decentralized robust economic dispatch framework combined distribution network and multi-microgrids," *Appl Energy*, vol. 253, p. 113588, Nov. 2019, doi: 10.1016/J.APENERGY.2019.113588

[29] H. Zhao, S. Guo, and H. Zhao, "Comprehensive assessment for battery energy storage systems based on fuzzy-MCDM considering risk preferences," *Energy*, vol. 168, pp. 450–461, Feb. 2019, doi: 10.1016/J.ENERGY.2018.11.129

[30] W. Gu et al., "Modeling, planning and optimal energy management of combined cooling, heating and power microgrid: A review," *International Journal of Electrical Power & Energy Systems*, vol. 54, pp. 26–37, Jan. 2014, doi: 10.1016/J.IJEPES.2013.06.028

[31] A. Mohsenzadeh, C. Pang, and M.R. Haghifam, "Determining optimal forming of flexible microgrids in the presence of demand response in smart distribution systems," *IEEE Syst J*, vol. 12, no. 4, pp. 3315–3323, 2018, doi: 10.1109/JSYST.2017.2739640

[32] K. Alqunun, T. Guesmi, and A. Farah, "Load shedding optimization for economic operation cost in a microgrid," *Electrical Engineering*, vol. 102, no. 2, pp. 779–791, 2020, doi: 10.1007/s00202-019-00909-3

[33] B. Ye, X. Shi, X. Wang, and H. Wu, "Optimisation configuration of hybrid AC/DC microgrid containing electric vehicles based on the NSGA-II algorithm," *The Journal of Engineering*, vol. 2019, no. 10, pp. 7229–7236, 2019, doi: 10.1049/joe.2018.5043

[34] A. Askarzadeh, "A memory-based genetic algorithm for optimization of power generation in a microgrid," *IEEE Trans Sustain Energy*, vol. 9, no. 3, pp. 1081–1089, 2018, doi: 10.1109/TSTE.2017.2765483

[35] F. Zhao, J. Yuan, and N. Wang, "Dynamic economic dispatch model of microgrid containing energy storage components based on a variant of NSGA-II algorithm," *Energies (Basel)*, vol. 12, no. 5, 2019, doi: 10.3390/en12050871

[36] N. Nikmehr and S. Najafi Ravadanegh, "Optimal power dispatch of multi-microgrids at future smart distribution grids," *IEEE Trans Smart Grid*, vol. 6, no. 4, pp. 1648–1657, 2015, doi: 10.1109/TSG.2015.2396992

[37] A. Younesi, H. Shayeghi, A. Safari, and P. Siano, "Assessing the resilience of multi microgrid based widespread power systems against natural disasters using Monte Carlo Simulation," *Energy*, vol. 207, p. 118220, Sep. 2020, doi: 10.1016/J.ENERGY.2020.118220

[38] B. Li, R. Roche, and A. Miraoui, "Microgrid sizing with combined evolutionary algorithm and MILP unit commitment," *Appl Energy*, vol. 188, pp. 547–562, Feb. 2017, doi: 10.1016/J.APENERGY.2016.12.038

[39] A. L. T. Ahamed, and S. Mohammed Sulthan, *Optimal Microgrid Battery Scheduling Using Simulated Annealing*. 2020. doi: 10.1109/PEREA51218.2020.9339727

[40] J. Radosavljević, M. Jevtić, and D. Klimenta, "Energy and operation management of a microgrid using particle swarm optimization," *Engineering Optimization*, vol. 48, no. 5, pp. 811–830, 2016, doi: 10.1080/0305215X.2015.1057135

[41] H. Jiang, S. Ning, and Q. Ge, "Multi-objective optimal dispatching of microgrid with large-scale electric vehicles," *IEEE Access*, vol. 7, pp. 145880–145888, 2019, doi: 10.1109/ACCESS.2019.2945597

[42] M. R. Ebrahimi and N. Amjady, "Contingency-constrained operation optimization of microgrid with wind and solar generations: A decision-driven stochastic adaptive-robust approach," *IET Renewable Power Generation*, vol. 15, no. 2, pp. 326–341, 2021, doi: 10.1049/rpg2.12026

[43] Adhiparasakthi Engineering College. Department of Electrical and Electronics Engineering, Adhiparasakthi Engineering College, and Institute of Electrical and Electronics Engineers, *2016 International Conference on Computation of Power, Energy, Information and Communication (ICCPEIC)*, 2016.

[44] B. Cao, W. Dong, Z. Lv, Y. Gu, S. Singh, and P. Kumar, "Hybrid microgrid many-objective sizing optimization with fuzzy decision," *IEEE Transactions on Fuzzy Systems*, vol. 28, no. 11, pp. 2702–2710, 2020, doi: 10.1109/TFUZZ.2020.3026140

[45] N. Yu, J. S. Kang, C. C. Chang, T. Y. Lee, and D. Y. Lee, "Robust economic optimization and environmental policy analysis for microgrid planning: An application to Taichung industrial park, Taiwan," *Energy*, vol. 113, pp. 671–682, Oct. 2016, doi: 10.1016/J.ENERGY.2016.07.066

[46] S. Mirjalili, S. M. Mirjalili, and A. Lewis, "Grey wolf optimizer," *Advances in Engineering Software*, vol. 69, pp. 46–61, Mar. 2014, doi: 10.1016/J.ADVENGSOFT.2013.12.007

[47] C. Green and S. Garimella, "Residential microgrid optimization using grey-box and black-box modeling methods," *Energy Build*, vol. 235, p. 110705, Mar. 2021, doi: 10.1016/J.ENBUILD.2020.110705

[48] J. P. Fossati, A. Galarza, A. Martín-Villate, J. M. Echeverría, and L. Fontán, "Optimal scheduling of a microgrid with a fuzzy logic controlled storage system," *International Journal of Electrical Power & Energy Systems*, vol. 68, pp. 61–70, Jun. 2015, doi: 10.1016/J.IJEPES.2014.12.032

4 Introduction to Optimization Techniques for Microgrid (continued)

P. Swati Patro and Sarat Kumar Sahoo
Parala Maharaja Engineering College, Berhampur, India

Nipon Ketjoy
Naresuan University, Phitsanulok, Thailand

Suman Lata Tripathi
Lovely Professional University, Phagwara, India

4.1 INTRODUCTION

The demand for energy has significantly expanded, and the rates of consumption have also increased accordingly due to technological improvements, population growth, and the urbanization of electricity. To fulfill the energy requirements brought on by a regular human lifestyle, conventional energy-producing resources based on fossil fuels are frequently utilized. The use of conventional energy sources harms the environment and contributes to climate change and global warming. It is seen as ineffective to solely rely on traditional energy sources to meet energy demands. As a result, renewable energy sources such as solar, wind, tidal, biomass, hydro, and geothermal are being used to meet the growing need for energy [1]. In conventional electricity networks, only energy sources made from fossil fuels received from power networks or utility companies are used to fulfill the bulk of energy demand. Because of the close proximity of the generation units and power plants to the availability of resources, the system is more difficult in terms of transmission. Because of this, the conventional electrical grid has a number of issues, including significant transmission losses and detrimental environmental effects. Due to the demand for a more sustainable solution and technological developments in the effectiveness and affordability of alternative electricity generation units, there has been considerable growth in the installation of concentrated renewable energy-generating resources that have been integrated into electricity systems over the last few decades [2]. Microgrids may run autonomously and locally by disconnecting from the regular grid, with their capacity to operate providing a grid asset for quicker system response and recovery when the primary grid

DOI: 10.1201/9781003481836-4

is offline. Microgrids can help increase grid resilience and alleviate grid disruptions. Microgrids contribute to a flexible and efficient electric grid by integrating rising renewable installations such as solar farms and electric cars. Furthermore, employing local energy sources to service local loads reduces energy losses in transmission and distribution. Boosting the efficiency of the electric delivery system by 2035, microgrids will be fundamental building blocks of the future power delivery system, supporting resilience, decarburization, and affordability. To make this a reality, we will focus on creating and verifying tools, processes, and technologies in each of the following strategic R&D areas: reducing the time and money required for microgrid implementation. In order to monitor, operate, and optimize large-scale grids of the future, establish microgrids as a foundation for future grids. Advance microgrid control and protection in order to react to changing grid circumstances while safeguarding the system and its consumers. Integrate microgrid planning, design, and operation models and tools to mix new and current capabilities to fulfill performance metrics and criteria. Microgrid effects and advantages are co-simulated by transmission and distribution to discover and confirm the value of microgrids. Microgrids were introduced into the existing electrical networks in order to minimize the major effects of integrating renewable energy. Microgrids with a supply of renewable energy may now transmit back to the primary electricity grid, eliminating the over-generation issue. Additionally, it is clear that the utilization of renewable energy sources has increased, especially in light of how producing renewable energy affects society. Microgrids are projected to improve power quality while also delivering financial, ecological, and technical benefits to both the general public and the electrical industry, ensuring system effectiveness [3]. When the microgrid is connected to the main grid, it can monitor and provide the difference between regional demand and emissions. It is important to provide assurance that power quality and stability with demand and generation must always be in balance when the system is in island mode. The system's capital cost and environmental cost are both significantly increased by the additional energy-producing sources. Every residence in a microgrid consumes various quantities of power over the day, and in the event of an alternative power-producing unit, such as wind or solar power, it is most likely to be in non-linear mode. A power system's ability to provide quick and necessary information on the distribution of electrical power generated by generating assets is critical for sustaining the system's energy. Energy management refers to the process of altering the components of an energy system. Hence, the optimization methods employed in the electrical management system for microgrids receive the majority of attention. In this chapter, we discuss microgrid energy management systems (EMS) and further analyze various optimization methods employed in EMS. The classification of various energy management strategies is also discussed, utilizing the different categorized methodologies. The different optimization approaches used in the EMS are also discussed in detail.

ABBREVIATIONS

DM   Demand management
DSO   Distribution system operator
EMS   Energy management system

ESS   Energy Storage System
EV   Electrical Vehicle
FL   Fuzzy logic
MG   Microgrid
MIP   Mixed-integer programming
PV   Photovoltaic
RES   Renewable Energy Source
UC   Unit commitment

4.2 MICROGRID ENERGY MANAGEMENT SYSTEM

Because of the unpredictability of renewable generation intermittency, power system reliability and stability problems have generated disadvantages for RES integration into the smart grid. To minimize MG drawbacks, a reliable battery management method with an adequate forecasting algorithm is needed. Many optimization methods have been offered to overcome this problem, such as successfully arranging alternate energy supplies or energy storage to provide consistency while simultaneously maximizing economic value by optimally deploying the produced power. "Energy management" is a phrase frequently used in the electrical engineering domain. A detailed examination of present operations was carried out to gain a full grasp of the various energy management tactics and optimization methodologies utilized for energy management. The concept of a fundamental microgrid with various renewable power sources, loads, and energy management technologies is discussed in this chapter. The level of trust in any control system's inputs is determined by a variety of variables. Some inputs may be altered by an uncertainty level due to the unpredictability of nature or unexpected human behavior. For example, a wind forecast may have a standard deviation error of 45% to 55% 24 hours ahead of the horizon and 25% one hour ahead. The deviation from the mean error for sun irradiation can range between 15% and 55% for 24 hours prior and 1 to 12% for 1 hour prior [6]. The unknown parameters to account for may vary in nature and need various representations depending on the microgrid under consideration. Here, we discuss the uncertainties in EMS for different kinds of microgrids, such as commercial and residential, electric mobility, phantom power plants, and multi-channel microgrids. A well-functioning EMS lowers system running costs and minimizes demand and supply mismatches. As a consequence, numerous MG energy management control systems have been created. Microgrids, as power management systems, are being created using organized, disorganized, distributed, and hierarchical management structures that employ various optimization methodologies. In an organized system, a single control unit receives commands from the MGs component and maintains them. Due to this, a centralized control system is usually practical and optimal. On the other hand, increasing the number of control and optimization variables occurs when all the data and processes are combined into a single entity. This makes things more difficult and calls for an immediate and effective response. Additionally, the control system raises a number of security and privacy concerns and is vulnerable to single-point failures. Disorganized control architectures are intended to address the shortcomings of organized control systems by gathering knowledge regionally and transferring control to each subsystem based on the overall aim. The unknown parameters used to account for uncertainty in MGs might vary in nature and require different representations depending on the microgrid's behavior. The following section provides current EMS concerns for different microgrids, such as commercial and residential unpredictability, phantom power plant ambiguity, electric mobility, and multi-channel migration difficulties. Phantom power plants, electric mobility, and multi-channel microgrids are all possibilities.

4.2.1 COMMERCIAL AND RESIDENTIAL INDUSTRIAL APPLICATIONS

Appliances, cooling and heating systems, solar power systems, systems for storing energy, and electric vehicles are among the most important components utilized in home and commercial applications in industries [5]. The major goal of the EMS in these applications is to minimize overall energy consumption, while retaining levels of convenience. The extent of autonomy from the grid is also explored, which may be increased by increasing personal consumption and self-reliance. Unpredictability influencing the price of power, customer demand for electricity, living temperatures in room, and PV array power are usually considered. Certain studies have integrated additional uncertainties such as house utilization, energy price, solar power production projection, and solar irradiance. While researchers include electric vehicles (EVs) for studying purposes, the ambiguity associated with the duration of charging connectivity and discharging time are also taken into account.

4.2.2 PHANTOM POWER PLANTS

Phantom power plants have three basic goals: maximize profit, minimize environmental impact, and maximize efficiency. When virtual power plants contain photovoltaic systems and wind turbines, the usual mistake in forecasting produced electricity and weather conditions is considered an uncertainty [6]. The flow of energy is governed by unpredictability like power consumption price, quick fluctuation, electricity consumption, and power storage. EV related factors such as charging length, capacity of power and the condition of charge, the number of autos connected, and the charging duration are all variables taken into account.

4.2.3 ELECTRIC MOBILITY

Some studies that employ EMSs to optimize parking lot operations incorporate EVs. The major aims in this situation are to minimize operation expenses, maximize the period when charging outlets are accessible, and reduce CO_2 emissions [7]. The uncertainties impacting the arrival and departure times, daily travel distances, driving behavior, and connection and disconnection times are all considered. Parking lot optimization is frequently done one day prior, and the timetable is modified with immediate effect, when the quantity of automobiles and pricing hours change. Despite the fact that reducing CO_2 emissions is critical, no research on the uncertainty associated with these variables are accessible yet in any studies. When EVs are linked to an AC microgrid, their battery packs may be considered as additional storage units capable of sustaining the grid's auxiliary services.

4.2.4 MULTI-CHANNEL MICROGRIDS

Multi-channel microgrids are composed of a number of supplier generators and microgrids, which are linked to the medium voltage grid and must conform to electrical system requirements. For this case, distributing system operators (DSOs) plan the flow of energy in order to increase profit while limiting operational risk. Typically, uncertainties affecting the generating power produced, pricing of energy and demanding of load are often considered [8]. During this study, the changing of frequency and voltage, electricity breakdowns, and generator interruption are all considered unpredictable. Long-term planning risks include future installed capacity and load demand. Uncertainties such as power reserves, fluctuation of CO_2 emissions, and reactions to ancillary services might be incorporated to improve energy management in multi-carrier microgrids. Furthermore, the degradation and lifetime of distinct microgrid are required for long-term planning.

4.3 CLASSIFICATION OF OPTIMIZATION TECHNIQUES IN ENERGY MANAGEMENT SYSTEMS

As shown in Figure 4.1, there are four main categories in which the optimization strategies used in microgrids energy management system (EMS) may be broadly categorized. Despite the fact that traditional strategies are routinely employed to handle the energy management problem, this chapter places a greater emphasis on unconventional methods. The various optimization methods applied to the EMS of MGs are thoroughly discussed in this chapter.

4.3.1 ARTIFICIAL INTELLIGENCE-BASED EMS

Artificial intelligent based also known as AI-Based EMS is classified in five categories as shown in Figure 4.2. The five primary categories of artificial intelligence-based energy management systems employed in microgrid system are described below. The subsections that follow provide a thorough description of how certain strategies are used.

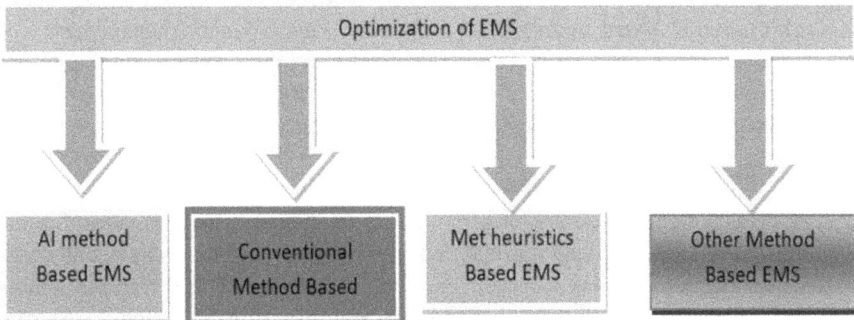

FIGURE 4.1 Different types of optimization techniques use in EMS.

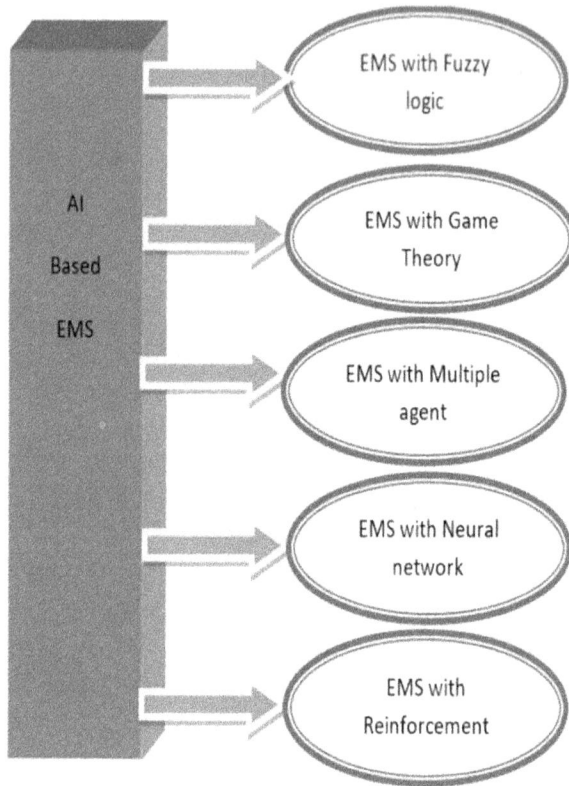

FIGURE 4.2   Different kinds of artificial based optimization techniques use in EMS.

4.3.1.1   EMS with Fuzzy Logic

Energy management system with fuzzy logic is a multiple purpose intelligent management of energy controller of a MG system is called as fuzzy logic-based EMS, which is developed to decrease operational costs and carbon emissions [3]. The optimization of the battery scheduling problem is the goal of a revolutionary neural network ensemble based metrological forecaster with a battery management system based on fuzzy logic (FL). With a decreased depth of battery discharge, the fuzzy logic-based battery scheduling scheme considerably reduces the cost of battery maintenance while extending the lifespan of the energy storage system [4].

4.3.1.2   EMS with Game Theory

Energy management systems with game theory is an area of mathematics that has been created to study the way rational decision-makers resolve conflicts, and cooperative organizations achieve a certain, explicitly stated goal. This technique demonstrated an ongoing decentralized demand-side administration in which a system is connecting to the grid house microgrid by using a new EV, energy storage system, and renewable energy source framework [9]. Each client linked to the microgrid estimates daily demand for electricity, rely on those values, the entire EMS conducts a non-cooperative hybrid strategy game until it finds an equilibrium find by to a predetermined utilization trend, with the purpose of minimizing the total cost of electricity [10].

4.3.1.3 EMS with Multiple Agents

Energy management system with a multiple agent system (MAS) is a computerized system composed of several communicating, intelligent agents. Because of the distributed character of the agent design, this optimization approach might be employed as a global control method in microgrids EMS. It uses a building management system to optimize the extraction and delivery of energy for heating and cooling, and power [11]. The goal of MAS-based structure of an independent microgrid is to sustain a constant voltage while achieving both environmental and financial advantages. Furthermore, a supply–demand imbalance is noticed as a result of the increased usage of RES.

4.3.1.4 EMS with Neural Network (NN)

Under imbalanced non-linear demand situations in both disconnected and connected configurations, neural network-based EMS provides a unique decentralized resilient power-sharing solution. A wavelet compression with a generalized neural network (NN) technique was used to create a smart energy management system consisting of variable loads, steady loads, and pumped hydroelectric storage. Generally NN-based forecasting is commonly used to predict stochastic renewable energy generation, while forecasting is an important component also used in sharing of power and load response scenarios, as well as management of energy algorithms based on NNs [12]. Furthermore, NNs are used in EMS plans that prioritize cost-effective dispatch and the incorporation of renewable energy sources.

4.3.1.5 EMS with Reinforcement

A dynamic programming-based action-dependent heuristic controller-based intelligent dynamic energy management system is known as reinforcement learning-based EMS. This optimization technique suggested a unique technique which is more dependable, eco-friendly, and the suggested approach's performance is compared to the flexible energy management plan based on decision chains. According to the anticipated wind power output a two-stage, a head reinforcement learning technique employing a Markov chain model is applied [13]. The RL method is also utilized to optimize the collaboration of numerous energy storage systems in a microgrid with an interrelated approach.

4.3.2 CONVENTIONAL EMS

Conventional or non-renewable energy sources are those that, once depleted, do not replace themselves within a set time frame, such as coal, gas, and oil. For a long time, these energy sources were widely employed to supply energy demands. Because the rate of use exceeds the rate of generation, these energy sources have been exhausted

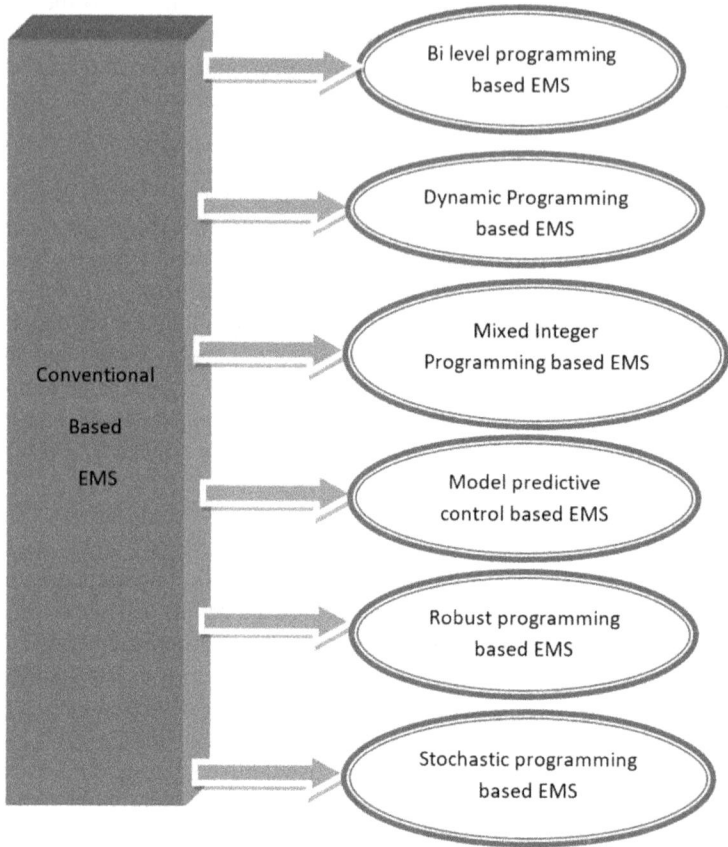

FIGURE 4.3   Classification of conventional optimization techniques use in EMS.

and will not be replenished. Though conventional energy sources such as natural gas, oil, coal, and nuclear are limited, they continue to dominate the energy industry. The six basic categories for conventional EMS employed in MG systems are as shown in Figure 4.3. In the subsections that follow, a thorough discussion of how the specific approach is used is provided [14].

4.3.2.1   Bilevel Programming (BP) EMS

Energy management systems with programming based on bilevel optimization and a new compressed air energy storage technology are used to solve the unit commitment issues. The solitary microgrid scenario provides an example of the need for sophisticated methods for evaluating economical assessment of the effects of regulations designed for reducing the carbon emissions as well as increasing renewable energy integration into MG [15]. A resilient planning model of the bilevel energy concept is built by developing an integrated integer second-level circular computing issue and solving it with a two-track resilient column limitation approach.

4.3.2.2 Dynamic Programming (DP) EMS

In order to reduce the overall operational costs of the microgrid, a dynamic programming (DP) technique is used. In addition to this, according to the planning approach, an actual time digital energy administration system based on windows, in which sliding is available, serves as an offline optimization issue. In terms of effectiveness, the recommended dynamic programming that integrated a time discrete controlling solution system performed better than the static dispatch methods [16]. A useful energy storage or battery management system utilizing the dynamic programming approach is used. The unique battery and energy storage system models, as well as the cost model based on restrictions, are the centerpiece of this technique.

4.3.2.3 Mixed-Integer Programming EMS (MIP-based EMS)

It is common practice to employ mixed-integer programming (MIP) methods for an MG's EMS. This is mostly due to the technique's simplicity and its minimal computing demands. This optimization technique is used in isolating energy management systems with a microgrid structure with a reserve system, a PV system, and gaseous micro turbine units. This optimization technique also considers a customized concept with ESS, photovoltaic, and ultra-capacitors. Alongside nearby electrical power management systems at the consumer's side that act as the microgrid, an overall energy management system based on the MIP concept is being taken into account. Mixed-integer linear modeling is used to regulate energy demand and output in addition to rolling horizon-based load forecasting. Adequate measurement and assessment of the RES and storage of energy systems, as well as a demand-side operating system of the microgrid using mixed-integer linear programming, are required in the residential MG configuration.

4.3.2.4 Model-Predictive Control (MPC) EMS

Another popular EMS solution that addresses the issue of unit commitment in MGs is the theory of model-predictive control. The mixed-integer non-linear problem was optimized with a model-predictive solution. The performance of the suggested model is used to handle MG; there is a demand-side governance issue. An improved energy oversight technique using model-predictive control, or MPC, is created using load shifting to redistribute power usage. Model-predictive control (MPC) is used to allocate power usage, reduce energy costs, and boost self-consumption capability [17]. If no feed-in tariff exists, a large but decreasing penalty coefficient might be introduced to the MPC cost function in order to raise the personal consumption ratio.

4.3.2.5 Robust Programming (RP) EMS

In the robust programming-based optimization strategy, a robust programming (RP) method was employed to offer a robust equivalent formulation to deal with the ambiguity in the energy production or energy consumption of an MG. This model forecasts energy use and RE generation using a MILP approach based on rolling horizon windows. The binary limitations are relaxed, and variable time steps are employed to solve the issue to minimize computing complexity. The decentralized MG model is created using agent-based modeling, and a robust programming-based optimization for the microgrid EMS is established.

4.3.2.6 Stochastic Programming (SP) EMS

A self-consumption-focused stochastic programming-based management system of energy that provides resiliency to unexpected demand variations while limiting power injection into the main grid. Due to the fact that industrial demand is usually predictable based on the scheduling of the machines used in the process (unpredictability in industrial demand is typically caused by variations in raw material quality), the industrial situation is analyzed. The impact of self-consumption laws is larger in businesses than in homes because the synchronicity between energy supply and demand is greater in industrial processes than in residential load centers, and the option of co-generation allows for the avoidance of the requirement for a battery storage system, which elevates investment prices to roughly 25% of total cost. Furthermore, there are certain environmental problems associated with the growing usage of batteries in non-conventional energy system applications [18]. The efficacy and reliability of the indicated stochastic programming (SP)-based microgrid energy routing method highlight the significance of this sort of optimization technique. Using this optimization method, the intelligent MG EMS seeks to balance the energy system's demand and supply while delivering high-quality power delivery.

4.3.3 META-HEURISTICS EMS

Energy management system (EMS) based on meta-heuristics program used in MG systems are essentially categorized into three fundamental classes, which are graphically depicted in Figure 4.4. The subsections that follow provide a full description of

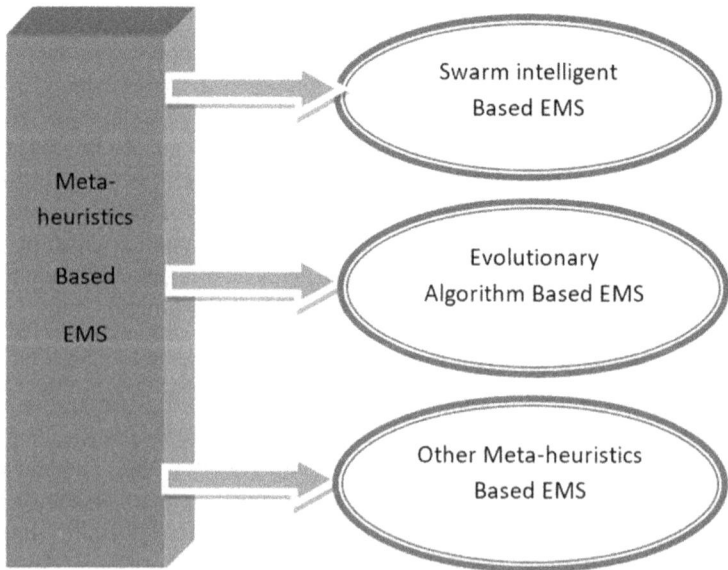

FIGURE 4.4 Types of meta-heuristics-based optimization techniques use in EMS.

the application of each approach. Algorithm for Swarm intelligent based and other meta-heuristic utilized in EMS are utilized in this optimization approach.

4.3.3.1 Swarm Intelligence EMS

Energy management systems, which use the technique of swarm intelligence, are influenced by natural swarm systems like insect territories, bird flocking, mammal herding, bird foraging, carp schooling, and so on. The optimization of particle swarm algorithms, also known as PSO algorithms, is popular, and in it, the algorithm is swarm-based, with the PSO addressing the EM approach with several objectives for optimizing the scheduling of battery energy storage systems to handle the power mis-handling issue. The PSO technique's rapid convergence rate and minimal processing requirements have been applied to a wide range of energy management issues [19].

4.3.3.2 Evolutionary Algorithm EMS

Algorithms influenced by biological evolution mechanisms, such as reproduction, variation, replication, and adoption, are examples of evolutionary algorithms. There are several uses for evolutionary algorithms, and there are a few evolutionary ones. Increased distributed energy output from energy generated from RES in smart grids has introduced stochastic intermittency into the system of MG management. Weather and load pattern forecasting accuracy is a vital component of the EMS. Short-period load projection is a difficult assignment due to the substantial non-linear nature of the historical data. It is also used to optimize unit commitment and battery scheduling problems' generation. One of the basic difficulties of MG EMS is the problem of bat-tery schedule optimization; to solve this, an algorithm-based coral reef optimization technique is used. The coral reef algorithm method differs from other optimization strategies in that it promotes mutual evolution in several exploring models through-out an entire population. This greatly aids handling of the scheduling problem. There are several different meta-heuristics methods utilized for optimizing the EMS of the MG, in addition to swarm optimization and evolutionary algorithm-based meta-heuristics strategies. A gravitational search algorithm (GSA) is employed with a probabilistic method to identify the ideal EM of the MG, with the goal of addressing the MG's energy and operation management challenges. Several distinct optimiza-tion solutions have been devised to identify the energy management dilemma; it is evident that algorithms have been intensively investigated in the study, including swarm-based and genetic algorithms. It is also suggested to conduct experimental assessments of the approach in order to harness the true impact of these potent algo-rithms to address energy management issues in microgrids. The fundamental draw-backs of all these approaches are that a great deal of the work has been restricted to simulation experiments and that the applications of these systems are currently quite limited in research.

4.3.4 OTHER EMS METHODS

Besides the already-listed classification of distinct optimization approaches used to handle energy, there are definitely a few issues with managing energy in microgrids with unique energy that are not being properly evaluated. Conventional unit commitment algorithms rely on past data regarding load demand and renewable energy output, making them unsuitable for circumstances where prior data is unavailable. Without using past data, the adaptive segmented situational learning method for unit commitment learns to determine the best possible unit commitment (UC) method while minimizing total operating costs [20]. While implementing an EMS for a DC distribution network, a Lagrangian primal-dual partially rdifferential algorithm approach is also applied to tackle the convex optimization problem under universal restrictions.

4.4 DISCUSSION

The previous sections discussed the usage of several optimization methodologies in MG EMS. This section discusses a more in-depth, insightful study of various optimization techniques based on the elements of supply of the EMS dependent on the usage and classification of the approach. This investigation revealed that the MILP is the most often employed optimization strategy for dealing with the EM issue in MGs. Furthermore, the PSO and GA were clearly the most commonly employed meta-heuristic algorithms. The employment of multiple agents and game-theoretic algorithms is increasing with the use of a more decentralized network. Multiple agent-based networks are more advantageous in a decentralized environment, and there is a need for a complete implementation of a decentralized modeling tool for the behavior of EMS in microgrids. The challenges of unit commitment and economical deployment have received more attention, taking into account the influence of the conclusions on the system's efficiency and cost. The top five optimization approaches used in this chapter are evaluated based on aspects such as clarity, productivity, dependability, flexibility, and projecting (UC, DM, and ED). Additionally, an efficient energy management strategy is necessary to accurately predict the amount of energy generated from renewable sources as a result of the rising usage of green energy-generating sources in the distributed system of microgrids. The difficulty of predicting algorithms arises owing to the stochastic nature of green energy supply, which may be related to or dependent on environmental factors peculiar to the distribution of energy resources.

4.5 CONCLUSION

Due to high energy consumption and the inclusion of intermittent renewable energy sources, an efficient energy management system is an essential component of microgrid design. With the widespread use of RES, the necessity for an optimal EMS approach has been demonstrated to be critical. The forecasting algorithm used in EMS has a considerable influence on its efficiency and reliability. In principle, the EMS aims to precisely estimate power production and load in order to successfully manage periods of high demand while lowering overall system costs. However, the sort of EM schemes chosen is also determined by aspects such as the EM scheme's complexity and flexibility in order to accomplish an outstanding degree of network reliability as well as effectiveness. This chapter conveys the various energy management system techniques in which it is found that the use of mix integer programming-based energy management solutions is extremely widely used, MAS-based platforms are going to be the best for resolving the complicated issues of DM and UC, and meta-heuristic algorithms like PSO, deep learning techniques, and neural networks have greater potential for estimating ED-based applications. It is additionally found that multiple agent-type algorithms and meta-heuristics-based algorithms are significantly less expensive while dealing with the difficulties of EMS problems. Finally, it is concluded that there is a need for future EMS to emphasize more greatly on the cooperative ecosystem established by microgrids and create highly precise projections along with planning algorithms will boost economic and cognitive benefits.

REFERENCES

[1] F. Katiraei, R. Iravani, N. Hatziargyriou, and A. Dimeas "Microgrids management: Controls and operation aspects of microgrids," *IEEE Power and Energy Magazine*, 6 (3), 54–65, 2008. https://doi.org/10.1109/MPE.2008.918702

[2] N.L. Panwar, S.C. Kaushik, and Surendra Kothari "Role of renewable energy sources in environmental protection A review," *Renewable and Sustainable Energy Reviews* 15 (3), 1513–1524, 2011. https://doi.org/10.1016/j.rser.2010.11.037

[3] A. Chaouachi, R.M. Kamel, R. Andoulsi, and K. Nagasaka "Multiobjective intelligent energy management for a microgrid," *IEEE Transactions on Industrial Electronics*, 60 (4), 1688–1699, 2013. https://doi.org/10.1109/TIE.2012.2188873

[4] M. Balamurugan, S.K. Sahoo, and S. Sukchai "Application of soft computing methods for grid connected PV system: A technological and status review," *Renewable and Sustainable Energy Reviews*, 75, 1493–1508, 2017. https://doi.org/10.1016/j.rser.2016.11.210

[5] Z. Chen, L. Wu, and Y. Fu. "Real-time price-based demand response management for residential appliances via stochastic optimization and robust optimization," *IEEE Trans Smart Grid*, 3, 1822–1831, 2013. https://doi.org/10.1109/TSG.2012.2212729

[6] N. Naval, and J.M. Yusta. "Virtual power plant models and electricity markets – A review," *Renewable and Sustainable Energy Reviews*, 149, 2021. https://doi.org/10.1016/j.rser.2021.111393

[7] L. Igualada, C. Corchero, M. Cruz-Zambrano, and F.J. Heredia. "Optimal energy management for a residential microgrid including a vehicle-to-grid system," *IEEE Trans. Smart Grid*, 5, 2163–2172, 2014. https://doi.org/10.1109/TSG.2014.2318836

[8] B. Zhang, Q. Li, L. Wang, W. Feng "Robust optimization for energy transactions in multi-microgrids under uncertainty," *Applied Energy*, 217, 346–360, 2018. https://doi.org/10.1016/j.apenergy.2018.02.121

[9] G.S. Thirunavukkarasu, M. Seyedmahmoudian, E. Jamei, B. Horan, S. Mekhilef, and A. Stojcevski. "Role of optimization techniques in microgrid energy management systems—A review," *Energy Strategy Reviews*, 43, 2022. https://doi.org/10.1016/j.esr.2022.100899

[10] F.A. Mohamed, and H.N. Koivo "Multiobjective optimization using modified game theory for online management of microgrid," *European Transactions on Electrical Power*, 21 (1), 839–854, 2010. https://doi.org/10.1002/etep.480

[11] P. Zhao, S. Suryanarayanan, and M.G. Simoes "An energy management system for building structures using a multi-agent decision-making control methodology," *IEEE Transactions on Industry Applications*, 49 (1), 322–330, 2012. https://doi.org/10.1109/TIA.2012.2229682

[12] M.E.G. Urias, E.N. Sanchez, and L.J. Ricalde "Electrical microgrid optimization via a new recurrent neural network", *IEEE Systems Journal*, 9 (3), 945–953, 2015. https://doi.org/10.1109/JSYST.2014.2305494

[13] G.K. Venayagamoorthy, R.K. Sharma, P.K. Gautam, and A. Ahmadi "Dynamic energy Management system for a smart microgrid," *IEEE Transactions on Neural Networks and Learning Systems*, 27 (8), 1643–1656, 2016. https://doi.org/10.1109/TNNLS.2016.2514358

[14] J. Zhang, K.-J. Li, M. Wang, W.-J. Lee, H. Gao, C. Zhang, and K. Li "A bilevel program for the planning of an islanded microgrid including CAES," *IEEE Transactions on Industry Applications*, 52 (4), 2768–2777, 2016. https://doi.org/10.1109/IAS.2015.7356783

[15] K. Rahbar, J. Xu, and R. Zhang "Real-time energy storage management for renewable integration in microgrid: An off-line optimization approach," *IEEE Transactions on Smart Grid*, 6 (1), 124–134, 2015. https://doi.org/10.1109/TSG.2014.2359004

[16] H. Kanchev, D. Lu, F. Colas, V. Lazarov, and B. Francois "Energy management and operational planning of a microgrid with a PV-based active generator for smart grid applications," *IEEE Transactions on Industrial Electronics*, 58 (10), 4583–4592, 2011. https://doi.org/10.1109/TIE.2011.2119451

[17] H. Yan, F. Zhuo, N. Lv, H. Yi, Z. Wang, and C. Liu "Model predictive control based energy management of a household microgrid," *IEEE 10th International Symposium on Power Electronics for Distributed Generation Systems (PEDG)*, Xi'an, China, pp. 365–369, 2019. https://doi.org/10.1109/PEDG.2019.8807657

[18] J. Barrientos, J.D. López, and F. Valencia "A novel stochastic-programming-based energy management system to promote self-consumption in industrial processes," *Energies*, 11, 441, 2018. https://doi.org/10.3390/en11020441

[19] P. Ray, C. Bhattacharjee, and K.R. Dhenuvakonda "Swarm intelligence-based energy management of electric vehicle charging station integrated with renewable energy sources," *International Journal of Energy Research*, a special issue research article, 46 (15), 21598–21618, 2022. https://doi.org/10.1002/er.7601

[20] H.-S. Lee, C. Tekin, M. Van Der Schaar, and J.-W. Lee "Adaptive contextual learning for unit commitment in microgrids with renewable energy sources," *IEEE Journal of Selected Topics in Signal Processing*, 12 (4), 688–702, 2018. https://doi.org/10.1109/JSTSP.2018.2849855

5 Energy Management Systems for Microgrids

Saqib Ali
NFC Institute of Engineering & Technology, Multan, Pakistan

Aamer Raza
Islamabad Electric Supply Company, Islamabad, Pakistan

Irfan Khan
Texas A&M University, Texas, USA

M. Kamran Liaqat Bhatti
NFC Institute of Engineering & Technology, Multan, Pakistan

Tahir Nadeem Malik
HITEC Taxila, Taxila, Pakistan

5.1 INTRODUCTION

Electric power systems play a pivotal role in the socioeconomic development of a human society. Residential comfort, commercial activities, industrial expansion, and agricultural progress depend on reliable and sustainable electric systems. Due to continuous load growth, utilities need to pour in meaningful amounts of capital and operational expenditure to maintain network sustainability, stability, reliability and adequate power generation, transmission and distribution resources at an acceptable level.

In conventional electricity networks, advances in stability and reliability require expansion of energy generation, transmission, and distribution capacities. Moreover, conventional electric systems do not support large amounts of renewable energy such as solar and wind concentrated on a single geographical location, to improve the sustainability (e.g., consumption cost and emission reduction). The reason for this is that conventional electric system has central generation located at distant locations from the load, therefore, an integration of large volume of unpredictable renewables may endanger the system stability. Due to all these issues, modern approaches like small-scale distributed generation, demand response, thermo-electric energy storages, and intelligent scheduling of loads may enhance system's stability, reliability, and sustainability.

DOI: 10.1201/9781003481836-5

In this regard, the term "microgrid" has been coined. A microgrid is a "a group of interconnected loads, energy storages, and distributed generation within clearly defined electrical boundaries, capable of working in either grid-connected or islanded mode" [1]. A microgrid energy management system is an artificial intelligence-based module that optimally controls the components such as energy sources (i.e., electricity from national grid, natural gas, renewable energy, heat and biogas), battery storage systems, and connected loads to provide low-carbon, affordable, and reliable electricity.

The concept of microgrid is not new; it dates back to around 1880 when Thomas Edison developed the first microgrid with a steam generator serving 82 customers in the New York City [2]. In the early part of the twentieth century, statewide regulated interconnected electric networks with central generation and distant loads emerged in the United States; however, developments in the area of the microgrid remained silent until 2007, when Geidl et al. proposed the future integrated energy networks [3]. Later on, the same concept was successfully applied to attain higher degree of reliability and sustainability in electric networks. According to Navigant Research, the microgrid market is expected to grow from 1.4 GW in 2015 to 7.6 GW in 2024 [4].

Practically speaking, a modern microgrid is a building or a portion of a power network working on distribution voltages such as 11 kV and/or 415 V in either islanded or grid-connected mode. The point where the microgrid physically links to the national is termed as the "point of common coupling" [5]. An intelligent energy management control module installed in the building wirelessly communicates with the microgrid's energy sources, storages, and load. In this regard, it receives operational preferences such as the required comfort and load scheduling from customer. Moreover, the module collects power tariff and weather-related information from the utility and the weather servers, respectively. Figures 5.1–5.3 describe the concept of individual microgrid control to N microgrids having building level micro energy hub controls under a utility-owned macro energy hub control module.

Category-wise, microgrids may be divided into residential, commercial, industrial, and agricultural buildings. Commercial entities may be shops, plazas, shopping malls, educational buildings, and hospitals, etc.

Keeping in mind the vitality of the concept of a microgrid, this chapter discusses its various aspects, identifies future research avenues, and provides simulation of a scenario for the readers. Sections 5.2–5.5 provide literature review on the developments achieved in the past.

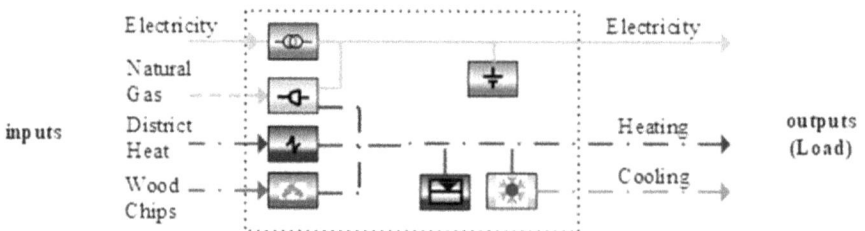

FIGURE 5.1   Basic concept of microgrid.

FIGURE 5.2 Mathematical description of energy hub or microgrid.

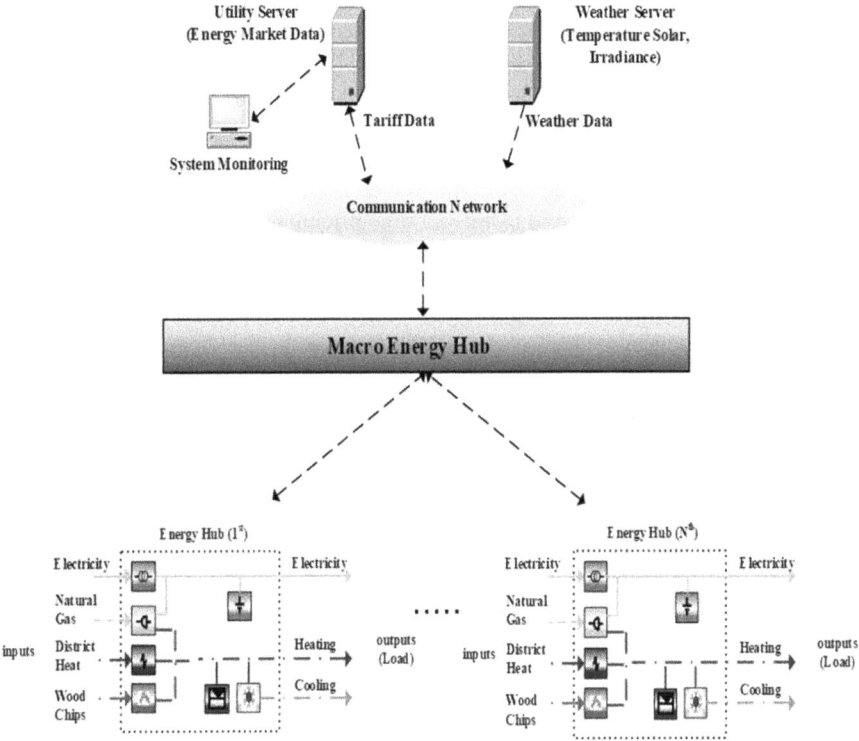

FIGURE 5.3 Two-layered control for utility and customers.

5.2 RESIDENTIAL MICROGRIDS

Globally, residential customers consume a generated significant proportion of electricity, therefore, optimal energy management in this sector may increase reliability and sustainability of electric networks. This section discusses the latest developments on residential microgrid energy management systems.

A mathematical framework has been presented that converts phase-balancing problem of a 400 V low voltage power distribution system into mixed integer quadratic optimization problem [6]. The proposed microgrid consists of a low tension 400 V secondary distributor, solar panels, individual residential buildings, and residential apartments. Simulation results show that energy losses at the distributor decrease by 41.74% by optimal handling of microgrid resources.

A group of residential microgrids has been proposed that are equipped with different types of load and electric vehicles [7]. Electric vehicles charge during off-peak and discharge in peak hours. Moreover, the electric charge in the vehicle battery plays the role of a reserve energy source during outage on national grid under extreme weather or natural disasters. Simulation results show that even after network failure, the present fleet of vehicles successfully feeds a load up to 6 hours.

A microgrid has been presented that consists of a group of residential buildings to optimally trade-off the customer comfort and consumption cost [8]. Buildings contain electric vehicles, solar photovoltaic panels, receive feed from the national grid, combine heat and power units, heat pump, battery, and home appliances. Moreover, the authors devised a building-to-building energy sharing strategy to increase profits. Results reveal that this application and combination decreases the power production cost by 44%.

A residential microgrid has been proposed, containing electric feed from national grid, solar panel, wind turbine, battery, and electric load [9]. Microgrid exchanges electric energy with national grid using real time pricing to reduce consumption cost. Results show that the proposed method reduces cost by 26%.

An optimal planning strategy has been presented for a residential microgrid to minimize user discomfort due to demand side management, electric energy cost, load fluctuations of microgrid, greenhouse gases, and net present value [10]. The proposed microgrid contains home appliances, solar panels, and a feed from the national grid. Simulation results using the proposed ABC algorithm shows that the energy cost reduces to $1405.18 from $ 1627.46 and emission lessens by 6699 kg per annum.

Shreyasi et al. [11] formulated a strategy to enhance resilience of an islanded residential microgrid with 100% solar energy. Results reveal that application of a battery storage system may avoid frequency instability during absences of solar energy.

This literature shows that the integration of different distributed generators, battery storages, electric vehicles, demand response techniques, and inter-building energy sharing mechanisms may increase comfort and resilience and reduce cost and emissions.

5.2.1 COMMERCIAL MICROGRIDS

Commercial buildings are quite large in size and load, and therefore, strategies to optimally handle the energy sources, load, and storage facilities may assist in realizing the reliable and sustainable electric system. Usually, the building owners have large, distributed generators to feed HVAC and lighting loads even under outages on the national grid. These generators may provide ancillary services such as spinning reserve to the national grid for voltage improvement and reduction in network congestion. Similarly, commercial microgrids are under frequent access to the customers and employees with a large number of electric vehicles arriving around the clock. Consequently, any strategy for optimal integration of these vehicles may reduce a building's energy cost and emissions, and increase resilience during natural disasters. This section provides a literature survey on the energy management system for commercial microgrids.

An islanded commercial microgrid for consumption cost reduction and improved reliability has been proposed [12]. The proposed microgrid contains batteries, wind turbines, solar panels, diesel generator, dumb and manageable loads, and energy input from the national grid. Results show that cost reduces by 18.23% with "loss of power supply probability" equaling 3.06%.

An eco-friendly and resilient office building microgrid has been proposed [13]. The microgrid contains 600 kW solar panels, 2.8 MWh battery and 200 kW natural gas-based generator as backup. Results show that CO_2 reduced by 287 tons per annum, and application of solar panels and batteries reduced the size of a natural gas-based backup generator by 100%.

Jubair et al. [14] devised commercial building microgrid control that calculates the optimal charging/discharging level of electric vehicles for energy cost reduction. Selected building for this study is a college located at the California University USA Microgrid contains HVAC, lights, national grid connection, solar panel, battery, and electric vehicles. Results show that energy cost reduced by 23%.

Sahbasadat [15] proposed a method to locate the optimal charging park and size of electric vehicle fleet to reduce energy costs. Simulation results show that a microgrid placed at the optimal location in the distribution system with optimum fleet size reduces the cost by 31.96%.

Seyd Ali et al. [16] proposed a commercial microgrid containing a fuel cell, solar panels, and battery. Bi-level energy management control improves system efficiency and the cost at the first and the second levels, respectively. Results show that the energy cost reduces, and efficiency improves by 3.47% and 80.44%, respectively.

Hegazy et al. [17] presented a commercial activity-based microgrid containing solar panels, solid oxide fuel cell, battery storage, and a connection with the national grid. Authors proposed four types of energy management controls using fuzzy logic, state machine control strategy, equivalent consumption minimization strategy and external energy maximization strategy (EEMS). Results show that the EEMS performs better than other techniques giving equipment efficiency as high as 84.91% with a 6.11% cost saving.

Sackey et al. [18] proposed a microgrid to supply power to an area in Ghana. The facility contains 94.8 kW solar panels, 231 kWh batteries, 158 kW diesel generator, and an external distribution grid. The initial capital was US $45,929.17 with payback period of 2 years and 11 months, producing total savings of US $322,563.21.

Literature reveals that optimal handling of load, energy sources, and storage in commercial buildings not only increases cost savings, resilience, and network stability during extreme events, but also reduces emissions to achieve environmental goals.

5.2.2 INDUSTRIAL MICROGRIDS

Modern-day progress of a country depends almost entirely on national industrialization. Pharmaceutical factories, textile mills, vehicles, aircraft, the mobile and computer manufacturing industry, furnaces, food processing units and cement factories are all examples of industrial microgrids. For inexpensive manufacturing, the concept of energy management may play central role in reducing the manufacturing price in industrial sector. In response to this, low manufacturing prices may create an investor-friendly atmosphere through the increase of exports and jobs, and the reduction of inflation rates and poverty. This section provides a literature review on industrial microgrid energy management systems.

Mohammad Reza et al. [19] proposed a grid-connected industrial microgrid with a diesel generator, solar photovoltaic panels, wind turbine, battery storage, and process load. Demand response techniques such as load shifting and peak load curtailment are utilized. To undo the adverse effects of stochastic nature of the solar irradiance, a risk averse attribute is added in energy management control using "information gap decision theory." Simulation results show that grid robustness increases by maintaining more energy in batteries during low solar hours.

Saqib Ali et al. [20] proposed an industrial microgrid aiming at to reduce energy cost and emission. Further, the authors presented a cybersecurity technique to block an unauthorized intrusion in the communication network of the devised energy management system. The microgrid participates in ancillary services such as selling spinning reserve to the external grid and contains a fuel cell, internal combustion engine, microturbine, solar panels, battery, and electric vehicle fleet. The external energy sources include biogas, electricity, and natural gas. Simulation results show that costs and emissions reduce from $ 6097.8 to $−704.05, and 6808.5 kg to 5302.8 kg, per day respectively.

A low-carbon energy management strategy was proposed for an industrial park to overcome high consumption costs and power shortages. The formulated microgrid contains solar panels, batteries, a connection with the outer grid, and industrial load [21]. Results show that introduction of the devised energy management strategy reduced CO_2 emissions, costs, and power shortages.

Shaila et al. [22] proposed a novel methodology for an islanded microgrid to enhance solar renewable energy hosting ability by application of storage batteries. The primary aim is to replace the diesel generator and national grid during faults. The microgrid contains 2.5 MW solar panels, 1.2 MVA battery storage and a 3 MVA diesel generator. Results show that application of proposed energy management method completely offsets the diesel generator and national grid during outages.

Mohammad Adel et al. [23] devised a mathematical framework to plan and optimize an industrial building microgrid. The proposed microgrid contains solar panels, wind turbines, diesel engines, battery storage, and a connection with the national grid. Results show that capital investment and operations and maintenance cost in grid-connected mode is US $9,397,003 as compared to US $12,750,864 for islanded mode after five years.

With increasing integration levels of fluctuating renewable energy, frequency stability has become as issue [24]. To address this, an "optimal fraction sliding mode control" for microgrid stability was proposed. In addition, the effect of electric vehicles and battery was also evaluated in terms of frequency stability against varying renewables. Considered scenario contains diesel engine, fuel cells, solar panels, wind turbine, battery, electric vehicle, and industrial and residential loads. Results show that with the application of the proposed control, energy sources, and storages, the post disturbance frequency settling time decreased by 20.2%.

The literature shows that optimal handling of industrial microgrids may increase network stability, improve voltages, enhance reliability and reduce CO_2 emissions and consumption costs.

5.3 MICROGRID PROTECTION

This section discusses protection related aspects.

5.3.1 PROTECTION CHALLENGES

Over time, the energy flows in the distribution networks have become complicated due to addition of building-owned dispatchable and non-dispatchable distributed generators and electric storages. In cases when the energy demand of a building is more than the installed distributed generators, microgrid imports power from the national grid and vice versa [25]. Besides this, there is quite a significant difference in fault current levels for grid-connected and islanded modes and high output impedance with consequent fault current limiting behavior of inverter-based generators posing problems for overcurrent, differential and distance relays [25].

A protection system is characterized by sensitivity, selectivity, and reliability. Sensitivity means the ability to detect even a minor abnormal situation above a preset threshold value, whereas selectivity isolates only the faulty zone of the network without affecting the healthy one. Finally, reliability of a protection system ensures the correct operation whenever desired [26]. Other mandatory parameters include the speed of operation, simplicity, and cost-effectiveness [27]. Major challenge in microgrid protection until now have included distinguishing the grid-connected and islanded modes and correct sensitivity and selectivity during faults [28, 29].

Literature shows that besides challenges mentioned earlier in this section, further modifications are required for microgrid protection systems. For example, an ability of "low-voltage ride through" needs to be added in conventional schemes to avoid complete blackouts during cascading tripping. Low-voltage ride through is an ability of renewable generators to remain operational in the system during faults and low-voltage conditions to avoid widespread network outages [30]. In islanded mode, frequency stability, voltage control and load management become a problem in the presence of a high level of non-dispatchable generation [31]. Besides this, renewables reduce system inertia leading to dynamic frequency instability during faults [25].

5.3.2 PROTECTION SCHEMES FOR MICROGRIDS

A comparison of various protections schemes is provided in Table 5.1 [25].

5.4 SUPPORTING TECHNOLOGIES

Realization of the concept of a microgrid requires supporting technologies. This section discusses some of the advances in these technologies.

TABLE 5.1
Comparison of Protection Strategies

No.	Scheme Type	Method	Advantages	Disadvantages
1.	Directional overcurrent	Current as threshold	Simple and cheap	Fixed settings, therefore not appropriate for dynamic networks
2.	Adaptive	Current as threshold	Appropriate for dynamic systems	Recalculation is required for fixed settings
3.	Differential	Difference of current	Effective for generators and turbines	Expensive
4.	Hybrid	Current as threshold	Combines benefits of different schemes	Confirmation of feasibility requires real time experimentation
5.	Fault current limiter	Resistor used with over-current protection	Caters both islanded and grid-connected modes	Fault current limiter location requires recalculation for any changes
6.	Over- and under- voltage	Voltage-based	Appropriate for islanded state	Cannot detect high impedance faults
7.	Traveling waves	Polarity and time of current	Potential scheme for microgrids	Lengthy calculations
8.	PMU based	Phasor difference	Low cost	Problem in synchronization
9.	Harmonic based	Fundamental frequency variation	Appropriate for certain faults	Unable to detect other faults
10.	Sequence components	Current-based	Senses changes in positive, negative, and zero sequence components	Unable to detect single-phase and unbalanced three-phase loads
11.	Wavelet transform	Difference of current	Suitable to identify signal patterns	Selection of the optimal active power
12.	Multi-agent system	Current-based	Incorporates IoT devices	Complicated

5.4.1 DEMAND RESPONSE

The demand response (DR) was defined as "actions by individual electric customers that reduce or shift their electricity usage in a given time period (peak hours) in response to a price signal, a financial incentive, or a grid reliability trigger" [32]. The DR is defined as

> [...] changes in electric usage by end-use customers, from their normal consumption patterns, in response to the changes in the price of electricity over time, or to the incentive payments designed to induce lower electricity use at times of high wholesale market prices, or when system reliability is jeopardized.

[33]

Figure 5.4 shows that there two ways to mitigate the negative effects of increase in energy usage: 1) spending capital to increase generation, transmission and distribution resources; and 2) enticing customers to shift energy usage on an incentive basis from peak to off-peak, installing building-owned generators, and reducing the consumption. Among the two, the second method may benefit both utility and customer at low cost.

5.4.2 DISTRIBUTED GENERATORS

A distributed generator coupled with any appropriate node of the distribution network or located in a building, may either be fossil fuel or renewable energy-based. Distributed generator can be defined as "electric power generation within the distribution networks or on the customer side of the network [34]". These generation units inject small manageable energy in the distribution network to reduce overloading and enable

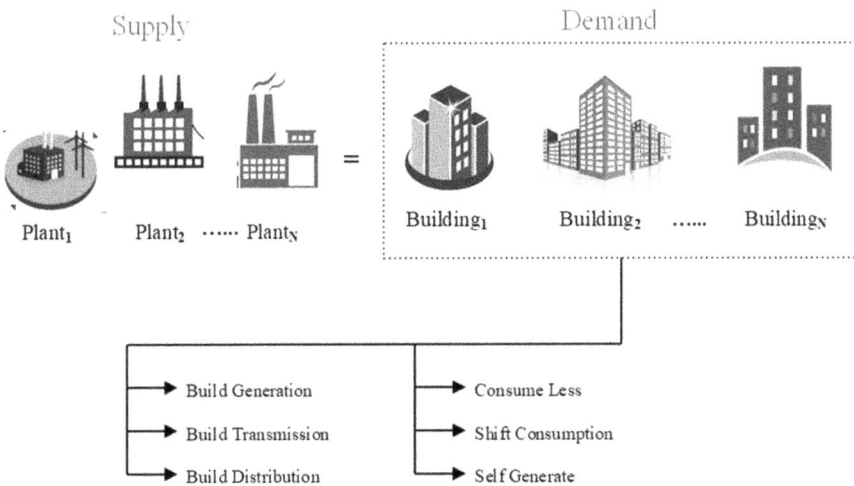

FIGURE 5.4 Elaboration of demand response.

bidirectional power exchanges between the grid and the building. Therefore, besides congestion management and bilateral energy trade, application of distributed generators may also improve network voltage, frequency stability, and resilience, as well as reduce consumption costs and emissions. In this regard, optimal sizing and siting of distributed generators may be investigated under liberalized energy market paradigm.

5.4.3 BATTERY STORAGES

Application of renewable energy has been reducing system inertia, voltage and frequency stability, and adequate reliability to keep the load running. Under such conditions, battery and electric vehicles may play a role as a reserve to mitigate these challenges [35–37]. Similarly, battery storages in microgrids may decrease emission, consumption costs and network overloading during peak hours [38]. Batteries have made it possible to feed 100% load through renewable energy sources.

However, the high manufacturing cost, durability and response time of batteries require further research endeavors [39, 40].

5.4.4 INFORMATION COMMUNICATION TECHNOLOGIES

Wired and wireless communication channels are important to transmit the data to an optimizer and then relay the dispatch signals back to the microgrid components. It is worth mentioning that any unnecessary delay in back-and-forth transmission of data to and from optimizer/control center may jeopardize system stability [41]. Therefore, there must be more research efforts to develop transmission medium with as low a latency as possible.

5.4.5 ADVANCED METERING INFRASTRUCTURES

Advanced metering infrastructure (AMI) is an interface between the customer and utility where energy consumed is accounted. The AMI supports bidirectional energy and data flow between utility and customer. Moreover, the AMI facilitates energy market by remote connect and disconnect; meter status monitoring; voltage and harmonic distortion monitoring; application of energy management systems; implementation of demand response techniques and real-time energy consumption pattern supervision [42].

In order to truly harness the benefits of the AMI, there is requirement for energy management modules capable of 1) mitigating harmonics; 2) reducing consumption cost, emission, and network losses; and 3) automatic VAR compensation. However, attainment of this level of grid modernization on both utility and customer sides may prove unnecessarily expensive for economically weak countries.

However, in case the AMI-aided energy management module is not incorporated in the distribution system, the advanced metering infrastructure will only provide benefits such as remote monitoring of 1) meter status; 2) meter reading; 3) voltage at customer premises without any automatic corrective measures; and 4) connect/disconnect. In reality, it would be important to mention that only these benefits will not turn into meaningful and rapid financial savings for both the utility and the customers.

5.4.6 Privacy and Security of a Microgrid

Secure information communication medium is mandatory for desired microgrid operations [43].

In cases where an unauthorized intruder accesses the microgrid, communications may change component settings and customer preferences. Such an event results in suboptimal operations and the system may even drift toward a blackout.

The literature survey reveals that application of energy management systems has been beneficial for all the stakeholders of the power system. To elaborate the benefits, a possible scenario of a residential building has been selected for modeling and simulation.

5.5 MODELING AND SIMULATION OF A RESIDENTIAL MICROGRID

The mathematical symbols of different variables and parameters are given below:

i	Number of homes
$P_{\mathrm{sr},i(t)}$	Output power from solar panels (kW)
$\eta_{\mathrm{sr},i}$	Efficiency of solar panels (%)
S_i	Panel area (m²)
$I(t)$	Solar irradiance (kW/m²)
$T_{\mathrm{ext}}(t)$	Atmospheric temperature (°C)
NG	Natural gas
$H_{\mathrm{cip},i}(t)$	Heat from combined cooling and heating power unit (CCHP) (kWh)
$P_{\mathrm{cip},i}(t)$	Output Power from CCHP (kW)
η_{th}	Thermal efficiencies (%)
$\eta_{\mathrm{e},i}$	Electrical efficiencies (%)
β	Conversion factor
$F_{\mathrm{NG},i}(t)$	Natural gas fed to the CCHP (m³)
$P_{\mathrm{chp},i}^{\mathrm{min}}$	Lower limit of output power from CCHP (kW)
$P_{\mathrm{chp},i}^{\mathrm{max}}$	Upper limit of output power from CCHP (kW)
$Hc_{\mathrm{hp}}^{\mathrm{min}}$	Lower limit on the CCHP produced heat (kWh)
$H_{\mathrm{chp},i}^{\mathrm{max}}$	Upper limit on the CCHP produced heat (kWh)
ramp$_i$	Ramp rate of generator unit (kW/h)
j	No. of shiftable appliances
$O_{i,j}(t)$	Binary variable
$\mathrm{IL}_{\mathrm{in},i(t)}$	Internal/appliance generated illumination (per unit)
$\rho(t)$	Linear function of electricity price
$\mathrm{IL}_{\mathrm{ou}}(t)$	Outer illumination (per unit)
$\mathrm{IL}_{\mathrm{req},i}(t)$	Required illumination (per unit)
$T_{\mathrm{ws},i}(t)$	Water storage temperature (°C)
$V_{\mathrm{cold},i}(t)$	Volume of entering cold water (liters)
$T_{\mathrm{cw},i}$	Temperature of entering cold water (°C)
V_i	Total water storage volume (liters)

w_s	Heat injected for water heating (kWh)
C_w	Specific heat capacity of water (kWh/°C)
$T_{in}, T_{in,i}(t)$	Internal room temperature (°C)
$H_{air,i}(t)$	Heat required to maintain air temperature (kWh)
$RH(t)$	Thermal resistance of building shell (°C/kW)
$T_{ext}(t)$	Atmospheric temperature (°C)
C_{air}	Specific heat capacity of air (kWh/°C)
$T_{ws,i}^{min}$	Min. water storage temperatures (°C)
$T_{ws,i}^{max}$	Max. water storage temperatures (°C)
$E_{phev,i}(t)$	Energy in the EV battery (kWh)
$P_{ch,phev,i}(t)$	Battery charging power (kW)
$P_{dch,phev,i}$	Battery discharging power (kW)
Δt	Duration of charging/discharging (h)
$\eta_{G2V,i}$	Grid to vehicle charging efficiency (%)
$\eta_{V2G,i}$	Vehicle to grid discharging efficiency (%)
$P_{ch,phev,i}(t)$	Battery charging power (kW)
$P_{charger,i}^{max}$	Vehicle battery charger rating (kW)
$P_{dch,hev,i}(t)$	Battery discharging power (kW)
$P_{charger,i}^{max}$	Vehicle battery charger rating (kW)
$SOV_{max,i}$	Upper limits on the state of charge (%)
$SOC_{min,i}$	Lower limits on the state of charge (%)
$P_{ch,phev,i}(t)$	Battery charging power (kW)
Cap_i	Capacity of vehicle battery (kW)
$E_{phev,i}(t)$	Energy level in battery (kWh)
$P_{li,i}$	Wattage of lighting load (kW)
$P_{EC}(t)$	Power fed to the electric chiller (kW)
$D_{cirt,i}(t)$	Critical load (kW)
$F_{NG,i}(t)$	Natural Gas (m³/h) fed to the building
$PRC_{NG}(t)$	Natural Gas price (cents)
$P_{net}(t)$	Power fed to or taken from the outer grid (kW)
PRC_{TOU}	Time of use price of electricity (cents)
$Cost_i$	Cost per day (cents/day)
$C_{th,i}$	Threshold fixed by the customer through micro EH control (cents)
$H_{AC,i}(t)he$	Fed to the absorption chiller (kWh)
$H_{in,i}(t)$	Heat injected into the thermal energy storage (kWh)
$H_{dr,i}(t)$	Heat drawn out of the thermal energy storage (kWh)
$COP_{AC,i}$	Coefficient of performance of absorption chiller (kWh)
$COP_{EC,i}$	Coefficient of performance of the electric chiller (%)
$H_{air}(t)$	Energy for maintaining air temperature (kWh)
$P_{to}(t)$	Total flowing power (kW)
P_{avg}	Average power (kW)
X_p^k	Flower/pollen p at kth iteration
S	Step-size
$L(\lambda)$	Levy flight step-size
$\Gamma(\lambda)$	Gamma function

X_l^k Pollen grains
σ Deviation
μ Mean of distribution function
α Value at risk

5.5.1   STRUCTURAL DESCRIPTION OF MICROGRID

Almost half of the load on electrical network relates to the residential sector. Therefore, in this section two residential microgrids connected with common electric and natural gas distribution networks, have been modeled for analysis as shown in Figure 5.5. Both microhubs contain "combined cooling and heating power (CCHP)" with "power generation unit (PGU)" and a "heat recovery unit (HRU)", "thermal energy storage (TES)", "water storage for heating and absorption chiller (AC) and electric chiller (EC)" for cooling. As can be seen in Figure 5.5, microhub-M1 has "plug-in hybrid electric vehicle (PHEV)" and micro hub-M2 does not have it. The reason for this is to investigate the effect of PHEV on energy cost, emission, and network losses. The PHEV discharges into the system during peak hours from 4:00 p.m. till midnight, recharges completely during off-peak hours from midnight till 8:00 a.m., whereas it remains out of the micro hub during 8:00 a.m. to 4:00 p.m. The PGU takes natural gas (NG) input to generate electricity, whereas heat produced

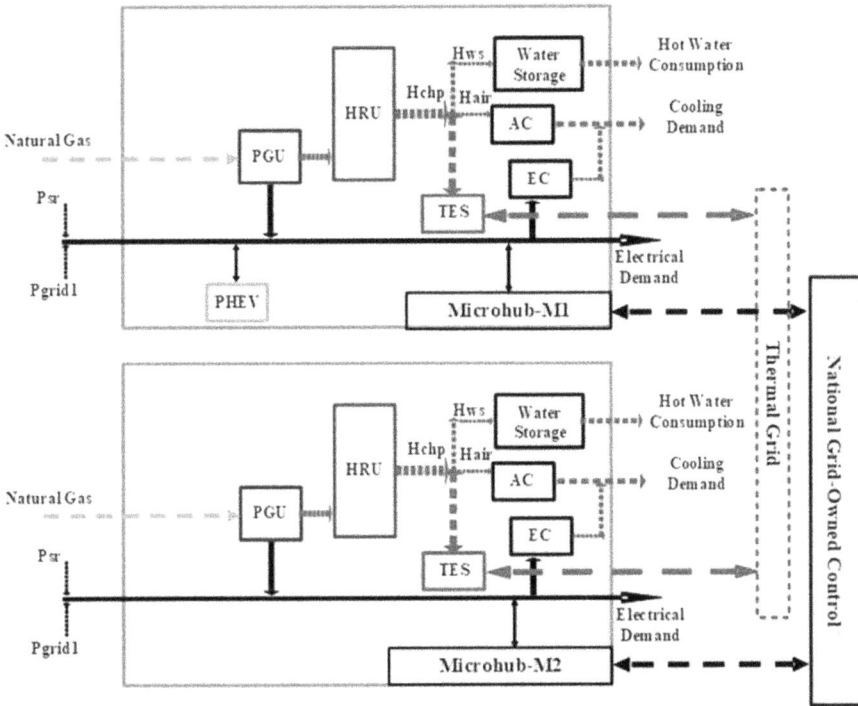

FIGURE 5.5   Overall concept [44].

during this process is captured by HRU. This heat is used for water heating and room chilling in AC. In case of inadequate heat for water heating and room chilling, extra heating is drawn from TES and/or imported from the central heat distribution system. An EC is used in combination with AC to enhance the micro hub's cooling efficiency [44]. Electricity, NG, and solar energy are the input carriers for both micro hubs.

In the modeling process, an effort has been made to incorporate all the attributes of a building located in any part of the globe. The microgrid considers: electricity flows; fossil fuel-based generator; renewable energy-based generators; electric vehicle storage; heat flows; heat storage connected with the community-based heating system; shiftable appliances (those the optimization layer moves from peak hours to off-peak hours) such as "iron, dishwasher, washing machine, dryer and pool pump; critical load like fridge, oven, toaster, stove, and computer; curtailable load such as lights and shed-able load; e.g., water heater, AC and the EC."

5.5.2 OPERATIONAL DESCRIPTION OF MICROGRID

The proposed residential microgrid energy management system functions in two layers: 1) microgrid-owned or micro energy hub control module residing in the customer premises optimally tracks energy consumption cost and greenhouse gas emission; and 2) national grid-owned or macro energy hub control module resides on pole or pad mounted transformer to optimally track the load. The national grid-owned control takes building level information through the micro hub module and the AMI and external data from utility and weather servers to generate optimal dispatch for all the components.

5.6 MATHEMATICAL MODELING OF MICROGRID COMPONENTS

Mathematical model of microgrid components is given below.

5.6.1 SOLAR PHOTOVOLTAIC POWER

The solar photovoltaic output power for ith home is given in Eq. (5.1):

$$P_{sr,i}(t) = \eta_{sr,i} . S_i . I(t) . \left(1 - 0.005 . \left(T_{ext}(t) - 25\right)\right) \tag{5.1}$$

To stochastically model the solar irradiance, hourly insolation data of 2013 for Islamabad is used to compute the "probability density function of normal distribution." The expression is given in Eq. (5.2):

$$h(I) = \frac{1}{\sigma\sqrt{2\pi}} e^{-(1-\mu)^2/2\sigma^2} \tag{5.2}$$

5.6.2 COMBINED COOLING AND HEATING POWER UNIT

Combined cooling and heating power with a 5 kW PGU and an HRU is used in the building as fossil fuel-based distributed generator. The HRU collects all the heat generated during the power generation process and utilizes it for rooming chilling and water heating. The unit functions on the natural gas.

For stochastic availability of natural gas network, the two state probabilities are in Eqs. (5.3)–(5.6):

$$p(\varphi_t = 1 | \varphi_{t_0} = 1) = \frac{\mu_i}{\mu_i + \lambda_i} + \frac{\lambda_i}{\lambda_i + \mu_i} \cdot e^{-(\lambda_i + \mu_i) \cdot (t - t_0)} \tag{5.3}$$

$$p(\varphi_t = 0 | \varphi_{t_0} = 1) = \frac{\lambda_i}{\lambda_i + \mu_i} \cdot \left(1 - e^{-(\lambda_i + \mu_i) \cdot (t - t_0)}\right) \tag{5.4}$$

$$p(\varphi_t = 1 | \varphi_{t_0} = 0) = \frac{\mu_i}{\lambda_i + \mu_i} \cdot \left(1 - e^{-(\lambda_i + \mu_i) \cdot (t - t_0)}\right) \tag{5.5}$$

$$p(\varphi_t = 0 | \varphi_{t_0} = 0) = \frac{\mu_i}{\mu_i + \lambda_i} + \frac{\lambda_i}{\lambda_i + \mu_i} \cdot e^{-(\lambda_i + \mu_i) \cdot (t - t_0)} \tag{5.6}$$

Heat generated and output power are related by Eq. (5.7):

$$H_{\mathrm{chp},i}(t) = P_{\mathrm{chp},i}(t) \cdot \frac{\eta_{\mathrm{th},i}}{\eta_{\mathrm{e},i}} \tag{5.7}$$

Fuel injected and output power are related by Eq. (5.8):

$$F_{\mathrm{chp},i}(t) = P_{\mathrm{chp},i}(t) \cdot \frac{\beta}{\eta_{\mathrm{e},i}} \tag{5.8}$$

Upper and lower limits of the PGU are presented in Eq. (5.9):

$$P_{\mathrm{chp},i}^{\min} \leq P_{\mathrm{chp},i}(t) \leq P_{\mathrm{chp},i}^{\max} \tag{5.9}$$

Upper and lower limits of heat generated are in Eq. (5.10):

$$H_{\mathrm{chp},i}^{\min} \leq H_{\mathrm{chp},i}(t) \leq H_{\mathrm{chp},i}^{\max} \tag{5.10}$$

Ramp rate limit of the PGU in terms of power generated is given in Eq. (5.11):

$$\left| P_{\mathrm{chp},i}(t) - P_{\mathrm{chp},i}(t-1) \right| \leq \mathrm{ramp}_i \tag{5.11}$$

Ramp rate limit of the PGU in terms of heat generated is given in Eq. (5.12):

$$H_{\mathrm{chp},i}(t) = H_{\mathrm{chp},i}(t-1) \leq \mathrm{ramp}_i \cdot \frac{\eta_{\mathrm{th},i}}{\eta_{\mathrm{e},i}} \tag{5.12}$$

5.6.3 FUEL CELL

The expression for fuel cell:

$$C_{i,f}(t) = B_{i,f} \cdot \frac{P_{i,f}(t)}{\eta_{i,f}} + C_{i,f}^{SU} + C_{i,f}^{SD} + C_{i,f}^{om} \cdot \overline{P}_{i,f} \qquad (5.12a)$$

This model and its parameter values are taken from [20]. Where $C_{i,f}(t), B_{i,f}.C_{i,f}^{SU}, C_{i,f}^{SD}, C_{i,f}^{om}, \overline{P}_{i,f}, P_{i,f}(t)$ and $\eta_{i,f}$ are fuel cell cost, fixed charges, start-up cost, shutdown cost, and operation and maintenance cost, respectively.

5.6.4 RESPONSIVE LOAD

Loads may be divided as "non-responsive and responsive." Responsive load may further be categorized as "curtailable; i.e., lights and shiftable such as dishwasher, dryer, washing machine, iron, and pool pump." All other loads like TV, fridge, laptop, toaster, oven, computer, and stoves are non-responsive or critical.

5.6.4.1 Shiftable Appliances

Optimization module intelligently shifts these appliances from peak hours to off-peaks hours for energy cost reduction. User-specified operational duration of a suitable appliance may be modeled as:

$$\sum_{t=bi,j}^{ei,j} on_{i,j}(t) = UT_{i,j} \qquad (5.13)$$

5.6.4.2 Curtailable Lighting Load

Lights are treated as a curtailable load. In this study, the curtailment factor is 1 during peak and 0 in off-peak hours. Moreover, external illumination is taken into account to reduce the energy consumption as:

$$IL_{in,i}(t) + IL_{out}(t) \geq (1 - 0.2.\rho(t)).IL_{req,i}(t) \qquad (5.14)$$

Probability density function of outer illumination given as:

$$h(IL_{out}) = \frac{1}{\sigma\sqrt{2\pi}} e^{-(1-\mu)^2/2\sigma^2} \qquad (5.15)$$

The parameters are calculated using hourly outer illumination (penetrating doors and windows) values for 2013 [45].

5.6.4.3 Flexible Thermal Load

Room chilling and water heating are served by either heat generated by a combined cooling and heating power unit and/or heat imported from the central heating system.

Tasdighi et al. provided utilization of hourly hot water, tank volume, and water inlet temperature [46]. Water temperature is given as:

$$T_{ws,i}(t+1) = \left[\frac{V_{cold,i}(t).(T_{cw,i}(t) - T_{ws,i}(t)) + V_i.T_{ws,i}(t)}{V_i} \right] + \left[\frac{H_{ws,i}(t)}{V_i.C_w} \right] \quad (5.16)$$

Room temperature is given as:

$$T_{in,i}(t+1) = T_{in,i}(t).e^{\frac{1}{R.C_{air}}} + \left[(-R.H_{air,i}(t) + T_{ext}(t)).\left(1 - e^{\frac{1}{R.C_{air}}} \right) \right] \quad (5.17)$$

Upper and lower temperature limits for stored water are:

$$T_{ws,i}^{min} \leq T_{ws,i}(t) \leq T_{ws,i}^{max} \quad (5.18)$$

Upper and lower room temperature limits are:

$$T_{in,i}^{min} \leq T_{in,i}(t) \leq T_{in,i}^{max} \quad (5.19)$$

5.6.4.4 Electric Vehicle
Energy stored in an electric vehicle:

$$E_{phev,i}(t+1) = E_{phev,i}(t) + \eta_{G2V,i}.P_{ch,phev,i}.\Delta t - \frac{1}{\eta_{V2G,i}}.P_{dch,phev,i}(t).\Delta t \quad (5.20)$$

Maximum charger capacity is modeled in Eqs. (5.21) and (5.22):

$$P_{ch,phev,i}(t) \leq P_{charger,i}^{max} \quad (5.21)$$

$$P_{dch,phev,i}(t) \leq P_{charger,i}^{max} \quad (5.22)$$

Lower and upper limits on state of charge are in Eq. (5.23):

$$SOC_{phev,min,i} \leq SOC_{phev,i}(t) \leq SOC_{phev,max,i} \quad (5.23)$$

Stored power in the electric vehicle battery is given in Eq. (5.24):

$$P_{ch,phev,i}(t).\eta_{G2V,i}.\Delta t \leq Cap_i - E_{phev,i}(t) \quad (5.24)$$

Energy available in PHEV battery is:

$$P_{dch,phev,i}(t).\frac{1}{\eta_{V2G,i}}.\Delta t \leq E_{phev,i}(t) \quad (5.25)$$

5.6.4.5 Power Balance Equation

Power balance equation without demand response equaling sources and loads is given in Eq. (5.26):

$$P_{net}(t) - \left(P_{ch,phev,i}(t).\eta_{G2V,i} - \frac{P_{dch,phev,i}}{\eta_{V2G,i}} \right) + P_{chp,i}(t) + P_{sr,i}(t)$$

$$= D_{total}(t) + \left(IL_{in,i}(t).P_{li,i} \right) + P_{EC}(t) \tag{5.26}$$

Power balance equation with demand response is in Eq. (5.27):

$$P_{net}(t) - \left(P_{ch,phev,i}(t).\eta_{G2V,i} - \frac{P_{dch,phev,i}}{\eta_{V2G,i}} \right) + P_{chp,i}(t) + P_{sr,i}(t)$$

$$= D_{crit,i}(t) + \left(IL_{in,i}(t).P_{li,i} \right) + \left(\sum_{j=1}^{N} on_{i,j}(t).P_{i,j} \right) + P_{EC}(t). \tag{5.27}$$

$P_{ne}(t)$ is the power imported from national grid.

5.6.4.6 Thermal Balances

Thermal balance is given as in Eq. (5.28):

$$H_{chp,i}(t) + H_{g,i}(t) = H_{ws,i}(t) + H_{AC,i}(t) + \left(-H_{in,i}(t).\eta_{in,i} + \frac{H_{dr,i}(t)}{\eta_{dr,i}} \right) \tag{5.28}$$

$H_{g}(t)$ is the heat imported from external thermal grid.

Heat input to absorption chiller is given in Eq. (5.29):

$$H_{AC,i}(t).COP_{AC,i} + E_{EC,i}(t).COP_{EC,i} = H_{air,i}(t) \tag{5.29}$$

5.6.4.7 Conditional Value at Risk

To improve the microgrid's resilience during emergencies, conditional value at risk (CVaR) is modeled in Eq. (5.30):

$$CVaR = \alpha + \frac{1}{p.(1-\beta)}.\sum_{s=1}^{p} f(x,y_s) \tag{5.30}$$

5.6.4.8 Objective Functions

Objective function for cost minimization is containing weighted CVaR is in Eq. (5.31):

$$Cost_i = min \left[\sum_{t=1}^{24} \left[P_{net,i}(t).PRC_{TOU}(t) + F_{NG,i}(t).PRC_{NG}(t) \right] + \omega.CVaR \right] \tag{5.31}$$

User-specified consumption cost threshold is given in Eq. (5.32):

$$\text{Cost}_i \le C_{\text{th},i} \tag{5.32}$$

Objective function for emission reduction containing weighted CVaR is in Eq. (5.33):

$$\text{Emission}_i = \min \sum_{t=1}^{24} \left[P_{\text{net},i}(t).\left(\mu_{\text{net}}^{CO_2} + \mu_{\text{net}}^{NO_x} + \mu_{\text{net}}^{SO_x} \right) \right.$$

$$\left. + F_{\text{chp},i}(t).\left(\mu_f^{CO_2} + \mu_f^{NO_x} + \mu_f^{SO_x} \right) \right] + \omega.\text{CVaR} \tag{5.33}$$

User-defined threshold on emission is given in Eq. (5.34):

$$\text{Emission}_i \le E_{\text{th},i} \tag{5.34}$$

Objective function for network load deviation is given in Eq. (5.35):

$$\text{Deviation} = \min \left[\sum_{t=1}^{24} \left| P_{\text{tot}}^2(t).R - P_{\text{avg}}^2 \right| + \omega.\text{CVAR} \right] \tag{5.35}$$

Total power over the horizon of 24 hours is given as in Eq. (5.36):

$$P_{\text{tot}}(t) = \sum_{i} \sum_{t=1}^{24} P_{\text{net},i}(t) \tag{5.36}$$

Average power in secondary distribution is given in Eq. (5.37):

$$P_{\text{avg}} = \frac{1}{24} \left[P_{\text{tot}}(t) \right] \tag{5.37}$$

5.7 METHODOLOGY

The proposed model is solved using "Mixed Integer Linear Programming (MILP) by CPLEX solver under GAMS on a Pentium IV, 2.6 GHz processor with 4 GB of RAM." All the appliance data and model parameters are taken from Ref. [47].

5.7.1 DESCRIPTION OF CASE STUDIES

Case 1 represents the current status of utility's distribution system with dumb/unintelligent charging of electric vehicle and with absence the demand response strategies. Customers switch the shiftable appliances at their own desire, without considering the national grid's off-peak and peak hours. Dumb charging irrespective of the peak or off-peak hours permits electric vehicle to connect with national grid for charging purpose only. Preferred room and water temperature control is "heat lead" [47] and "curtailment factor" [47] is omitted.

TABLE 5.2
Cases Studies

Case Studies	Dumb Charging	DR Program			TES	Smart EV	Biogas
		Load Shifting	Load Curtailment	Flexible Thermal Load			
Case 1	✓						
Case 2	✓	✓	✓	✓			
Case 3		✓	✓	✓	✓	✓	
Case 4		✓	✓	✓	✓	✓	
Case 5		✓	✓	✓	✓	✓	

In Case 2, "dumb charging," "load shifting," and "load curtailment and flexible thermal load" are considered.

In Case 3, smart electric vehicle charging/discharging and thermal energy storage are considered. Under smart interaction with the grid, vehicle charges in off-peak from midnight to 8:00 a.m. and discharges into the microgrid during peak hours from 4:00 p.m. to midnight.

Case 4 considers all the aspects of Case 3 while solving the bi-objective problem for both cost and emission reduction, using "Pareto Front Sets" [48].

Case 5 incorporates both cost and emission as constraints while minimizing network load deviations. This description is summarized in Table 5.2.

5.7.2 SIMULATION RESULTS

Scenario 14 with an extremely low occurrence probability is designated for analysis and shown in Table 5.3. It can be seen that electric, natural gas network, and central thermal grids switch off due to load shedding or fault at 2:00 a.m., 8:00 a.m., and 3:00 p.m. during a day. Table 5.4 shows the simulation results.

Case 1 relates to the "heat-led control" of combined cooling and heating power. It doesn't permit any change in water and room temperatures and tracks the preferred settings stringently. Preferred room and water temperatures are 22°C and 70°C respectively. Such a harsh temperature regulation decreases the output power of combined cooling and heating power unit, thus, diminishes the sold power, and the earned

TABLE 5.3
Scenario 14 with Occurrence Probability of 1.63%

Electric network	1	0	1	1	1	1	1	1	1	1	1	1	1	1	1	1	1	1	1	1	1	1	1	1
Natural gas network	1	1	1	1	1	1	1	1	0	1	1	1	1	1	1	1	1	1	1	1	1	1	1	1
Thermal grid	1	1	1	1	1	1	1	1	1	1	1	1	1	1	1	0	1	1	1	1	1	1	1	1

revenue. Case 1 evidently determines that sharp temperature controls raise the energy cost and shrinks the revenue, as shown in Table 5.4. Exclusion of demand response techniques also affects the cost. Another prominent feature is that the energy cost of Home 1 remains high while earning a small revenue in comparison to Home 2, due to dull electric vehicle presence in the former microgrid.

Case 2 considers the demand response programs. The allowed temperature variation range for water is 60°C to 80°C and 21°C to 14°C for room. It is eminent that in Case 2, output power from combined cooling and heating power unit is high in comparison to Case 1, with high sold power, low cost, and greater revenues. The surge in sold power and output from combined cooling and heating power is because of the presence of the "flexible thermal load" instead of heat-led control. The allowed temperature change permits the combined cooling and heating power unit to produce higher power. The fall in cost in this case compared to Case 1 is due to the inclusion of "load shifting" and "illumination curtailment." Analysis proves that with the inclusion of demand response techniques, building-owned control shifts the load from peak to off-peak for cost reduction. However, such a load shifting may rebound the peak load on the network during off-peak hours thus causing another complicated problem for utility owners which may increase losses, instability, and trigger load shedding.

Case 3 comprises thermal energy storage in both buildings and a smart electric vehicle in Home 1 only. Thermal energy storage is a building-owned heat storage, whereas electric vehicle is a mobile electricity storage. Table 5.4 shows that energy cost and network load fluctuations reduce compared to Cases 1 and 2. Moreover, a decrease in power imported and an increase in earned revenue, power sold, and output power from combined cooling and heating power unit are also witnessed.

Case 4 addresses a bi-objective framework with cost and emissions taking environment as a stake holder. Table 5.4 shows a drop in power purchased from national grid for a reduction in emission. This decrease is due to the high "emission rates"

TABLE 5.4
Simulation Results

Case Studies	Home	Energy Cost (cents/day)	CCHP (kW)	Purchased Power (kW)	Sold Power (kW)	Revenue (cents/ day)	Deviation (kW)	Emission (kg/day)
Case 1	1	195.394	16.230	24.564	10.670	116.560	17.723	
	2	167.349	13.578	19.040	10.730	114.923		
Case 2	1	168.574	16.045	23.450	15.190	169.250	19.347	
	2	134.092	14.923	22.560	15.780	186.560		
Case 3	1	124.504	18.219	24.257	16.429	258.257	17.550	24.779
	2	129.401	19.420	21.310	17.602	266.691		22.001
Case 4	1	126.064	18.161	18.071	9.878	227.118	12.982	16.917
	2	131.383	15.554	15.904	8.192	173.056		14.948
Case 5	1	243.992	27.716	11.479	12.647	143.732	10.057	13.021
	2	229.615	21.234	12.188	9.0752	104.583		13.270
	2	171.363	24.327	13.230	13.894	152.144		14.491

[47] of purchased energy from the external grid. Likewise, sold power also declines. Subsequently, this decreases in grid trade and lessens the deviation.

Case 5 addresses load deviation on the low-tension distributor through externally placed national grid-owned control. Whenever the outer control activates, the building level control simply behaves like a buffer and acts as a medium of information transmission to macro hub control. The customer fixes the cost (≤260 cents) and emissions (≤22 kg/day for a house) constraints which are communicated to the macro hub control to dispatch appliances for load variation reduction. As can be seen, load deviations reduce from 19.347 kW to 10.057 kW, thereby ensuring the reliability and loss reduction. However, the costs of both buildings increase.

With this analysis, it can be inferred that both customer and utility stakes cannot be attained at the same time. For this purpose, low-cost fuel (5 kW) is injected, and the results are given in Table 5.5. Simulations show a marked decrease in energy prices compared with Case 5 in Table 5.4. Consequently, it can be inferred that fuel cells or any low-priced energy sources help in attaining bilateral stakes simultaneously.

Figures 5.6–5.8 show room and water temperatures, hourly building load after demand response and network load deviation at low-tension network for Home 1 in Case 5 with fuel cell. Figure 5.9 shows the state of battery charge in risk neutral and risk-averse cases under Case 5.

TABLE 5.5

Results of Case 5 with Fuel Cell

Case Studies	Home	Energy Cost (cents/day)	CCHP (kW)	Purchased Power (kW)	Sold Power (kW)	Revenue (cents/day)	Deviation (kW)	Emission (kg/day)
Case 5	1	171.363	24.327	13.230	13.894	152.144	11.5	14.491
	2	178.98	22.765	14.67	14.87	145.67		14.578

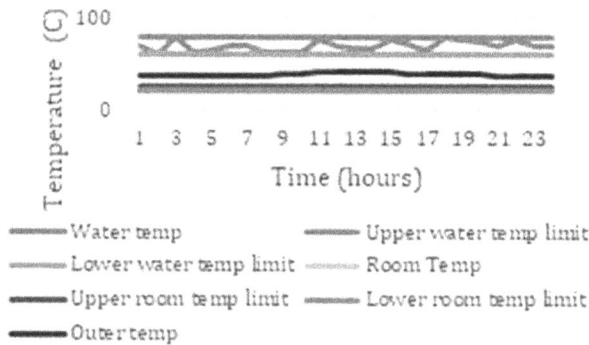

FIGURE 5.6   Water and room temperatures for Home 1.

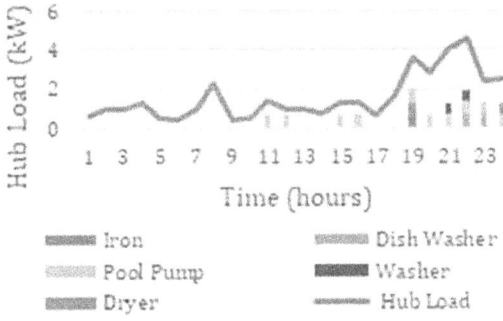

FIGURE 5.7   Hourly building load of Home 1.

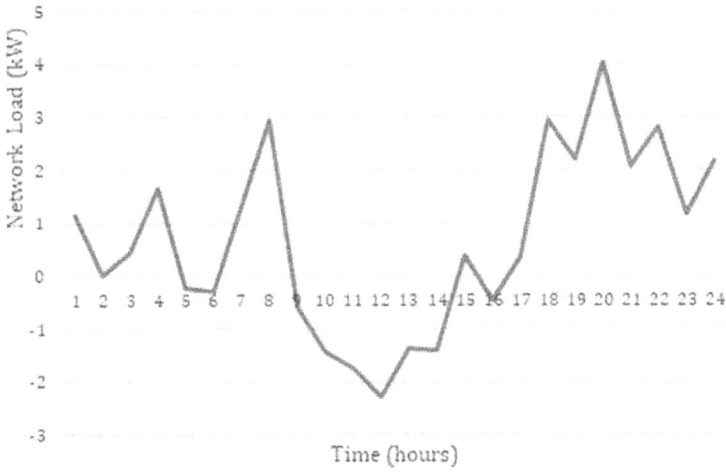

FIGURE 5.8   State of battery charge.

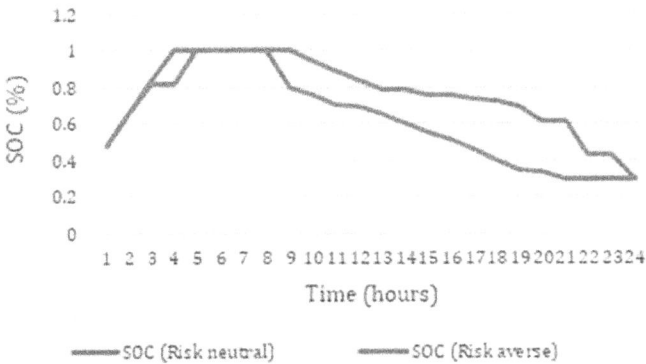

FIGURE 5.9   Network load deviation.

It can be seen that with the addition of conditional value at risk, the charge retaining capability of battery increases; this charge can serve any load during outage on a carrier, thereby improving the resilience of microgrid during natural disasters.

REFERENCES

[1] S&C Electric Company, "Microgrids: An Old Idea with New Potential," tech. rep., S&C Electric Company.

[2] Schewe, P. F., *The grid: A journey through the heart of our electrified world.* Washington, D.C.: The National Academies Press, Feb 2007.

[3] Geidl, Martin. Integrated modeling and optimization of multi-carrier energy systems. Diss. ETH Zurich, 2007.

[4] "Market Data: Microgrids, Campus/Institutional, Commercial & Industrial, Community, Community Resilience, Military, Utility Distribution, and Remote Microgrid Deployments: Global Capacity and Revenue Forecasts," 2016.

[5] Su, W. and Wang, J., "Energy management systems in microgrid operations," *The Electricity Journal* 25 (2012): 45–60.

[6] Garces, Alejandro, "A mixed-integer quadratic formulation of the phase-balancing problem in residential microgrids." *Applied Sciences* 11.5 (2021): 1972.

[7] Simental, O. Q., P. Mandal, and E. Galvan. "Enhancing distribution grid resilience to power outages using electric vehicles in residential microgrids." *2021 North American Power Symposium*, 2021.

[8] Nikkhah, Saman, et al. "A community-based building-to-building strategy for multiobjective energy management of residential microgrids." *2021 12th International Renewable Engineering Conference (IREC).* IEEE, 2021.

[9] Khezri, Rahmat, et al. "Optimal sizing of grid-tied residential microgrids under real-time pricing." *2021 IEEE Energy Conversion Congress and Exposition (ECCE).* IEEE, 2021.

[10] Habib, Habib Ur Rahman, et al. "Optimal planning of residential microgrids based on multiple demand response programs using ABC algorithm." *IEEE Access* 10 (2022): 116564116626.

[11] Som, Shreyasi, et al. "bess reserve-based frequency support during emergency in Islanded residential microgrids." *IEEE Transactions on Sustainable Energy* 14 (2023). 10.1109/TSTE.2023.3244002

[12] Çetinbaş, Ipek, Bünyamin Tamyürek, and Mehmet Demirtaş. "Sizing optimization and design of an autonomous AC microgrid for commercial loads using Harris Hawks optimization algorithm." *Energy Conversion and Management* 245 (2021): 114562.

[13] Sepúlveda-Mora, Sergio, and Steven Hegedus. "Design of a Resilient and Eco-friendly Microgrid for a Commercial Building." *Aibi Revista De investigación, administración E ingeniería* 9.1 (2021): 8–18.

[14] Yusuf, Jubair, et al. "A Centralized Optimization Approach for Bidirectional PEV Impacts Analysis in a Commercial Building-Integrated Microgrid." arXiv preprint arXiv:2104.03498 (2021).

[15] Rajamand, Sahbasadat. "Optimal electrical vehicle charging park in residential-commercial microgrid based on probability of vehicles routing and cost minimization approach." *International Journal of Energy Research* 45.6 (2021): 8593–8605.

[16] Ferahtia, Seydali, et al. "Optimal techno-economic multi-level energy management of renewable-based DC microgrid for commercial buildings applications." *Applied Energy* 327 (2022): 120022.

[17] Rezk, Hegazy, et al. "A comparison of different renewable-based DC microgrid energy management strategies for commercial buildings applications." *Sustainability* 14.24 (2022): 16656.

[18] Sackey, David Mensah, et al. "Techno-economic analysis of a microgrid design for a commercial health facility in Ghana-Case study of Zipline Sefwi-Wiawso." *Scientific African* 19 (2023): e 01552.

[19] Daneshvar, Mohammad Reza, et al. "A novel techno-economic risk-averse strategy for optimal scheduling of renewable-based industrial microgrid." *Sustainable Cities and Society* 70 (2021): 102879.

[20] Ali, Saqib, and Tahir Nadeem Malik. "Two-cored energy management system for an industrial microgrid." *Mehran University Research Journal of Engineering & Technology* 40.4 (2021): 793–808.

[21] Guo, Juntao, et al. "Low-carbon robust predictive dispatch strategy of the photovoltaic microgrid in industrial parks." *Frontiers in Energy Research* 10 (2022): 900503.

[22] Arif, Shaila, et al. "Enhancement of solar PV hosting capacity in a remote industrial microgrid: A methodical techno-economic approach." *Sustainability* 14.14 (2022): 8921.

[23] Ahmed, Mohammad Adel, et al. "Techno-economic optimal planning of an industrial microgrid considering integrated energy resources." *Frontiers in Energy Research* 11 (2023).

[24] Swain, Dipak R., et al. "Optimal fractional sliding mode control for the frequency stability of a hybrid industrial microgrid." *AIMS Electronics and Electrical Engineering* 7.1 (2023): 14–37.

[25] Uzair, Muhammad, et al. "Challenges, advances and future trends in AC microgrid protection: With a focus on intelligent learning methods." *Renewable and Sustainable Energy Reviews* 178 (2023): 113228.

[26] Hewitson, Leslie, Mark Brown, and Ramesh Balakrishnan. *Practical power system protection*. Elsevier, 2004.

[27] Blackburn, J. Lewis, and Thomas J. Domin. *Protective relaying: principles and applications*. CRC Press, 2015.

[28] Pabst, P. Challenges of microgrid deployment, 2017. Challenges of Microgrid Deployment – IEEE Smart Grid.

[29] Razibul, Islam Md, and Hossam A. Gabbar. "Analysis of microgrid protection strategies." *2012 International Conference on Smart Grid (SGE)*. IEEE, 2012.

[30] Nelson, R. J., H. Ma, and N. M. Goldenbaum. "Fault ride-through capabilities of Siemens full-converter wind turbines." *2011 IEEE Power and Energy Society General Meeting*. IEEE, 2011.

[31] Eto, Joseph H., et al. "The certs microgrid concept, as demonstrated at the certs/aep microgrid test bed." US Department of Energy, Berkeley 53 (2018).

[32] Zhang, Qin, and Juan Li. "Demand response in electricity markets: A review." *2012 9th International Conference on the European Energy Market*. IEEE, 2012.

[33] Siano, Pierluigi. "Demand response and smart grids—A survey." *Renewable and Sustainable Energy Reviews* 30 (2014): 461–478.

[34] Ackermann, Thomas, Göran Andersson, and Lennart Söder. "Distributed generation: A definition." *Electric Power Systems Research* 57.3 (2001): 195–204.

[35] Zia, Muhammad Fahad, et al. "Energy management system for a hybrid PV-Wind-TidalBattery-based islanded DC microgrid: Modeling and experimental validation." *Renewable and Sustainable Energy Reviews* 159 (2022): 112093.

[36] Lipu, MS Hossain, et al. "A review of controllers and optimizations based scheduling operation for battery energy storage system towards decarbonization in microgrid: Challenges and future directions." *Journal of Cleaner Production* 360 (2022): 132188.

[37] Wei, Yifan, et al. "Toward more realistic microgrid optimization: Experiment and high-efficient model of Li-ion battery degradation under dynamic conditions." *eTransportation* 14 (2022): 100200.

[38] Bhatt, Ankit, and Weerakorn Ongsakul. "Optimal techno-economic feasibility study of netzero carbon emission microgrid integrating second-life battery energy storage system." *Energy Conversion and Management* 266 (2022): 115825.

[39] Ding, Lei, et al. "Low-cost mass manufacturing technique for the shutdown-functionalized lithium-ion battery separator based on Al_2O_3 coating online construction during the β-iPP cavitation process." *ACS Applied Materials & Interfaces* 14.5 (2022): 6714–6728.

[40] Yun, Lingxiang, Lin Li, and Shuaiyin Ma. "Demand response for manufacturing systems considering the implications of fast-charging battery powered material handling equipment." *Applied Energy* 310 (2022): 118550.

[41] Saleh, Mahmoud, et al. "Impact of information and communication technology limitations on microgrid operation." *Energies* 12.15 (2019): 2926.

[42] Firoozabadi, Mehdi Savaghebi, et al. "Voltage harmonics monitoring in a microgrid based on advanced metering infrastructure (AMI)." *Seminario Anual de Automática, Electrónica Industrial e Instrumentación 2015.* 2015.

[43] Ghosh, Prasanta K., et al. "Applications of blockchain methodologies for microgrid energy transactions while maintaining user privacy and data security–A review." *2022 IEEE 10th International Conference on Smart Energy Grid Engineering (SEGE).* IEEE, 2022.

[44] Raza, Aamir, and Tahir Nadeem Malik. "Biogas supported bi-level macro energy hub management system for residential customers." *Journal of Renewable and Sustainable Energy* 10.2 (2018): 025501.

[45] See http://weather.uwaterloo.ca/data.html#select for UW Weather Station, "Data Archives – Incoming Shortwave Radiations, 2009"; See http://oeb.gov.on.ca/OEB/For+Consumers/Underestanding+Your+Bill+Rates+and+Prices/Electricity+in+ontario forElectricity, "Prices in Ontario".

[46] Tasdighi, M., H. Ghasemi, and A. Rahimi-Kian, "Residential microgrid scheduling based on smart meters data and temperature dependent thermal load modeling," *IEEE Transactions on Smart Grid* 5.1 (2014): 349–357.

[47] Brahman, Faeze, Masoud Honarmand, and Shahram Jadid. "Optimal electrical and thermal energy management of a residential energy hub, integrating demand response and energy storage system." *Energy and Buildings* 90 (2015): 65–75.

[48] Yalcin, G. D. and N. Erginel, "Determining weights in multi-objective linear programming under fuzziness," *Proceedings of WCE.* 2011, p. 2.

6 Energy Storage Systems in Microgrid Operation

Sonu Kumar
Koneru Lakshmaiah Education Foundation, Vaddeswaram, India
K.S.R.M. College of Engineering, Kadapa, India
Manipur International University, Manipur, India

Y. Lalitha Kameswari
Koneru Lakshmaiah Education Foundation, Vaddeswaram, India
MLR Institute of Technology, Hyderabad, India

S. Koteswara Rao
Koneru Lakshmaiah Education Foundation, Vaddeswaram, India

B. Pragathi
DVR & Dr. HS MIC College of Technology, Kanchikacherla, India

6.1 INTRODUCTION

Energy storage systems play a vital role in the operation of microgrids, which are local power systems operating within distribution networks. These systems have gained popularity due to their ability to reduce environmental impacts, increase energy efficiency, provide ride-through capability during grid outages, and integrate renewable energy sources effectively. Microgrids combine renewable energy generation and storage technologies to meet regional energy demand and enhance local energy security. Renewable sources like wind, solar, and hydro energy are cost-effective and contribute to improved energy supply in remote communities. Energy storage in microgrids ensures stability, enables bidirectional power flows, and addresses challenges such as low inertia, load perturbations, and uncertainties. Microgrids can be designed using DC or AC, and advanced control, protection, and optimization methods are being explored to enhance reliability and efficiency [1]. Various energy storage technologies, such as pumped hydro, compressed air energy, flywheels, fuel cells, biomass, fossil fuels, batteries, supercapacitors, and heat pumps, are utilized in microgrids [2]. The choice of energy storage technology depends on factors like

DOI: 10.1201/9781003481836-6

discharge time, application requirements, and advantages and disadvantages associated with each technology [2]. As microgrids continue to evolve and renewable energy adoption expands, energy storage systems will remain crucial for their efficient operation and integration of sustainable power sources.

6.1.1 ENERGY STORAGE SYSTEMS

Energy Storage Systems (ESS) play a crucial role in the operation of microgrids by providing various benefits such as energy management, grid stability, and integration of renewable energy sources. ESS are key components in microgrid operation, enabling efficient energy management, and enhancing grid stability. Microgrids are localized power systems that can operate independently or in connection with the main grid. They consist of various components, including generation sources, loads and control systems. Traditional grid infrastructure faces challenges such as grid instability, limited capacity for renewable energy integration, and vulnerability to power outages. Energy storage systems address these challenges by storing excess energy during periods of low demand and releasing it during peak demand, providing backup power, and improving overall grid reliability. Energy storage systems offer numerous benefits in microgrid operation. They provide flexibility in energy management, allowing for efficient load balancing and demand response. ESS can also support the integration of renewable energy sources by smoothing out their intermittency and optimizing their utilization. Additionally, they enhance grid stability by providing frequency regulation, voltage support, and grid backup during outages. There are various energy storage technologies utilized in microgrid applications. Battery-based systems, such as lithium-ion and lead–acid batteries, are widely used due to their high energy density and fast response. Mechanical energy storage systems, including pumped hydro and flywheels, store energy in kinetic or potential form. Thermal energy storage systems utilize materials with high heat capacity to store and release energy.

In order to overcome the difficulties of contemporary energy management and make the transition to a more sustainable and decentralized energy grid, energy storage devices are essential. These systems give power systems flexibility, stability, and efficiency by allowing the capture, storage, and subsequent release of energy for a variety of uses. Energy storage systems help to optimize energy use, improve grid resilience, and encourage the integration of renewable energy sources by storing excess energy during times of low demand and releasing it during times of peak demand or when intermittent renewable sources are not available. We shall examine the foundations and significance of energy storage systems in the current energy landscape in this introduction.

The majority of conventional energy systems have relied on fossil fuel-based power plants, which can continually provide electricity to meet varying energy demands. But the increasing use of renewable energy sources like solar and wind creates a particular issue. These sources are sporadic, weather-dependent, and may produce excess energy at times while their production may not immediately correspond to the energy demand. By capturing and storing extra renewable energy for later use, energy storage technologies provide a way to close this gap. Energy storage systems come in a variety of forms, each with a unique set of benefits and uses. One of the most popular and adaptable types of energy storage are batteries. They are capable of repeated charging and discharging and electrochemical energy storage. For instance, lithium-ion batteries are commonly utilized in domestic energy storage systems and electric vehicles because of their high energy density, efficiency, and long cycle life.

Pumped hydro storage, which makes use of the gravitational potential energy of water, is another method of energy storage. During times of low demand, extra electricity is used to pump water uphill to a reservoir. The water is released and flows downhill through turbines to produce energy as demand for electricity rises. The massive scale capacity, long-duration storage capabilities, and high efficiency of pumped hydro storage systems are well known. Other energy storage technologies include thermal energy storage (TES), which stores energy in the form of heat or cold in substances like molten salt or phase change materials, and compressed air energy storage (CAES), which compresses air and stores it in underground caverns for later use in power generation. Depending on elements like energy capacity, response time, and location-specific needs, these technologies have distinct qualities and are suited for a variety of applications.

Energy storage devices offer advantages beyond regulating swings in renewable energy sources. By offering grid ancillary services like frequency control, voltage control, and grid resynchronization, they also improve grid stability. Energy storage devices can also increase the reliability of the power supply, especially in locations with weak grid infrastructure or during times of high demand. They also make it possible to integrate microgrids and off-grid systems, which helps communities become more resilient and self-sufficient.

6.2 MICROGRID

The biggest problem with renewable energy has always been how to store power and then use it when it's needed. CHP plants and diesel-powered generator sets are always available, but they are not as cheap as renewable sources of energy. Microgrids are a good way to store energy because they use both types of power generation and include batteries and a control system to put everything together in a smart way. Alexander Patt, who is in charge of MTU's microgrid development team, says, "Microgrids are a concept for the future of power generation. They combine renewable energy sources that are cheap and good for the environment with the reliability of our gensets."

Microgrids are becoming more popular because solar panels are getting cheaper and batteries last longer. Batteries usually store energy during times of high production, when there is more than enough to go around. This makes up for the fact that solar panels can't make energy at night or when there's no wind. Batteries can also control peak consumption when the AC grid has been overworked to the point where it can't handle any more.

A microgrid is a small power system that can work connected to the larger grid or on its own, without being connected to the larger grid. Microgrids can be small and only power a few buildings, or they can be big enough to power an entire neighborhood,

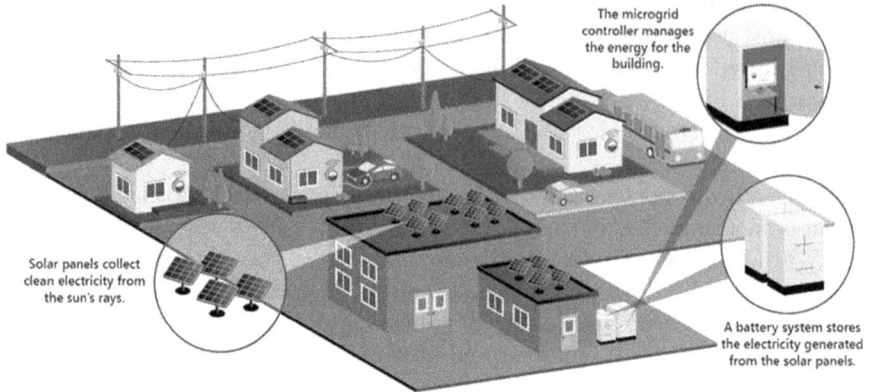

FIGURE 6.1   Micro grid controller [1].

college campus, or military base. Today, many microgrids are built around existing combined-heat-and-power plants on college campuses or in factories. But microgrids are increasingly based on energy storage systems and renewable energy sources like solar, wind, and small hydro, which are usually backed up by a generator that runs on fossil fuels. Microgrid energy storage options have become very promising and are often talked about in the energy community.

A microgrid is made up of five main parts: 1) energy sources such as generators as well as storage; 2) energy loads (sinks); 3) connection / disconnection from a power system (large); 4) regulating the microgrid; and 5) appropriate safety-assurance systems (protection). A typical microgrid is represented in Figure 6.1.

The energy sources must be able to do some important things that are usually done by the larger grid. These include: 1) after a full outage, they were able to start up the microgrid on their own; 2) when some loads turn on, they need big bursts of power, which the microgrid must be able to provide; and 3) keeping the voltage at your outlets within certain ranges.

6.2.1   Advantages of Microgrids

Microgrids offer numerous advantages in the context of energy management and distribution. Here are some key advantages of microgrids:

1. **Increased Energy Resilience**: Microgrids are designed to operate independently or in conjunction with the main grid, which enhances energy resilience. In the event of a power outage or disruption in the main grid, microgrids can continue to generate and distribute electricity to connected buildings or communities. This capability is particularly beneficial in areas prone to natural disasters or where grid infrastructure is unreliable.
2. **Integration of Renewable Energy**: Microgrids facilitate the integration of renewable energy sources, such as solar panels and wind turbines, at a localized level. By generating clean energy locally, microgrids reduce

dependence on fossil fuels and contribute to the overall reduction of greenhouse gas emissions. This promotes sustainability and helps achieve renewable energy targets.

3. **Energy Efficiency**: Microgrids enable localized energy generation, which reduces transmission and distribution losses that typically occur in centralized grid systems. By generating electricity closer to the point of consumption, microgrids minimize energy losses and improve overall energy efficiency. This translates into cost savings and a more environmentally friendly energy system.

4. **Grid Independence and Autonomy**: Microgrids provide the opportunity for grid independence and autonomy, especially in remote or isolated areas. Instead of relying solely on the main grid, communities or facilities can establish their microgrid infrastructure to meet their energy needs. This independence allows for greater control over energy supply, pricing, and reliability.

5. **Demand Response and Load Management**: Microgrids enable efficient demand response and load management strategies. By incorporating advanced energy management systems and smart grid technologies, microgrids can dynamically adjust electricity consumption based on demand and supply conditions. This flexibility helps optimize energy usage, reduce peak demand, and alleviate stress on the grid during periods of high electricity consumption.

6. **Cost Savings and Economic Benefits**: Microgrids can offer cost savings and economic benefits to communities, businesses, and individuals. By generating electricity locally and leveraging renewable energy sources, microgrid users can reduce their reliance on expensive grid electricity. Additionally, microgrids create opportunities for local job creation, investment, and economic development, particularly in the renewable energy sector.

7. **Voltage and Frequency Stability**: Microgrids can maintain stable voltage and frequency levels within their localized area. This stability is crucial for sensitive equipment, such as medical devices or industrial machinery, which may be adversely affected by voltage fluctuations. Microgrids ensure a reliable and stable power supply, contributing to the overall quality and reliability of electricity supply.

8. **Scalability and Modular Design**: Microgrids offer scalability and modular design, allowing for easy expansion or modification of the system based on evolving energy needs. Whether it is adding more renewable energy sources, accommodating increased demand, or integrating new technologies, microgrids provide flexibility and adaptability to meet changing energy requirements.

The main benefit of a microgrid is that it is more reliable. Since the power sources are close to the loads, the microgrid is less likely to lose power because of storms or other natural disasters. Most microgrids that are used commercially today were put in place to improve reliability.

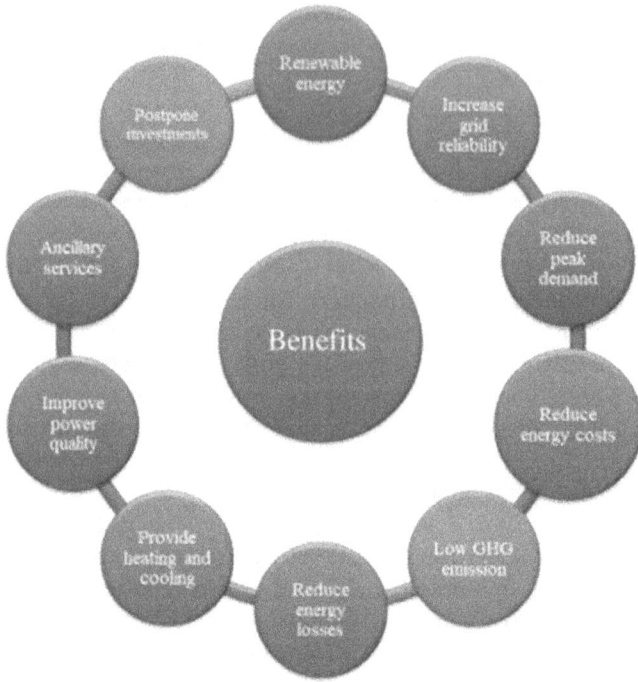

FIGURE 6.2 Benefits of micro grid [2].

In the long run, microgrids may be cheaper. If microgrid equipment is mass-produced, energy storage and renewable energy technology improves, and design and operation are standardized, microgrids may become affordable. The "steam plant" stated above, hydropower, or powerful solar resources can be used. Figure 6.2 represents major benefits of microgrid. Local generation eliminates transmission losses. DC electricity may boost microgrid energy efficiency and system management.

6.3   BATTERY ENERGY STORAGE SYSTEMS

A common kind of energy storage is the battery, which transforms chemical energy into electrical energy. They are critical elements of different electronic devices, electric vehicles (EVs), renewable energy systems, and portable electronics. Batteries are a popular option for energy storage in a variety of applications due to their many benefits. Batteries are incredibly portable due to their small size and light weight. For powering portable electronic gadgets like cell phones, laptops, and cameras, this quality is essential. Furthermore, the portability of batteries makes it possible for electric vehicles to be developed, allowing them to store and supply energy for transportation. Many contemporary batteries can be recharged, enabling multiple uses, and obviating the need for frequent battery replacements. When compared to single-use disposable batteries, rechargeable batteries, such as lithium-ion batteries, are more affordable and environmentally beneficial because they can be charged and discharged several times. Batteries are able to store a substantial quantity of energy in a relatively small volume or weight due to their high energy density. This quality is essential for applications with weight and space restrictions, such as portable electronics and electric cars. Longer runtimes and increased device utilization are made possible by high energy density batteries. Batteries are a good source of electrical energy that is converted from chemical energy in an efficient manner. Batteries are appropriate for instant power applications like starting a car or powering electronic gadgets with high power requirements since the conversion process is reasonably quick and can give power quickly.

Rechargeable batteries help to maintain the environment by eliminating the use of disposable batteries and waste. They assist in lowering the amount of hazardous waste disposed of that is related to single-use batteries, which can have a detrimental effect on the environment. Additionally, improvements in battery technology have concentrated on enhancing their environmental performance, including the use of more environmentally friendly materials, and enhancing their capacity for recycling.

Batteries are essential to grid energy storage systems because they allow them to store excess electricity during times of low demand and release it during times of high demand. Applications like this improve grid stability, smooth out swings in renewable energy generation, and balance the supply and demand of electricity. Grid-scale battery storage devices facilitate the grid's integration of renewable energy sources, assisting in the shift to a cleaner and more dependable energy system. In terms of capacity and setup, batteries offer scalability and flexibility. To provide the desired voltage, capacity, and output of power, they can simply be connected in series or parallel. Whether for small-scale applications or large-scale grid storage initiatives, this flexibility enables the design of battery systems that correspond to particular energy storage requirements.

The use of safety features in battery technology helps to avoid accidents and property damage. Short-circuit prevention devices, overcharge and over discharge protection, and thermal management systems are some of these features. Safety concerns are extremely important, particularly in situations where batteries are subjected to extreme circumstances or where failure can result in negative effects.

As a whole, batteries provide many benefits, including portability, rechargeability, high energy density, efficient energy conversion, environmental friendliness, grid energy storage capabilities, scalability, and safety features. Batteries are an essential part of many sectors because to their unique properties, which have helped progress portable devices, electric vehicles, the integration of renewable energy sources, and grid stability. In order to broaden the applications and advantages of batteries, ongoing research and development efforts are directed at improving their performance, longevity, and environmental sustainability. Figure 6.3 represents different kinds of batteries.

Batteries are one way that energy can be stored in microgrids. Battery energy storage systems (BESS) can be lithium-ion, magnesium, or another type, to store energy that has been generated. In backup power applications, residential energy storage usually meets the energy needs in case the grid goes down. If there is a problem with

| Lithium-Ion Battery | Ni-Cd Battery | Zinc Carbon Battery | Alkaline Battery |

| Ni-MH Battery | Coin Cell Battery | Lead-acid Battery | Sealed Battery |

FIGURE 6.3   Battery types [3].

the grid, energy can be taken from the batteries up to the point where they can no longer be used. There are also different sizes of batteries. Before buying storage batteries, the owner of the microgrid should figure out how much energy the system makes. Once the maximum output is known, batteries with the same amount of power can be bought and added to the system. If the microgrid goes down, solar and wind can be used to charge the batteries, making the backup power available for more hours per day.

6.3.1   ENERGY STORAGE

Energy storage is a flexible, adaptable, and distributed source of energy that can help a microgrid in a big way. Improving storage technologies offers many benefits besides more renewable energy to be used to make electricity. One of these benefits is a grid that is more efficient and not affected by problems.

Storing energy means saving energy that is made now for later use. Energy is usually stored in a battery or collector. Some energy storage systems are short-term, while others are long-term. Applications for residential energy storage are driven by the need for backup power. Right now, this demand is driven more by customer appeal than by economics, because the cost can't be justified on a dollar per kilowatt-hour basis. Customers like these are willing to pay more for the newest technology.

Batteries are key for battery energy storage systems (BESS), which are energy storage devices. When demand is low or renewable energy production is high, they store electrical energy, and release it when demand is high or renewable energy production is low. Due to their adaptability, scalability, and potential to facilitate the grid integration of renewable energy sources, BESS have attracted a lot of interest in recent years.

The elements and operating principles of battery energy storage systems are described in the below section. Battery energy storage system components include:

1. **Batteries**: The main part of a BESS is its battery bank, which is made up of a number of separate batteries connected in series to produce the required power output and storage capacity. Batteries of many sorts, such as lithium-ion, lead–acid, flow batteries, and sodium-ion batteries, can be utilized in BESS. The most popular battery type is lithium-ion because of its high energy density, efficiency, and long cycle life.
2. **Power Electronics**: Power electronics are used in the power conversion system to enable the bidirectional flow of electricity between the batteries and the grid. It consists of transformers, converters, and inverters, amongst other components. Voltage regulation, synchronization with the grid, and effective energy transfer are all guaranteed by the power conversion system.
3. **Energy Management System**: The brain of the BESS is the energy management system (EMS). It keeps an eye on and manages the charging, discharging, and battery operation procedures. Based on current energy costs, grid demand, the availability of renewable energy sources, and other variables, the EMS optimizes the battery use. It provides effective functioning, extends the battery life, and makes it possible to provide different grid services.

6.3.2　How Battery Energy Storage Systems Operate

Charging: Extra electricity is used to charge the batteries during times of low demand or high renewable energy production. The power conversion system converts the AC (alternating current) electricity from the grid or renewable sources into DC (direct current) electricity to charge the batteries. The EMS controls the charging procedure while taking into account variables including battery capacity, voltage restrictions, and charging effectiveness.

Discharging: When there is a need for additional power on the grid or renewable energy generation is low, the batteries are discharged. To supply the grid, the power conversion system transforms the DC electricity stored in the batteries into AC electricity. The EMS controls the discharging process to ensure the batteries' performance, regulate voltage and frequency, and respond to grid demand or specific grid services.

Grid Services: Battery Energy Storage Systems can provide various grid services, enhancing grid stability and flexibility. Some common grid services include a) frequency regulation: BESS can respond rapidly to fluctuations in grid frequency by either absorbing excess power or injecting power into the grid, helping maintain a stable frequency; b) peak shaving: BESS discharges during periods of high electricity demand, reducing the strain on the grid and avoiding the need for expensive peaker plants; c) load shifting: BESS can shift energy consumption from peak hours to off-peak hours, taking advantage of lower electricity prices and reducing peak demand; d) Ancillary Services: BESS can provide ancillary services like voltage control, reactive power support, and grid resynchronization during grid disturbances; and e) Renewable Energy Smoothing: BESS helps mitigate the intermittent nature of renewable energy sources by storing excess renewable energy and releasing it during low generation periods, ensuring a consistent and reliable power supply.

6.3.3 Advantages of Battery Energy Storage Systems

1. Flexibility: BESS are highly flexible and can be deployed in various applications, from residential and commercial settings to utility-scale installations. They can be easily scaled up or down based on energy requirements.
2. Fast Response: BESS have fast response times, allowing them to provide instantaneous power when needed. They can respond within milliseconds to frequency
3. Keep up good power and dependability
4. Provide customer services, such as cost control, flexibility, and convenience
5. Improve transmission and distribution (T&D) stability
6. Make better use of assets and put off upgrades
7. Raise the value of variable renewable generation

In terms of energy storage and grid management, battery energy storage systems (BESS) have a number of benefits. Here are a few of BESS's main benefits:

a. **Grid Stability and Reliability**: By offering quick reaction capabilities, BESS contributes to grid stability. They may swiftly inject or absorb energy to balance variations in electrical supply and demand, assisting in keeping the frequency and voltage of the system within allowable bounds. This lowers the possibility of blackouts or disruptions and increases grid dependability.
b. **Integration of Renewable Energy**: The grid's ability to integrate renewable energy sources like solar and wind depends heavily on BESS. Although the production of renewable energy is sporadic and variable, BESS can store extra renewable energy during times of high production and release it when demand outstrips supply or renewable energy production is low. This promotes stability in the electricity supply, lessens renewable energy curtailment, and smooths out swings.
c. **Peak Load Management**: By releasing stored energy during moments of high demand, BESS facilitates peak load management. This lessens the load on the grid and helps prevent the need for more expensive peaker plants or other forms of extra power producing capacity. BESS improves grid efficiency and save costs by diverting demand from peak times.

    Ancillary services, including frequency regulation, are provided by BESS to the grid as valued ancillary services. They can quickly inject or absorb power in response to fluctuations in grid frequency, assisting in the maintenance of a stable frequency. Other auxiliary services that BESS can offer include voltage control, reactive power assistance, and grid resynchronization in times of grid disturbance. These services improve the dependability and stability of the grid. BESS permits energy time-shifting, which entails holding excess energy during times of low demand or low electricity prices and releasing it during times of high demand or high electricity costs. This makes it possible for customers to save money on energy expenditures, maximize energy use, and benefit from off-peak electricity rates.

d. Grid resilience and backup power: BESS can offer backup power during emergencies or grid failures. They improve grid resilience and reliability by ensuring an uninterrupted power supply to vital facilities, including hospitals, data centers, and telecommunications networks. BESS can also be used in microgrid setups to build islanded systems that function separately from the main grid in times of emergency.

e. Environmental Benefits: By simplifying the integration of renewable energy and reducing reliance on fossil fuel-based power generation, BESS contributes to environmental sustainability. BESS aids in lowering greenhouse gas emissions and battling climate change by facilitating greater penetration of renewable energy. By providing charging stations and balancing the demand for electricity from EVs, BESS may also aid in the transition to electric vehicles.

BESS provides scalability and adaptability, enabling flexible deployment in a variety of applications and locales. Depending on the needed energy storage, they may simply be scaled up or down by adding or deleting battery modules. BESS is appropriate for a variety of applications because to its scalability, including utility-scale installations as well as settings in the home and workplace. Figure 6.4 represents a battery energy storage system set-up.

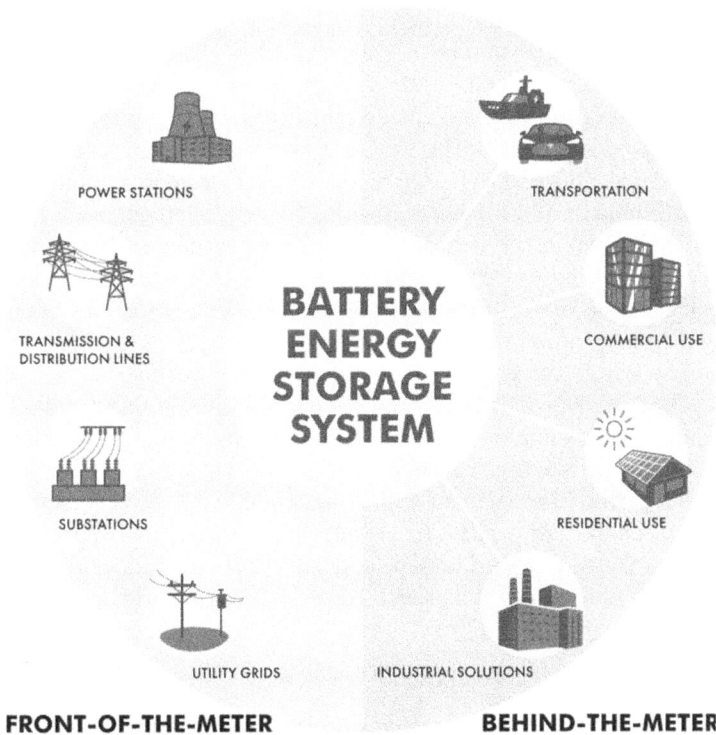

FRONT-OF-THE-METER   **BEHIND-THE-METER**

FIGURE 6.4   Battery energy storage system [4].

Battery energy storage systems have benefits including grid stability, integrating renewable energy, managing peak loads, frequency regulation, backup power, having environmental advantages, scalability, and modularity. BESS is an important option for improving energy management, boosting grid dependability, and assisting the shift to a clean and sustainable energy future because of these benefits.

6.3.4　Energy Storage Systems and Their Role in the Energy Sector

Energy storage systems (ESS) are technologies that store energy for later use. They play a crucial role in the energy sector by addressing the challenges of variable energy supply, grid stability, and demand management. ESS are capable of storing excess energy during times of low demand and releasing it during peak demand periods or when the energy supply is limited.

The primary role of energy storage systems in the energy sector can be summarized as follows:

1. **Load Balancing and Demand Management**: ESS helps balance the supply and demand of electricity. During periods of low demand, when energy generation exceeds consumption, the excess energy can be stored in the storage system. This stored energy can then be discharged when demand increases, ensuring a stable and reliable power supply.
2. **Integration of Renewable Energy**: Renewable energy sources such as solar and wind power are intermittent and can fluctuate based on weather conditions. ESS helps smooth out these fluctuations by storing excess energy generated during peak production periods and supplying it when renewable energy generation is low. This enables better integration of renewables into the grid and reduces the reliance on fossil fuel-based power plants.
3. **Grid Stability and Frequency Regulation**: ESS can provide fast response times in regulating the frequency and voltage of the grid. They can absorb excess power during periods of high frequency and release stored energy during periods of low frequency, helping maintain grid stability and reducing the risk of power outages.
4. **Backup Power and Grid Resilience**: In case of power outages or emergencies, ESS can provide backup power to critical infrastructure, homes, and businesses. This improves grid resilience and ensures uninterrupted power supply during unforeseen events.
5. **Peak Demand Management and Cost Savings**: ESS can help reduce peak demand on the grid, which often results in higher electricity costs. By discharging stored energy during peak demand periods, ESS can help lower electricity bills for consumers and reduce strain on the grid infrastructure.

6.3.5　Introduction to Microgrids and Energy Storage Systems

Microgrids are twenty-first-century energy ecosystems. A controlled local electricity system for a college campus, medical complex, commercial center, or community is called a microgrid. Under typical conditions, it can link to the main grid at a common coupling point and operate in sync. However, during emergencies or grid outages, a microgrid can disconnect from the main grid and operate autonomously, utilizing its local energy generation units. These energy generation units typically consist of renewable sources like solar panels, wind turbines, combined-heat-and-power systems, and may also include energy storage systems such as batteries, along with electric vehicle charging stations [5].

Microgrids offer several advantages and contribute to enhancing flexibility, reliability, and resiliency in the energy sector. They provide access to green and safe energy, participate in demand response programs, optimize costs, and help balance the grid. The concept of microgrids encompasses various aspects, including the structure (AC, DC, or hybrid) and control scheme (centralized, decentralized, or distributed). These aspects influence the operation and management of a microgrid [5].

One crucial component of microgrids is energy storage systems. The integration of energy storage allows microgrids to overcome the intermittency and unpredictability of renewable energy sources. Energy storage systems play a vital role in stabilizing and ensuring a reliable supply of electricity within microgrids. They enable the storage of excess energy during periods of low demand or high renewable energy generation and provide it when needed, such as during peak demand or when renewable sources are not generating sufficient power. Commonly used energy storage technologies in microgrids include lithium-ion batteries, lead–acid batteries, and sodium–sulfur batteries [6].

Furthermore, microgrids employ EMS in conjunction with smart control systems to effectively manage energy supply and demand. The EMS integrates hardware and software components to monitor, control, and optimize the flow of energy within the microgrid. It ensures efficient utilization of available energy resources, coordinates energy generation and storage, and enables demand response capabilities. The combination of energy storage systems and advanced EMS plays a vital role in enhancing the overall performance and grid integration of microgrids [6]. Microgrids are localized energy systems that provide a range of benefits, including energy security, resiliency, and cost savings. They integrate renewable energy sources, energy storage systems, and advanced control technologies to ensure a reliable and sustainable energy supply. The combination of microgrids and energy storage systems is a promising solution for addressing the evolving energy landscape and achieving a more sustainable future [5, 6].

6.3.6 BENEFITS OF ENERGY STORAGE SYSTEMS IN MICROGRID OPERATION

Energy storage systems play a crucial role in the operation of microgrids, offering several benefits for efficient and reliable energy management. One of the primary advantages is their ability to address power imbalances and ensure grid stability, particularly with the increasing penetration of renewable energy sources [7]. Energy storage systems enable microgrids to provide ancillary services such as frequency regulation, peak shaving, and energy arbitrage, contributing to grid reliability and power quality [7]. By storing excess energy generated during periods of low demand, these systems ensure a stable and reliable supply of electricity during times of high demand or when renewable sources are not producing enough energy [7].

The deployment of energy storage in microgrids improves energy self-sufficiency and reduces reliance on the main power grid, enhancing energy security [7]. Additionally, energy storage systems support the integration of renewable energy sources by mitigating their intermittency, enabling smoother power production, and reducing carbon emissions [6, 7]. They enhance the operational flexibility of microgrids, allowing for effective load balancing, demand response participation, and cost optimization [7]. The choice of energy storage technology depends on factors such as cost, technical benefits, cycle life, ease of deployment, energy and power density, and operational constraints, ensuring that the most suitable solution is selected for each microgrid application [7]. Overall, energy storage systems in microgrids provide a multitude of advantages, contributing to grid stability, energy reliability, renewable energy integration, and operational flexibility.

6.3.7 Challenges and Solutions for Energy Storage in Microgrids

Implementing energy storage systems in microgrids presents certain challenges, but they can be overcome with appropriate solutions. One of the key challenges is microgrid stability, ensuring that the system maintains a consistent and reliable power supply. Energy storage systems play a vital role in addressing this challenge by providing grid support, voltage and frequency regulation, and seamless transition between grid-connected and islanded modes of operation [1]. Another significant challenge is effective power and energy management in microgrids, particularly when integrating renewable energy sources (RES). Energy storage systems enable efficient management by storing excess energy generated from renewables during periods of low demand and supplying it during high-demand periods or when RES output is low [2]. Reliability and power quality are crucial considerations in microgrids, and energy storage systems help maintain a stable and consistent power supply by smoothing out variations in renewable energy generation and providing backup power during grid outages [1, 2]. Furthermore, cost, technical feasibility, cycle life, ease of deployment, energy and power density, and operational constraints are factors that need to be carefully considered when selecting energy storage technologies for microgrids [2]. Overcoming these challenges involves comprehensive research, innovative design, and continuous improvements in energy storage technologies, ensuring their effective integration into microgrid systems and enabling the realization of their full potential.

6.3.8 Modeling and Simulation of Energy Storage Systems in Microgrids

Modeling and simulation of energy storage systems in microgrids is a crucial aspect of designing and analyzing their performance. By employing reliable models, it becomes possible to accurately represent the behavior and characteristics of energy storage systems in microgrid applications. This enables engineers and researchers to assess their impact on microgrid operation and optimize their design. Different modeling approaches have been developed to capture the behavior of energy storage systems, particularly battery energy storage systems (BESS), in microgrids.

These models can be analytical or electrical, and they are often based on experimental measurements [2]. The derived models are integrated into comprehensive methodologies and computational tools for robust microgrid design, such as the Poli. NRG tool developed by Politecnico di Milano [2]. Through these modeling and simulation techniques, it becomes possible to evaluate the performance, reliability, and cost-effectiveness of different energy storage system models, and their impact on microgrid design can be assessed [2]. Ultimately, modeling and simulation provide valuable insights for optimizing the integration and operation of energy storage systems in microgrids, facilitating the efficient utilization of renewable energy sources, and improving overall system performance.

6.3.9 CONTROL AND OPTIMIZATION STRATEGIES FOR ENERGY STORAGE IN MICROGRID OPERATION

Control and optimization strategies play a crucial role in the effective operation of energy storage systems in microgrids. These strategies aim to ensure optimal utilization of energy storage resources, enhance system performance, and support grid stability and reliability [8, 9]. Through advanced control algorithms and optimization methods, the energy flow and storage capacity of the microgrid can be efficiently managed and coordinated with various energy sources and loads. Control strategies involve the real-time monitoring and adjustment of energy storage parameters such as state of charge (SOC), charging and discharging rates, and voltage regulation [10]. Optimization methods, on the other hand, utilize mathematical algorithms and models to optimize the dispatch and scheduling of energy storage, considering factors such as energy prices, demand patterns, and renewable energy availability. These approaches can be applied at different time scales, ranging from short-term load leveling to long-term capacity planning [11]. By employing control and optimization strategies, microgrid operators can maximize the economic benefits of energy storage, enhance grid resilience, and support the integration of renewable energy sources, ultimately contributing to a more sustainable and efficient energy system.

6.3.10 INTEGRATION OF RENEWABLE ENERGY SOURCES WITH ENERGY STORAGE IN MICROGRIDS

The integration of renewable energy sources with energy storage in microgrids is a key strategy for achieving a sustainable and reliable power system. Renewable energy sources such as solar and wind power are variable in nature, which can pose challenges for grid stability and supply-demand balance. By combining these renewable sources with energy storage systems, the intermittency and variability of renewable generation can be mitigated, enabling a more consistent and controllable power supply. Energy storage systems, such as batteries or hybrid storage solutions, can store excess renewable energy during periods of high generation and release it when demand exceeds supply, thus maintaining grid stability [12]. Additionally, energy storage can provide ancillary services such as frequency regulation and voltage support, enhancing the overall performance and reliability of the microgrid [13]. Real-time energy storage management techniques and optimization algorithms play a crucial role in effectively coordinating the charging and discharging of energy storage systems to maximize the utilization of renewable energy and ensure efficient grid operation [14]. The integration of renewable energy sources with energy storage in microgrids is a promising approach toward achieving a sustainable and resilient energy system by reducing reliance on fossil fuels and promoting the utilization of clean and abundant renewable resources.

6.3.11 ENERGY MANAGEMENT AND DISPATCH STRATEGIES IN MICROGRIDS WITH ENERGY STORAGE

Energy management and dispatch strategies play a crucial role in optimizing the operation of microgrids with energy storage. These strategies aim to ensure efficient utilization of available energy resources, maintain grid stability, and minimize operating costs. An optimal energy management strategy for microgrids with energy storage can involve various considerations, such as state of charge (SOC) of the storage system, renewable energy generation, load demand, and demand response (DR) [15]. By integrating these factors, an energy management strategy can determine the optimal scheduling and dispatch of energy resources to meet the demand while maximizing the utilization of renewable energy and minimizing the reliance on non-renewable sources. Optimization algorithms, such as particle swarm optimization (PSO), can be applied to solve the energy management problem and achieve optimal results [15]. These strategies also consider factors like carbon emissions, energy consumption penalties, and the life cycle of hybrid energy storage systems (HESS) to enhance the overall economic and reliable performance of microgrids [15]. By implementing efficient energy management and dispatch strategies, microgrids with energy storage can achieve improved renewable energy consumption rates, reduced loss of load, and enhanced grid resilience.

6.3.12 ADVANCED TECHNOLOGIES AND INNOVATIONS IN ENERGY STORAGE FOR MICROGRID OPERATION

Advanced technologies and innovations in energy storage are driving significant advancements in microgrid operation. With the increasing integration of renewable energy sources in power systems, energy storage systems have become essential for managing power imbalances, ensuring grid stability, and providing ancillary services [16]. These advanced technologies offer a range of benefits, including frequency regulation, peak shaving, energy arbitrage, and backup power supply, which enhance the reliability and quality of power supply in microgrids [16]. Various energy storage technologies are being explored for microgrid applications, each with its own advantages and limitations. A critical review of these technologies, focusing on mature technologies, provides insights into their feasibility in terms of cost, technical benefits, cycle life, ease of deployment, and operational constraints [16]. The selection of appropriate energy storage solutions depends on specific requirements and circumstances, considering factors such as response time, autonomy period, energy density, and self-discharge rates [16]. The integration of energy storage systems with renewable energy resources in microgrids helps address challenges related to intermittency and improves overall grid efficiency and reliability [17]. Additionally, hybridization of multiple energy storage systems can further enhance storage capacity, lifespan, and efficiency [17]. The concept of microgrids, which incorporate distributed generations, energy storage systems, and electrical loads, offers numerous benefits such as reduced greenhouse gas emissions, decentralized energy supply, and demand response [17]. These advancements in energy storage technologies are paving the way for more sustainable, resilient, and efficient microgrid operations, contributing to the transition toward a decarbonized and reliable energy future.

6.3.13 CASE STUDIES AND BEST PRACTICES OF ENERGY STORAGE SYSTEMS IN MICROGRID OPERATION

Microgrids are local power systems that operate independently or in coordination with the main grid, incorporating renewable energy sources (RESs) and energy storage systems (ESSs). These systems have gained significance in addressing the challenges posed by the increasing global demand for electrical energy and the need to reduce carbon emissions. The integration of RESs, such as photovoltaic (PV) and wind power, into microgrids helps meet electricity demand while reducing reliance on fossil fuels [18]. Energy storage systems play a crucial role in mitigating the inherent variability and intermittency of RESs, ensuring grid stability, and enabling effective utilization of renewable energy [18]. The deployment of energy storage in microgrids is expected to grow significantly, with global energy storage demand predicted to increase by 15 times by 2030 [18]. Effective control and coordination between RESs and ESSs remain a challenge, and microgrids provide a framework for achieving this integration [18]. By operating in both grid-connected and islanding modes, microgrids can support local loads, provide auxiliary services to the main grid, and ensure system stability [18]. The typical structure of a microgrid includes RESs, ESSs, and loads, enabling the efficient utilization of renewable energy and maximizing the benefits for end-users [18]. In summary, microgrids with energy storage systems offer a promising solution for integrating renewable energy sources, addressing grid challenges, and enhancing the overall sustainability of the power system [18].

6.3.14 IMPORTANCE OF ENERGY STORAGE SYSTEMS IN MICROGRID OPERATION

Energy storage systems are essential to the operation of microgrids because they offer a number of advantages and improve the system's dependability, stability, and efficiency. Microgrids frequently include renewable energy sources, including solar cells and wind turbines, which have a natural tendency to fluctuate depending on the weather. When the demand for energy is high or the supply of renewable energy is low, energy storage devices can store extra energy produced during certain times and release it. This contributes to the microgrid's supply-demand balance, maintaining a steady and dependable supply of power. Peak load management and load shifting are made possible by energy storage systems in microgrids by storing excess energy during off-peak hours and supplying it during periods of high demand. By doing this, you can avoid paying exorbitant peak electricity rates and lessen your reliance on external power sources during peak hours. Microgrids can optimize their energy use and lower their overall energy costs by effectively regulating peak demands.

Microgrids' energy storage technologies can offer ancillary services that help maintain grid stability. They can react swiftly to changes in supply and demand, aiding in the microgrid's regulation of frequency and voltage levels. Additionally, capable of participating in frequency control at the grid level, energy storage devices can offer stability services to the wider utility grid. One of the key characteristics of microgrids is their capacity to function without the assistance of the primary utility grid in times of crisis. By providing backup power, energy storage devices are essential for enabling islanding capabilities. The stored energy in the storage devices can be used to keep the microgrid's important loads powered during grid interruptions, assuring continuous operation. Microgrid energy storage systems can benefit from time-of-use pricing models. They can charge when prices and demand for electricity are low and discharge when those conditions are high. Operators of microgrids can reduce their dependency on the grid during periods of high energy prices and optimize energy expenses as a result.

Energy storage devices can contribute to the improvement of power quality in microgrids. They can support reactive power, control voltage, and reduce variations brought on by sporadic renewable energy sources. This produces a power source that is more dependable and steadier, lowering the possibility of voltage sags, flickering, and other power quality problems.

Microgrids frequently integrate different DERs, such as electric vehicle charging stations, combined-heat-and-power (CHP) systems, and renewable energy sources. By balancing supply and demand, controlling intermittency, and offering backup power when necessary, energy storage systems can integrate and optimize the operation of these DERs.

Energy storage technologies are crucial to the operation of microgrids. They offer advantages include smoothing out sporadic renewable energy production, shifting loads, managing peak loads, supporting and stabilizing the system, being grid independent, managing time-of-use, enhancing power quality, and integrating distributed energy resources. These benefits help microgrids be more reliable, effective, and resilient overall, which makes them an essential component of contemporary energy systems. Energy storage systems are essential to the functioning of microgrids because they provide a variety of applications that improve the system's functionality, dependability, and efficiency. Figure 6.5 represents grid energy storage technologies. The following are some major uses for energy storage devices in microgrids.

Energy storage devices assist microgrids in managing peak loads by storing excess energy during times of low demand and releasing it during times of high demand. This is known as peak shaving. This minimizes the need for costly peaker plants, lessens reliance on the primary grid during peak hours, and improves energy efficiency. Microgrids can reduce costs and boost grid efficiency by moving load away from peak times.

Energy storage systems offer frequency regulating services to microgrids, ensuring grid stability. They may quickly inject or absorb power in response to frequency variations, assisting in the maintenance of a stable frequency. This application improves grid stability, lowers the possibility of power fluctuations, and makes sure

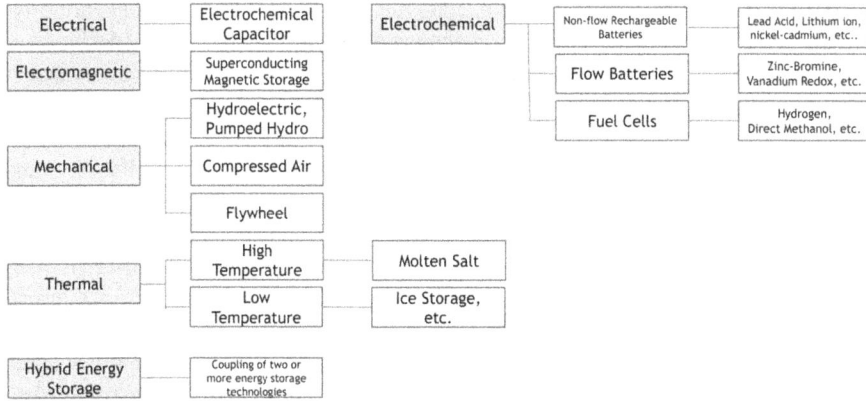

FIGURE 6.5 Grid energy storage technologies.

that the microgrid always has a steady supply of electricity. Energy storage devices can assist in regulating voltage levels inside microgrids and can also help enhance power quality. They assist reactive power and lessen voltage swings brought on by sporadic renewable energy sources. Energy storage systems contribute to a dependable and high-quality power supply by maintaining steady voltage levels.

Microgrids frequently incorporate renewable energy resources, such as solar panels and wind turbines. By holding excess energy during times of high generation and releasing it during times of low generation or high demand, energy storage devices make it possible to integrate these intermittent sources. This guarantees a steady and uninterrupted supply of electricity, lessens dependency on the primary grid, and maximizes the use of renewable resources.

Energy storage devices allow microgrids to function without the assistance of the main grid in times of crisis. important loads receive backup power from them, assuring an uninterrupted supply of electricity to crucial buildings like hospitals, emergency response centers, and important infrastructure. This increases the microgrid's resilience and increases the overall reliability of the grid. Microgrids can benefit from time-of-use pricing structures thanks to energy storage devices. They can charge when prices and demand for electricity are low and discharge when those conditions are high. With less reliance on the grid at times of high energy prices, microgrids can better manage their energy expenditures.

6.4 FUTURE TRENDS AND OUTLOOK FOR ENERGY STORAGE IN MICROGRIDS

Energy storage plays a crucial role in modern power systems, particularly in the context of microgrids. As the penetration of renewable energy sources increases, energy storage technologies become essential for addressing power imbalances, ensuring grid stability, and providing backup supply and resilience [16]. Looking into the future, several trends and outlooks can be identified in the realm of energy storage for microgrids. First, the advancement and adoption of distributed energy resources

(DERs) in microgrids, such as solar panels, fuel cells, and battery storage, are driving the need for sophisticated energy management systems. Microgrids, designed as self-sustaining power networks, require intelligent automation and smart management to optimize the integration and operation of DERs. The utilization of information technology, including the IoT, enables real-time data collection and analysis, facilitating decision-making processes that optimize energy usage, reliability, and cost-effectiveness.

Second, there is a growing focus on energy storage technologies with improved response time (power density) and autonomy period (energy density). Different operational requirements exist for various applications, such as long-term storage or primary reserves. In long-term storage, high cycling rates may not be necessary, but high energy density and low self-discharge rates are desirable. On the other hand, primary reserves require faster response times for shorter durations. Hence, the selection of energy storage technologies should be tailored to specific use cases [16]. Third, the market for energy storage in microgrids is projected to witness significant growth in the coming years. According to Technavio, the global energy storage market for microgrids is expected to grow by US $1,361.05 million from 2022 to 2027, with a compound annual growth rate (CAGR) of 18.9%. North America is anticipated to account for a substantial portion of this growth. This emphasizes the increasing recognition of the benefits of energy storage in microgrid applications and the demand for sustainable and reliable power solutions.

Furthermore, the integration of renewable natural gas, such as biogas, in microgrids is gaining traction as a means of reducing carbon footprints and cutting energy costs. Biogas can be produced from organic materials found in landfills, sewage treatment, or bio-digesters, and can be used to generate electricity. Several companies have implemented biogas-based microgrids to achieve cleaner and more cost-effective power supply. The future trends and outlook for energy storage in microgrids are characterized by the integration of advanced energy management systems, the development of energy storage technologies optimized for specific use cases, the projected market growth, and the adoption of renewable natural gas. These factors contribute to the continued advancement and deployment of energy storage in microgrid applications, enabling greater reliability, resilience, and sustainable energy solutions.

6.5 CONCLUSION

In order to assist the overall grid operation, energy storage devices in microgrids might offer ancillary services. These services include grid resynchronization during disturbances, frequency regulation, and voltage support. Microgrids with energy storage devices can help the larger utility grid maintain stability and dependability by taking part in grid-level functions. The integration of renewable energy, grid independence, time-of-use management, peak shaving, load management, frequency regulation, voltage support, and provision of ancillary services are just a few of the uses for energy storage systems in microgrids. These applications improve microgrids' functionality, dependability, and efficiency as a whole, making them an important part of contemporary energy systems.

REFERENCES

[1] Microgrid Energy System Introduction, (accessed 10 June 2023), https://www. franklinwh.com/blog/microgrid-energy-system-introduction

[2] Marco Aurelio Lenzi Castro, (2020). Chapter 9 – Urban Microgrids: Benefits, Challenges, and Business Models. In *The Regulation and Policy of Latin American Energy Transitions*, Elsevier, 153–172, ISBN 9780128195215, https://doi.org/10.1016/ B978-0-12-819521-5.00009-7

[3] Different Types of Batteries and their Applications, (accessed 10 June 2023), https:// www.theengineerspost.com/types-of-batteries/

[4] Andrey Solovev, Anna Petrova, Efficient Energy Management and Energy Saving with a BESS (Battery Energy Storage System), (accessed 10 June 2023), https://www. integrasources.com/blog/energy-management-and-energy-saving-bess/

[5] Shahbazitabar, M., Abdi, H., Nourianfar, H., Anvari-Moghaddam, A., Mohammadi-Ivatloo, B., Hatziargyriou, N. (2021). An Introduction to Microgrids, Concepts, Definition, and Classifications. In Anvari-Moghaddam, A., Abdi, H., Mohammadi-Ivatloo, B., Hatziargyriou, N. *Microgrids. Power Systems*. Springer. https://doi. org/10.1007/978-3-030-59750-4_1

[6] Luke Talltree, Microgrids 101: An Introduction to Microgrids, (accessed 10 June 2023), https://www.gridpoint.com/blog/what-are-microgrids/

[7] Oliveira, D.Q., Saavedra, O.R., Santos-Pereira, (2021). A Critical Review of Energy Storage Technologies for Microgrids. *Energy System*. https://doi.org/10.1007/ s12667-021-00464-6

[8] Nazaripouya, H., Chung, Y., Akhil, A., (2019). Energy Storage in Microgrids: Challenges, Applications and Research Need, *International Journal of Energy and Smart Grid*, 3(2), 60–70. https://doi.org/10.23884/IJESG.2018.3.2.02

[9] Moncecchi, M., Brivio, C., Mandelli, S., Merlo, M. (2020). Battery Energy Storage Systems in Microgrids: Modeling and Design Criteria. *Energies*, 13, 2006. https://doi. org/10.3390/en13082006

[10] Malysz, P., Sirouspour, S., Emadi, A., (2014). An Optimal Energy Storage Control Strategy for Grid-connected Microgrids, *IEEE Transactions on Smart Grid*, 5(4), 1785–1796. https://doi.org/10.1109/TSG.2014.2302396

[11] Al-Ismail, F. S., (2021). DC Microgrid Planning, Operation, and Control: A Comprehensive Review, *IEEE Access*, 9, 36154–36172. https://doi.org/10.1109/ ACCESS.2021.3062840

[12] Etxeberria, A., Vechiu, I., Camblong, H., Vinassa, J. M., (2010). Hybrid Energy Storage Systems for Renewable Energy Sources Integration in Microgrids: A Review, *2010 Conference Proceedings IPEC*, 532–537. https://doi.org/10.1109/IPECON. 2010.5697053

[13] Anglani, N., Di Salvo, S. R., Oriti, G., Julian, A. L., (2020). Renewable Energy Sources and Storage Integration in Offshore Microgrids, *IEEE International Conference on Environment and Electrical Engineering and 2020 IEEE Industrial and Commercial Power Systems Europe (EEEIC / I&CPS Europe)*, 1–6 https://doi.org/10.1109/EEEIC/ ICPSEurope49358.2020.9160760

[14] Rahbar, K., Xu, J., Zhang, R., (2015). Real-Time Energy Storage Management for Renewable Integration in Microgrid: An Off-Line Optimization Approach, *IEEE Transactions on Smart Grid*, 6(1), 124–134. https://doi.org/10.1109/TSG.2014.2359004

[15] Chen, H., Gao, L., Zhang, Z. (2021). Optimal Energy Management Strategy for an Islanded Microgrid with Hybrid Energy Storage. *Journal of Electrical Engineering & Technology*, 16, 1313–1325. https://doi.org/10.1007/s42835-021-00683-y

[16] Oliveira, D.Q., Saavedra, O.R., Santos-Pereira, K. (2021). A Critical Review of Energy Storage Technologies for Microgrids. *Energy System.* https://doi.org/10.1007/s12667-021-00464-6

[17] Choudhury, Subhashree, (2022). Review of Energy Storage System Technologies Integration to Microgrid: Types, Control Strategies, Issues, and Future Prospects, *Journal of Energy Storage*, 48, 103966. https://doi.org/10.1016/j.est.2022.103966

[18] Xin Lin, Ramon Zamora, (2022). Controls of Hybrid Energy Storage Systems in Microgrids: Critical Review, Case Study and Future Trends, *Journal of Energy Storage*, 47, 103884. https://doi.org/10.1016/j.est.2021.103884

7 Intelligence-based Charging System

Prashant Kumar, Adit Srivastava,
Prince Rajpoot, and Vikas Patel
Rajkiya Engineering College, Ambedkar Nagar, India

Ritika Yaduvanshi
Mahamaya College of Agricultural Engineering and
Technology, Ambedkar Nagar, India

Shivendu Mishra
Rajkiya Engineering College, Ambedkar Nagar, India

7.1 INTRODUCTION

It is essential to have a reliable charging method for batteries due to the rising demand and widespread use of batteries in EVs and electronic devices and to store renewable energy as backup power. Traditional charging methods frequently use a fixed charging rate that does not consider the battery's state of charge or temperature and may continue to supply a constant charging rate even after the battery is fully charged, resulting in energy waste and battery degradation. Furthermore, traditional charging methods may need more charge to fully charge the battery, leading to overcharging or undercharging. Using a charging rate that is not optimized for the battery can damage it and reduce its lifespan and performance over time, leaving traditional methods nearly inept. On the other hand, modern intelligent control methods can adjust the charging rate based on the battery's SOC, temperature, and other factors using Cromlech algorithms that can address these issues and provide a more efficient charging experience while extending the battery's lifespan and improving its performance. A summary of the critical concepts of intelligent control for battery charging systems is given in this chapter. It covers recent works, different ideas and algorithms used in intelligent control systems, including their advantages and limitations. The chapter also discusses the challenges associated with enforcing intelligent control systems and the unexplored directions of exploration in this field.

DOI: 10.1201/9781003481836-7

7.1.1 Intelligent Charging Systems

Customers' preferences shift toward electrified cars as batteries improve and urbanization grows. The long-term viability of electric automobiles, on the other hand, depends on an entirely novel architecture that will assist the vehicles: rooftop photovoltaic systems, house battery backup, and smart home technology such as "intelligent charging systems." Intelligent charging shows a situation in which a charging station or car with access to an exterior API, like a utility, acquires commands to start charging, slows down the charging rate, or stops charging following the infrastructure's requirements [1]. As a result, it permits adaptive utilization of the charging facility and charging activity to offer a specific demand reaction utility or an additional benefit to the facility. The intelligent control system comprises various approaches and algorithms that allow the charging system to function appropriately. These methods include current control, voltage control, temperature administration, and battery status estimation. The computerized control system can guarantee that the power source is being charged adequately while safeguarding against excessive charging, overcharging, and overheating by monitoring and managing the variables mentioned above.

Smart charging requires considerable consumer understanding. Suppose this process is isolated to autonomous electric vehicles. In that case, it is a handy software control to take advantage of the surplus of off-peak electricity supply and to vary the charging periods between different vehicles at a given charging station. In that case, presumably, there would be enough of these independent EVs to be available 24/7 while being charged at the cheapest times of the 24-hour clock or by vehicle. However, consumers do not have this capacity, so if this "smart" charging is based on realistic public education, the actual energy optimization results will drastically differ.

7.1.2 Importance of Intelligent Control in Battery Charging Systems

Intelligent control systems are intended to optimize the charging process, guaranteeing that the power source is charged swiftly, effectively, and securely. Moreover, intelligent control can optimize the charging parameters, such as voltage and current, based on the battery characteristics and charging method. By optimizing the charging parameters, intelligent control can reduce the charging time, minimize the energy consumption, and prevent overcharging and undercharging, which can damage the battery. Batteries are dynamic systems that change over time due to aging, temperature, and usage. Intelligent control can adapt to evolving battery characteristics and adjust the charging parameters accordingly. By adapting to the changing battery characteristics, intelligent control can maintain battery performance and prolong the battery lifespan, resulting in improved energy efficiency. By improving energy efficiency, intelligent control can reduce the environmental impact and promote sustainable development [2–7].

The remaining portion of this chapter is organized into the following sections: Section 7.2 introduces popular state-of-the-art methods to implement automated charging, followed by a literature survey for the same techniques in Section 7.3. Section 7.4 elaborates on the challenges state-of-the-art methods face and their future possibilities. At the end of the article, some rewarding factors of these approaches are given for convenience in Section 7.5, followed by a conclusion in Section 7.6.

7.2 METHODS TO IMPLEMENT INTELLIGENT CONTROL FOR BATTERY CHARGING SYSTEMS

Intelligent control can be implemented in battery charging systems using various techniques, depending on the specific requirements and constraints of the system. For example, different types of batteries have other characteristics that affect the charging method used. For example, lithium–ion batteries need constant current and voltage charging, while lead–acid batteries need constant voltage charging. The charging techniques also depend on the battery state of charge (SOC), temperature, and charging time [3, 4, 8]. Figure 7.1 shows the various factors considered for smart charging of Lithium -ion batteries.

Some fundamental approaches to implementing intelligent charging in batteries follow.

7.2.1 FUZZY LOGIC CONTROL

Fuzzy logic is a mathematical approach that allows for reasoning with uncertain or vague information. In intelligent battery charging systems, fuzzy logic can determine the optimal charging current and voltage based on several variables, including battery type, temperature, and charge level. Fuzzy logic controllers can be designed using software tools such as MATLAB, Simulink, or Mathworks and then implemented on a microcontroller or a programmable logic controller (PLC) to control the charging process.

7.2.2 NEURAL NETWORK CONTROL

Artificial intelligence that mimics complex relationships between inputs and outputs involves neural networks. Regarding battery charging systems, neural networks can

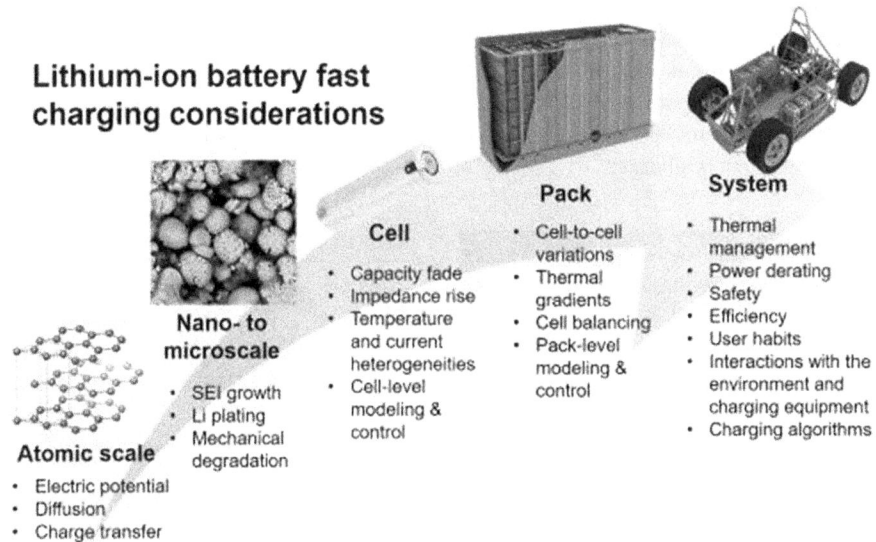

FIGURE 7.1 Factors to regulate for smart charging [9].

be trained to forecast the best charging current and voltage using historical time-series data and various input variables, including battery temperature, charge level, and charging time. The microcontroller or PLC that manages the charging procedure can incorporate the neural network logic.

7.2.3 ADAPTIVE CONTROL

Adaptive control modifies the controller's parameters according to system behavior changes. In battery charging systems, adaptive control can adjust the charging voltage and current based on changes in the battery's charge level, temperature, and other parameters. The adaptive control algorithm can be implemented on a microcontroller or a PLC that controls the charging process.

7.2.4 MODEL-PREDICTIVE CONTROL

Model-predictive control is a type of control that employs the system's mathematical model to forecast its future behavior and optimize the control inputs. In battery charging systems, model-predictive control can be used to optimize the charging current and voltage based on the predicted behavior of the battery. Although the model-predictive approach is very tedious for real-world implementation due to uncertain noise in real-time charging scenarios, the model-predictive control algorithm can be implemented on a microcontroller or a PLC to regulate the battery's charging process.

7.2.5 HYBRID INTELLIGENT CONTROL

Hybrid intelligent control combines multiple control methods to simultaneously take advantage of several features of respective models, such as fuzzy logic, neural networks, and adaptive control, to achieve optimal control performance. In battery charging systems, hybrid intelligent control can enhance charging current and voltage based on factors such as battery temperature, state of charge, and charging time. The hybrid intelligent control algorithm can be implemented on a microcontroller or a PLC that controls the charging process. Figure 7.2 shows fully automatic and intelligent battery charger.

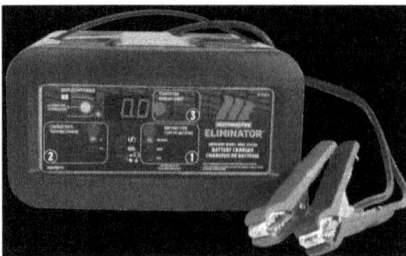
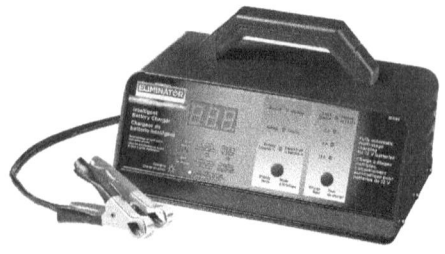

(a)   (b)

FIGURE 7.2   (a) Fully automatic smart battery charger [10], (b) Intelligent battery charger [11].

7.3 LITERATURE SURVEY

Fuzzy control techniques were first used in lead–acid battery charge systems in 1993 to get a better charge current and improve battery-charging performance. The system has good performance traits, like an optimal charging rate that can be achieved without knowing the battery model attributes [8, 12]. The controller of an electric vehicle can accommodate the proposed charging system. It performs poorly because fuzzy logic lacks a mathematical foundation and is sensitive to noise and uncertainties. An intelligent two-step charging method fuzzy controller for a lead–acid battery charger was investigated in 1999, and the depolarization technique built into the battery charger improves the effectiveness of the charging process by lowering battery internal resistance [13]. Moreover, to boost efficiency and enhance current practices, in 2009, a fuzzy inference process was used in a power management system (PMS) to control a buck converter as a battery charging system, and the results of the process were confirmed by computer modeling to show control potentialities [14].

Traditional lead–acid battery chargers suffer from drawbacks like low reliability, poor charging efficiency, and a lack of charging protection when using a single closed-loop equivalent control strategy. Thus, an intelligent lead–acid battery charger based on SCM double closed-loop controls was suggested in 2013 to address these drawbacks. This model's viability is also tested in an experimental setting [15]. A dynamic response model for charging systems based on adaptive control was developed in 2014. According to this model, the battery pack charging system's stability and dynamic performance have been substantially enhanced by decreasing the impact of charger input voltage fluctuations and changes in the battery pack's characteristics throughout the charging process [16].

A hybrid system innovative management method for battery lifetime increment using state of charge estimation in PV batteries is proposed in 2020, overcoming all these particular model approaches. In off-grid PV-battery hybrid systems, the proposed battery management system control strategy significantly reduces charging and deep-discharging disturbances, meeting the battery lifetime objective. Experimental measurements of functioning battery life, performance, and state of charge (SOC) are also made in this model [17] to assess model effectiveness. Triwijaya further argued in 2021 that the system's dependability must be considered in addition to the charging time. The battery and control system are not harmed by the quick charging time, as predicted by the model "Automatic Design of Battery Charging System Power Supply from Photovoltaic Sources Based on Voltage." According to the model's experimental results, the new controller charging period is significantly shortened. The proposed controller also minimizes battery overcharging and has high accuracy [18].

7.4 CHALLENGES AND FUTURE SCOPE

Several difficulties and research junctures exist in implementing intelligent control for battery charging systems. Below are some of the problems that need to be overcome to implement intelligent control of battery charging systems, along with the potential solutions in each of their respective dimensions.

7.4.1 Accuracy and Reliability

Intelligent control techniques rely on accurate and reliable data to provide optimal control outputs. However, sensor noise, environmental conditions, and system variability can all impact data accuracy and reliability. Future research should develop more accurate and dependable sensing and data processing techniques, such as machine learning and signal processing [19], to address this issue.

7.4.2 Battery Degradation and Capacity Loss

Batteries are dynamic systems that deteriorate with use and are affected by several variables, including temperature, usage, and cycling [20]. It is necessary for intelligent control techniques to adjust their outputs in response to changing battery characteristics. Future research can concentrate on creating methods for adaptive control that can calculate the battery's state of health (SOH) and modify the charging parameters to maximize battery lifespan to address this problem.

7.4.3 Hardware and Implementation

Intelligent control techniques require suitable hardware platforms and software algorithms to be implemented. The hardware platform should be capable of processing and transmitting the data in real time, while the software algorithm should be optimized for the hardware platform. To address this challenge, future research can focus on developing hardware and software co-design techniques that can optimize the performance and efficiency of the intelligent control system [21].

7.4.4 Scalability and Interoperability

Intelligent control techniques must be adaptable to various battery types, charging procedures, and applications. The smart control system must be flexible enough to respond to shifting needs and deliver the best possible control results [22]. Future research can concentrate on creating standardized interfaces and protocols that can help improve the versatility and connectivity of the intelligent control system to address this problem.

7.5 COMFORTING ELEMENTS PROMOTING WIDESPREAD ACCEPTANCE OF THE INTELLIGENCE-BASED STRATEGY

Among the benefits of intelligent control over conventional battery charging techniques are.

7.5.1 Improved Charging Efficacy

Intelligent control enhances the charging process, resulting in quicker battery filling. This means that, compared to traditional charging techniques, batteries can be recharged more rapidly and with fewer watts of energy.

7.5.2 Prolonged Battery Life

Intelligent control may extend battery life by preventing overcharging and undercharging. Overcharging can cause battery damage, while undercharging can result in a shorter total battery life.

7.5.3 Reduced Energy Costs

By optimizing the charging process, intelligent control can reduce the energy required to charge the battery, decreasing long-term utility costs.

7.5.4 Increased Safety

Intelligent control can help avoid charging battery potential risks like overheating and overcharging, thus guaranteeing the charging procedure is healthy for both the battery and the user.

7.5.5 Improved User Experience

Using intelligent control, the charging procedure for batteries can be streamlined and more straightforward for customers, resulting in enhanced user satisfaction and a lower chance of human error.

The following table shows the relative comparison of the pros and cons and the limitations of the topic in question (Table 7.1).

TABLE 7.1

Tabular Comparison of Various Prospects of Automated Control Charging Systems

Advantages	Disadvantages	Limitations
Longer battery life and faster charging times result from the intelligent control system's ability to optimize the charging procedure based on the specific needs and state of the battery [23].	The charging method may become complex due to the intelligent control system, which requires supplementary software and hardware [24].	The technique might only work with some battery types, which would restrict its utility in some circumstances.
The equipment can track the battery's electrical current, voltage, and temperature to prevent excessive charging and decrease the danger of harm to the battery or potential fire hazards [24].	Due to the additional software and hardware required, autonomous battery charging systems may be priced higher than routine charging systems [24].	A sophisticated control system may require recurring maintenance, calibration, or software updates to achieve optimal operation, which might raise the overall cost.

(Continued)

TABLE 7.1 (Continued)

Advantages	Disadvantages	Limitations
The appliance can charge several battery types since it can adjust the voltage at which it charges and the current to suit the needs of the battery.	The automated control system, which might be more challenging to use than a traditional charging framework, might require users to be trained [24].	Unlike regular charging systems, the automated control system may be more challenging to engineer, execute, and debug, necessitating specialized expertise and abilities.
Users can check the current condition of their batteries and conduct charging operations using the system's intuitive interfaces.	Issues in the procedure for charging may occur when the automated control system does not operate at its best [25].	The automated control method's performance depends on the application in question and energy category, which restricts its implementation in some situations.
A battery-powered charging system with automated administration may be significantly more economical and energy-effective than conventional recharging systems.	Because it depends on information technology, a smart battery charging solution might not work effectively if the technology behind it breaks down [24].	The technology might be challenging to scale, which would restrict its applicability to applications on a massive scale.

7.6 CONCLUSION

Numerous sectors, including automotive, renewable energy, and consumer electronics businesses, use intelligent battery charging systems more often. These systems refine the charging method and boost battery performance using evolutionary algorithms, resulting in more profound charging efficiency, reduced charging times, and prolonged battery life. While numerous intelligent charging algorithms, such as adaptive neuro-fuzzy inference systems, genetic algorithms, and neural networks, can offer better charging performance when compared to ordinary charging methods, their efficacy can vary based on the particular application and context. More study is required to boost these systems' performance and find the best algorithms for specific applications. Overall, intelligent automated battery charging systems might significantly improve battery performance, which might render numerous rewards.

REFERENCES

1. Wu, Y., Wang, Z., Huangfu, Y., Ravey, A., Chrenko, D. and Gao, F., 2022. Hierarchical operation of electric vehicle charging station in smart grid integration applications—An overview. *International Journal of Electrical Power & Energy Systems*, *139*, p. 108005.
2. Alkawsi, G., Baashar, Y., Abbas, U.D., Alkahtani, A.A. and Tiong, S.K., 2021. Review of renewable energy-based charging infrastructure for electric vehicles. *Applied Sciences*, *11*(9), p. 3847.
3. Chen, C., Wei, Z. and Knoll, A.C., 2021. Charging optimization for Li-ion battery in electric vehicles: A review. *IEEE Transactions on Transportation Electrification*, *8*(3), pp. 3068–3089.

4. Song, C., Kim, K., Sung, D., Kim, K., Yang, H., Lee, H., Cho, G.Y. and Cha, S.W., 2021. A review of optimal energy management strategies using machine learning techniques for hybrid electric vehicles. *International Journal of Automotive Technology*, 22, pp. 1437–1452.

5. Lai, C.S., Lai, L.L., and Lai, Q.H., 2021. Blockchain Applications in Microgrid Clusters. *Smart Grids and Big Data Analytics for Smart Cities*, pp. 265–305. DOI: 10.1007/978-3-030-52155-4_3

6. Huang, Y. and Yang, H., 2012, March. A Method based on K-Means and fuzzy algorithm for industrial load identification. In *2012 Asia-Pacific Power and Energy Engineering Conference* (pp. 1–4). IEEE.

7. Schuller, A., Dietz, B., Flath, C.M. and Weinhardt, C., 2014. Charging strategies for battery electric vehicles: Economic benchmark and V2G potential. *IEEE Transactions on Power Systems*, 29(5), pp. 2014–2022.

8. Poorani, S., Kumar, K.U. and Renganarayanan, S., 2003, April. Intelligent controller design for electric vehicle. In *The 57th IEEE Semiannual Vehicular Technology Conference, 2003. VTC 2003-Spring.* (Vol. 4, pp. 2447–2450). IEEE.

9. Tomaszewska, A., Chu, Z., Feng, X., O'Kane, S., Liu, X., Chen, J., Ji, C., Endler, E., Li, R., Liu, L. and Li, Y., 2019. Lithium-ion battery fast charging: A review. *ETransportation*, 1, p. 100011.

10. "Intelligent battery charger". https://www.canadiantire.ca/en/pdp/motomaster-eliminator-intelligent-battery-charger-12-8-2a-0111518p.html. Accessed on April 1, 2023.

11. "Smart battery charger". https://www.canadiantire.ca/en/pdp/motomaster-eliminator-workshop-series-smart-battery-charger-fully-automatic-15-3-amp-6v-12v-0111978p.html?rrec=true. Accessed on April 2, 2023.

12. Liang, Y.C. and Ng, T.K., 1993. Design of battery charging system with fuzzy logic controller. *International Journal of Electronics Theoretical and Experimental*, 75(1), pp. 75–86.

13. Bandara, G.E.M.D.C., Ivanov, R. and Gishin, S., 1999, October. Intelligent fuzzy controller for a lead–acid battery charger. In *IEEE SMC'99 Conference Proceedings. 1999 IEEE International Conference on Systems, Man, and Cybernetics (Cat. No. 99CH37028)* (Vol. 6, pp. 185–189). IEEE.

14. Malekjamshidi, Z. and Jafari, M., 2009, December. Design, simulation and implementation of an intelligent battery charging system. In *2009 Second International Conference on Computer and Electrical Engineering* (Vol. 2, pp. 242–246). IEEE.

15. Luo, W., Yang, Y., Li, H. and Jiang, Y., 2013, August. Design of intelligent battery charger based on SCM double closed-loop control. In *2013 IEEE International Conference on Mechatronics and Automation* (pp. 1413–1418). IEEE.

16. Haoyu, L., Yong, G., Lei, Z. and Zhenwei, L., 2014, September. Adaptive control of charging system based on online parameters identification. In *2014 IEEE 36th International Telecommunications Energy Conference (INTELEC)* (pp. 1–6). IEEE.

17. Qays, M.O., Buswig, Y., Basri, H., Hossain, M.L., Abu-Siada, A., Rahman, M.M. and Muyeen, S.M., 2020. An intelligent controlling method for battery lifetime increment using state of charge estimation in pv-battery hybrid system. *Applied Sciences*, 10(24), p. 8799.

18. Triwijaya, S., Pradipta, A. and Wati, T., 2021, November. Automatic design of battery charging system power supply from photovoltaic sources based on voltage. In *Journal of Physics: Conference Series* (Vol. 2117, No. 1, p. 012011). IOP Publishing.

19. Jiang, J., Tang, H., Mohamed, M.S., Luo, S. and Chen, J., 2020. Augmented tikhonov regularization method for dynamic load identification. *Applied Sciences*, 10(18), p. 6348.

20. Ren, Z. and Du, C., 2023. A review of machine learning state-of-charge and state-of-health estimation algorithms for lithium-ion batteries. *Energy Reports*, 9, pp. 2993–3021.

21. Sadeghian, O., Oshnoei, A., Mohammadi-Ivatloo, B., Vahidinasab, V. and Anvari-Moghaddam, A., 2022. A comprehensive review on electric vehicles smart charging: Solutions, strategies, technologies, and challenges. *Journal of Energy Storage, 54*, p. 105241.

22. Liu, Y., Duan, Q., Xu, J., Li, H., Sun, J. and Wang, Q., 2020. Experimental study on a novel safety strategy of lithium-ion battery integrating fire suppression and rapid cooling. *Journal of Energy Storage, 28*, p. 101185.

23. Lipu, M.H., Hannan, M.A., Karim, T.F., Hussain, A., Saad, M.H.M., Ayob, A., Miah, M.S. and Mahlia, T.I., 2021. Intelligent algorithms and control strategies for battery management system in electric vehicles: Progress, challenges and future outlook. *Journal of Cleaner Production, 292*, p. 126044.

24. Lipu, M.H., Hannan, M.A., Hussain, A., Ayob, A., Saad, M.H., Karim, T.F. and How, D.N., 2020. Data-driven state of charge estimation of lithium-ion batteries: Algorithms, implementation factors, limitations and future trends. *Journal of Cleaner Production, 277*, p. 124110.

25. Schmutzler, J., Andersen, C.A. and Wietfeld, C., 2013. Evaluation of OCPP and IEC 61850 for smart charging electric vehicles. *World Electric Vehicle Journal, 6*(4), pp. 863–874.

8 Impacts of Energy Storage Systems on PV Prosumer Based on Household Load Profiles

Bevara Srikanth and Sarat Kumar Sahoo
Parala Maharaja Engineering College, Berhampur, India

Franco Fernando Yanine
Universidad Finis Terrae, Providencia, Santiago, Chile

8.1 INTRODUCTION

Every living entity on earth needs solar energy, which keeps our planet comfortable. However, the concentration of CO_2 is growing gradually, which produces additional heat on earth and results in climatic changes. Due to the huge level of CO_2 emissions due to the fossil fuels use in the energy industry, it is inevitable that we shall need to substitute conventional power generation with renewable generation [1]. Among the available renewable sources, solar energy with photovoltaic (PV) technology stands out globally [2]. The increase in residential rooftop installations shows the spreading of PV technology around the world [3]. In this regard, the consumer, who produces and consumes the energy, plays a vital role in the renewable transition from conventional power generation [4]. The PV systems produces the power is irregular and varies with time. Similarly, household power consumption differs throughout the day; there will therefore be a mismatch between the production and load profiles [5].

Even though renewable production is advantageous, a huge injection of PV power into the utility may cause voltage fluctuations. In order to overcome these fluctuations and mismatches between the load and production profiles, a storage system is necessary. Using the storage system, household consumers can consume most of the power produced and maximize their own consumption [5]. A battery energy storage system (BESS) is widely used as installation prices are decreasing. Lead–acid and Li-ion batteries are the variants of battery technology available on the market. Because of their high energy density and efficiency, Li-ion batteries are preferred to lead–acid batteries, even though Li-ion faces certain safety issues [1]. Frequent charge and discharge cycles may deteriorate the battery's performance. Super capacitors combined

DOI: 10.1201/9781003481836-8

with batteries as a storage system are the solution to overcome the aging effect of batteries [6]. But the storage system size is challenging for PV prosumers. In this chapter, prosumer importance in PV technology development, the role of energy storage systems, and methods to choose the size of storage systems are discussed as in the preceding sections.

8.2   PROSUMER IMPORTANCE IN THE PV TECHNOLOGY DEVELOPMENT

Alvin Toffler is the author of the 1980 book "The Third Wave," and he specified the term "prosumer" based on his observation of market saturation accompanied by typical items. Don Tapscott and Williams revived the concept of "prosumer" in numerous ways by which a consumer can be part of the technical domain. Earlier, these domains were taken care of by professionals [7]. The limited definition of "prosumer" is "a consumer who can also produce." Basically, a prosumer is a participant in the energy market today, and will be a dynamic participant tomorrow. A new definition emerges in an inspirational way: the consumer is not restricted to only the production and storage of energy but is involved in creating wealth through electricity as a commercial supplier of energy [8].

The prosumer physically contains innovative equipment of the smart grid, such as smart meters, information, and communication technologies (ICT), sources of energy, storage devices, and loads.

The functions of a prosumer are production, storage, consumption, and market participation, which are related to interaction with others around the globe, as shown in Figure 8.1 [9].

PV technology is best emerging technologies that had the ability to raise its share in the future energy market. Residential prosumers who adopt photovoltaic technology are known as PV prosumers.

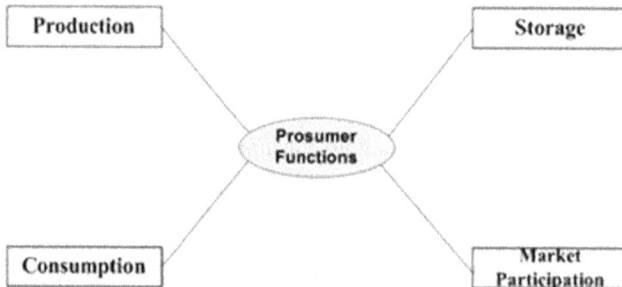

FIGURE 8.1   Functions of prosumer.

8.3 CLASSIFICATION OF PROSUMER

Prosumers are classified based on their relationship with traditional utilities [10–13].

8.3.1 GRID DEFECTION

Prosumers have the right to break up the agreement with the utility system such that all the residential electric requirements are supplied by photovoltaic systems and storage devices. They move toward off-grid mode.

8.3.2 SELF-CONSUMPTION

Prosumers acquire power continuously from the utility; however, they decrease the power bought from the grid by obtaining most of their residential needs from photovoltaic systems.

8.3.3 COMMERCIAL PRODUCTION OF ELECTRICITY

Prosumers are able to sell a huge percentage of the power from their PV generation to the grid, even though they continue to buy the power as well.

8.4 MERITS OF PROSUMER

If PV is chosen for power generation, consumers' contribution to the energy sector's development will be enhanced. The merits from the prosumer perspective and the PV industry are represented by three categories.

8.4.1 ECONOMIC MERITS

Due to the PV installation, the energy sector does not need to be dependent on fossil fuels alone, as they are increasing day by day [14]. The utility is also benefited by the reduction in the cost of distribution and transmission and the losses associated with them as the PV energy is consumed locally. On the other side, a PV consumer can make profit by vending the excess power to the utility or consuming it for residential loads, which allows him to be in an affordable region [15]. Prosumers may be awarded subsidies for the installation of PV. Based on the system size and the load demand of the household, both self-sufficiency and self-consumption can be obtained with higher values by household owner. Self-sufficiency measures what percentage of energy generated by PV can meet household energy requirements. More self-sufficiency means less consumption from the grid [15].

8.4.2 ENVIRONMENTAL MERITS

Household PV systems release fewer emissions from their operation, and they do not need cooling systems. PV owners are the best participants in the development of a viable environment by transforming energy production from traditional to renewable. Prosumers are the drivers of environmental sustainability and make a big impact on a greenhouse gas-free atmosphere. But the process of manufacturing PV cell components produces greenhouse gas emissions, which should also be considered [16].

8.4.3 SOCIETAL MERITS

PV household deployment allows prosumers to contribute to societal benefits through the development of industry in an innovative way. The prosumers will gain awareness of the power consumption profiles and thereby strive to increase their own consumption [17, 18].

8.5 ROLE OF ENERGY STORAGE SYSTEM

8.5.1 WHY ESS?

As solar PV generation is an intermittent energy source, it is not possible to produce power during peak demand. In order to get the greatest advantage from this source and make self-consumption and self-sufficiency as possible, an ESS is used. The ESS refers to energy storage systems that reserve energy off-peak and deliver it during peak hours. The usage of ESS could also lead to other merits like peak shaving, leveling of loads, and control of voltage and frequency when connected to the grid [19, 20]. The oscillations in load demand can be decreased by peak shaving. The ESS effect on the load curve is illustrated in Figure 8.2 [21].

The fragmentary feature of solar PV produces the electric energy discontinuously, due to which there will be oscillations in the grid voltage and frequency. When the PV generates maximum power output, the ESS keeps it in reserve for future use and contributes to the grid when it needs the power, as illustrated in Figure 8.3. These functions of ESS keep the levels of voltage and frequency on the grid constant [22].

The ESS are categorized based upon their distinct discharge times and capabilities, for which they have multiple applications. The various types of ESS present are

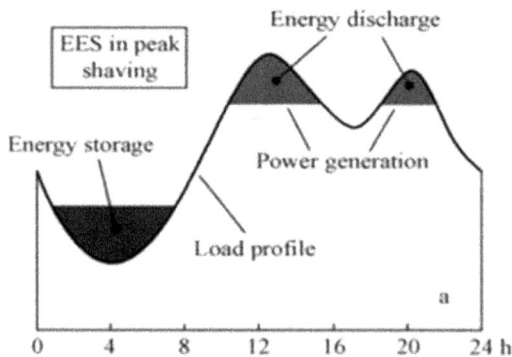

FIGURE 8.2 Peak shaving illustration [21].

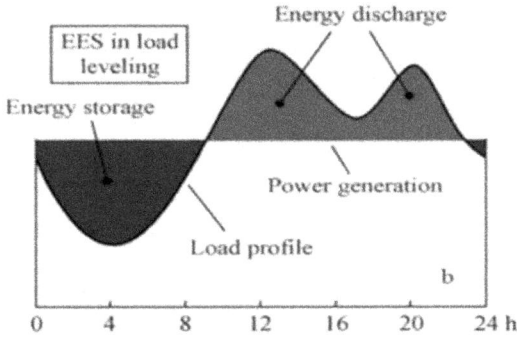

FIGURE 8.3 Leveling of load illustration [21].

like batteries, pumped hydroelectric storage, flywheels, etc. Among the various ESS, battery energy storage systems (BESS) are found as the prime choice of storage systems to reserve the irregular solar PV energy. The BESS is so advantageous due to its compact size and cost-effective considerations. Solar PV supported by BESS will increase the self-consumption of household consumers and bring a localized power generation market to individual households with PV renewable generation [23–25].

In recent times, the electricity generated from photovoltaic systems with battery support has increased self-consumption dynamically in developed countries [26, 27].

8.5.2 PV-Battery Household Configuration

A PV system connected to the grid is depicted in Figure 8.4 [28].

Initially, PV panels absorb the irradiance from the sun and transform it into electric energy (DC). The DC energy is given to the inverter, which transforms it into AC. The AC power is either utilized by household demand or transferred to the utility

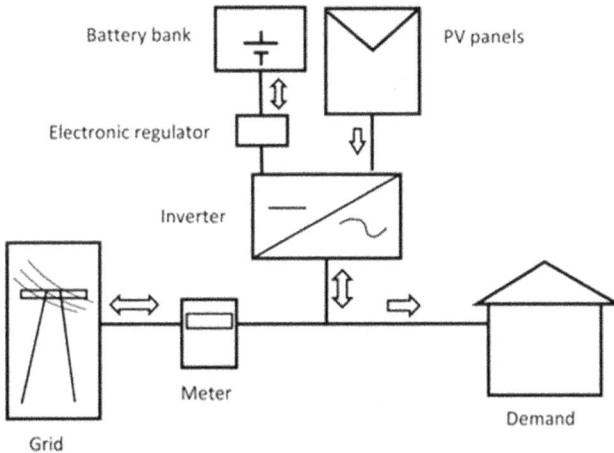

FIGURE 8.4 Representation of household PV-Battery configuration [28].

grid. A battery is used to store the energy and to operate discharge and charge cycles based on the requirements. The electronic regulator is placed so as to avoid heavy battery discharge and charge operations, which results in an increase in battery life [29, 30].

The PV-supported battery system can be attached either to the AC circuitry of the consumer home, which is known as AC coupling, or to the DC circuitry, which is called DC coupling. Additional losses are generated due to the existence of one more inverter in the AC coupling system; however, they increase flexibility and decrease dependence on the PV inverter. Higher efficiencies are obtained in DC coupling systems [31]. Battery performance can be maximized through the supervisory system, which controls the battery discharge and charge cycles [32].

8.5.3 FACTORS INFLUENCING BATTERY PERFORMANCE

Table 8.1 depicts the factors which affect the degradation of battery life cycle and performance.

Most of the batteries used in photovoltaic systems are made from lead–acid or lithium-ion (Li-ion) combinations. Out of the two combinations, lead–acid is preferred due to its economic cost [35]. But lead–acid batteries are facing issues of less lifetime anticipation and fewer efficiencies when compared to Li-ion batteries.

Emerging battery technology has acquired a huge market share in recent times in the form of lithium-ion batteries. The advantages of Li-ion include its high efficiency, the ratio of energy-to-weight, and numerous discharge and charge patterns. But one of the barriers faced by the Li-ion batteries is that the price per kWh for the high energy storage is very high. In spite of a reduction of up to 65% in cost per kWh during 2009–2013, even then it is overpriced compared with remaining battery technologies like sodium sulfur and lead–acid batteries [39].

The safety factor is the other barrier faced by Li-ion batteries. The decomposition of metal oxide electrons at high temperatures causes a thermal runway, which tends

TABLE 8.1
Factors Influencing Battery Performance [33]

Factor	Description
Depth of discharge (DOD)	It indicates the depth of the level of discharge of the battery. Zero percentage DOD shows the battery is charged and 100% DOD indicates 100% discharged battery [33].
Stage of charge (SOC)	It indicates the level of availability expressed as a percentage of battery's full capacity. The battery degradation rate increases with a high level of SOC [34].
Ambient/cell temperature	The battery degradation rate will increase with the change in ambient temperature. 25°C–35°C is the safe operating region of ambient temperature for the battery [35, 36].
Charge/discharge voltage and rate	Peak current rates and temperatures are evolved due to peak discharge and charge cycles of the battery, which deteriorate battery performance [37, 38].

to flare up cells. Higher discharging, charging, and current charging are the causes of this phenomenon. A battery management system is utilized to overcome these issues through supervisory control of battery discharge and charge cycles in the area of safety and to achieve high performance [40].

The high discharge currents in Li-ion batteries produce enormous heat, which tends to a decrement in efficiency with the respective energy conversion from chemical to electrical. Furthermore, there is a rise in internal resistance and deterioration of capacity due to these high temperatures. The upswing of high currents ends the choice of Li-ion batteries for the purpose of storage of energy [41, 42].

Apart from batteries, capacitors can discharge high currents exponentially, which is useful in applications where the loads draw high currents, like, for example, to prevent the initial stiction at the start of an electric motor, with devices with high inrush currents, and also to synchronize the growing demands in digital circuits [43].

If the capacitor is arranged in parallel with the battery at the output end, most of the current is supplied by the capacitor rather than the battery. This arrangement prevents the battery from experiencing huge cycles of discharge and further improves the life cycle and reliability of the battery. Nevertheless, there should be a compromise in the volume and weight of the system [44].

The characteristics of different storage systems differ from each other, such as ultra-capacitors, which have low energy density and high-power density, and lithium-ion batteries, which have low power density and high energy density. But super capacitors are advantageous due to their high charge and discharge flow, which are usually more than 10,000. The switching mode of operation between discharge and charge is possible frequently [45].

8.6 METHODS TO CHOOSE THE SIZE OF STORAGE SYSTEMS

In order to increase the self-consumption of PV systems supported by battery-super capacitor configurations, the size of the PV systems needs to be optimally determined. The methods to determine the PV household system size are discussed in the subsequent sections.

8.6.1 TECHNO ECONOMIC MODEL

The workings of this model are based on variations in parameters such as the PV system size and storage device specification. The entire cost of the power supply and the respective parameters associated with economics and technologies need to be determined for every system configuration. The optimal price of the system combination can be identified by arranging the results with the respective whole power supply cost during the lifespan of the system in descending order [47, 48].

Due to the aging of batteries, regulatory and economic schemes are taken into consideration, and the production, demand, and charging cycles pose high resolution. A simulation-based algorithm model is employed due to the above reasons. This model is applicable to household PV systems whose size is limited to 10 kW, even though the size could be increased further. The model structure is illustrated in Figure 8.5.

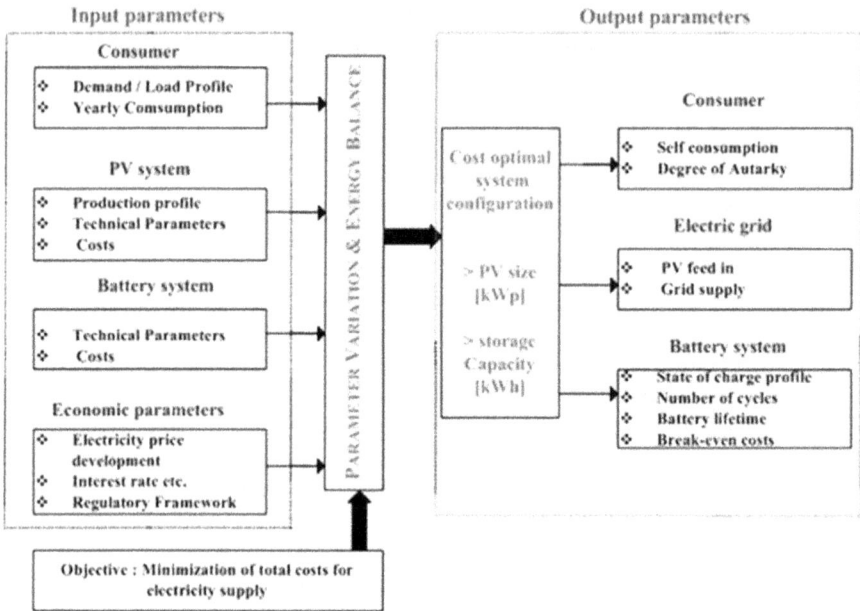

FIGURE 8.5   Representation of BaPSi model [46].

For model calculations, the parameters required from the input side are the data collection of PV profiles, consumer, battery, and economic constraints. The BaPSi model is framed for the PV-Battery system implementation based on the energy balance concept of demand and power supply. For every step, the energy balance will be determined.

High resolutions are preferred for production and load profiles when short-term fluctuations occur in the power supply. Priority is given to direct self-consumption of PV systems that are generated locally. The battery replaces direct consumption if it fails to deliver. If the PV generation is more than sufficient, the battery will be charged, and it is discharged if the generation of PV does not match the load demand [49].

The power is drawn from the utility to meet the power demand of households when batteries and direct consumption are not able to meet the demand. The battery utilization is represented in the flow chart in Figure 8.6 (source: IEK-STE 2015). Increasing self-consumption is the aim of this approach [46].

8.6.2   ECONOMIC MODEL

The criteria based on the evaluation of principal investments are required for the economic analysis. Payback Period (PBP), Net Present Value (NPV), Discounted Profitability Ratio (DPR), and Internal Rate of Return (IRR) are the parameters considered for this model. The aforementioned analysis considers both starting investments and futuristic costs for determining the solution. Every parameter is formed

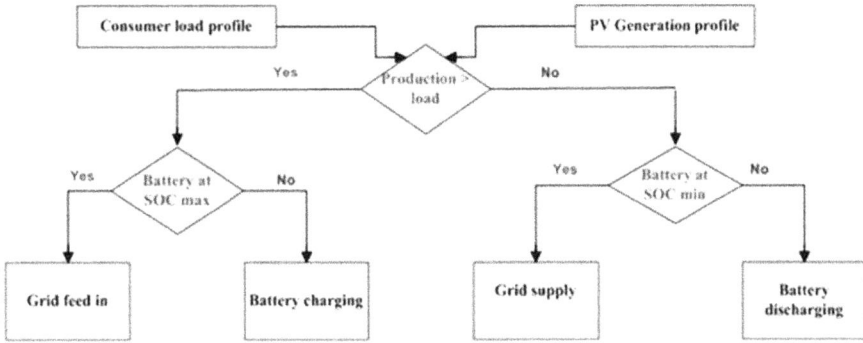

FIGURE 8.6 Representation of household PV-battery configuration [46].

as a function of the rating of the PV plant and the storage, which are represented by $W_{ST,max}$ and $P_{PV,max}$ respectively, and they can be determined for any equipment rating. To verify if the below-mentioned solution is performing to determine the minimum cost or not, all the parameters related to optimum ratings are determined. i.e., parameters are taken as functions of $W_{ST,max}^{opt}$ and $P_{PV,max}^{opt}$ [50].

This model is analyzed by choosing the life of the PV system as 20 years and 10 years as the storage period.

8.6.2.1 Payback Period

This is the period required to recover the capital money disbursed in the investments. It can be calculated from Equations (8.1), (8.2), (8.3), and (8.4) [50].

$$\text{Payback Period} = \min\left[n : \sum_{j=1}^{n} \text{FCFO}_j - \text{FCFO}_0 \geq 0 \right] \tag{8.1}$$

Where FCFO_j is cash flow since year j and $j = 1, 2, 3, \ldots 20$.

$$\text{FCFO}_j = \left(C_{NET,Y,j}(0,0) - C_{NET,Y,j}\langle P_{PV,max}^{opt}, W_{ST,max}^{opt} \rangle \right) - \begin{bmatrix} 0 & j \neq 10 \\ C_{ST}W_{ST,max}^{opt} & j = 10 \end{bmatrix} \tag{8.2}$$

And FCFO_0 is cash flow since year 0; i.e., initial investment.

$$\text{FCFO}_0 = C_{PV}P_{PV,max}^{opt} + C_{ST}W_{ST,max}^{opt} \tag{8.3}$$

$$\sum_{j=1}^{n}\left\{ C_{NET,Y,j}(0,0) - C_{NET,Y,j}\left(P_{PV,max}^{opt}, W_{ST,max}^{opt} \right) \right\}$$
$$- \left[C_{PV}P_{PV,max} + \begin{bmatrix} C_{ST}W_{ST,max}^{opt} & n < 10 \\ 2C_{ST}W_{ST,max}^{opt} & n \geq 10 \end{bmatrix} \right] \geq 0 \tag{8.4}$$

Where

C_{ST} refers to the storage system cost per kWh

C_{PV} refers to the cost of PV per peak kW

$C_{NET,Y,j}$ refers to the jth year annual cost of energy

8.6.2.2 Net Present Value

Cash flows corresponding to surplus or deficit are measured by the NPV in terms of present values. The NPV is defined as the sum of FCFO during all years at a discounted interest rate (WACC). The NPV is given in the Equation (8.5) [50].

$$\text{Net present value} = \sum_{j=1}^{20} \frac{FCFO_j}{(1+WACC)^{j-0.5}} - FCFO_0 \tag{8.5}$$

Where

WACC refers to the weighted average cost of capital

8.6.2.3 Discounted Profitability Ratio

It gives the return on investment over the lifetime of the operation, expressed in percentage. It is the ratio of NPV to starting investment. It is represented by Equation (8.6) [50].

$$\text{Discounted Profitability Ratio } (\%) = \frac{NPV}{FCFO_0} \times 100 \tag{8.6}$$

8.6.2.4 Internal Rate of Return

It is the rate of return at which the value of NPV becomes zero. In other words, it is defined as the rate at which the investment reaches breakeven. IRR measures whether an investment is desirable or not by which a project is selected [50].

The above parameters should be integrated for the analysis.

8.7 CONCLUSION

In this chapter, the necessity of renewable energy sources compared with fossil fuels and the superiority of photovoltaic technology (PV) are explained. The prosumer concept is defined, and the functions and classifications of prosumers are represented. With this representation, it is evident that the consumer is one of the consumers who will become a stakeholder in the energy sector for the betterment of society. Due to the intermittent nature of PV, grid-connected PV systems cannot produce enough power to meet peak load demand. Energy storage systems are the solution for irregular PV power producers. In BESS, lithium-ion batteries are preferred over lead–acid batteries because of their high efficiency and life cycle. Despite the merits of Li-ion batteries over lead–acid batteries, they are not capable of delivering high pulses of current continuously. The battery's performance will deteriorate if it tries to supply power for loads that draw high currents frequently. Super capacitors, which have the feature of high-power density and can supply peak pulses, should be connected in shunt with the battery. This configuration minimizes the battery stress and, in addition, avoids the problem of battery replacement.

However, it is quite challenging to calculate the PV household system size. The literature provides two successful models, namely technical and economical, for the evaluation of the PV system size. Both models are discussed in this chapter and can be applied from the perspective of the application.

REFERENCES

[1] Mertens, K. (2018). *Photovoltaics: Fundamentals, Technology, and Practice*. 2nd Edition. John Wiley & Sons Ltd., ISBN: 978-1-119-40133-9.

[2] Palm, A. (2016). Local factors driving the diffusion of solar photovoltaics in Sweden: A case study of five municipalities in an early market. In *Energy Research & Social Science* (Vol. 14, pp. 1–12). Elsevier BV.

[3] Aune, M., Godbolt, Å. L., Sørensen, K. H., Ryghaug, M., Karlstrøm, H., & Næss, R. (2016). Concerned consumption. Global warming changing household domestication of energy. In *Energy Policy* (Vol. 98, pp. 290–297). Elsevier BV.

[4] Suna, D., Haas, R., & Lopez-Polo, A. (2008). Country specific added value analysis of PV systems. In *Proceedings of ISES World Congress 2007* (Vol. I–V, pp. 161–166). Springer Berlin Heidelberg.

[5] Luthander, R., Widén, J., Nilsson, D., & Palm, J. (2015). Photovoltaic self-consumption in buildings: A review. In *Applied Energy* (Vol. 142, pp. 80–94). Elsevier BV.

[6] Wang, G., Ciobotaru, M., & Agelidis, V. G. (2014). Power smoothing of large solar PV plant using hybrid energy storage. In *IEEE Transactions on Sustainable Energy* (Vol. 5, Issue 3, pp. 834–842). Institute of Electrical and Electronics Engineers (IEEE).

[7] Rodríguez-Molina, J., Martínez-Núñez, M., Martínez, J.-F., & Pérez-Aguiar, W. (2014). Business models in the smart grid: Challenges, opportunities and proposals for prosumer profitability. In *Energies* (Vol. 7, Issue 9, pp. 6142–6171). MDPI AG.

[8] Goncalves Da Silva, P., Ilic, D., & Karnouskos, S. (2014). The impact of smart grid prosumer grouping on forecasting accuracy and its benefits for local electricity market trading. In *IEEE Transactions on Smart Grid* (Vol. 5, Issue 1, pp. 402–410). Institute of Electrical and Electronics Engineers (IEEE).

[9] Grijalva, S., & Tariq, M. U. (2011). Prosumer-based smart grid architecture enables a flat, sustainable electricity industry. In *ISGT 2011. 2011 IEEE PES Innovative Smart Grid Technologies (ISGT)*. IEEE.

[10] Espe, E., Potdar, V., & Chang, E. (2018). Prosumer communities and relationships in smart grids: A literature review, evolution and future directions. In *Energies* (Vol. 11, Issue 10, p. 2528). MDPI AG.

[11] Quoilin, S., Kavvadias, K., Mercier, A., Pappone, I., & Zucker, A. (2016). Quantifying self-consumption linked to solar home battery systems: Statistical analysis and economic assessment. In *Applied Energy* (Vol. 182, pp. 58–67). Elsevier BV.

[12] Wuebben, D., & Peters, J. (2022). Communicating the values and benefits of home solar prosumerism. In *Energies* (Vol. 15, Issue 2, p. 596). MDPI AG.

[13] Schulte, E., Scheller, F., Sloot, D., & Bruckner, T. (2022). A meta-analysis of residential PV adoption: the important role of perceived benefits, intentions and antecedents in solar energy acceptance. In *Energy Research & Social Science* (Vol. 84, p. 102339). Elsevier BV.

[14] Freitas, C. J. P., & da Silva, P. P. (2015). European Union emissions trading scheme impact on the Spanish electricity price during phase II and phase III implementation. In *Utilities Policy* (Vol. 33, pp. 54–62). Elsevier BV.

[15] Radhi, H. (2012). Trade-off between environmental and economic implications of PV systems integrated into the UAE residential sector. In *Renewable and Sustainable Energy Reviews* (Vol. 16, Issue 5, pp. 2468–2474). Elsevier BV.

[16] Okoye, C. O., & Oranekwu-Okoye, B. C. (2018). Economic feasibility of solar PV system for rural electrification in Sub-Sahara Africa. In *Renewable and Sustainable Energy Reviews* (Vol. 82, pp. 2537–2547). Elsevier BV.

[17] Best, R., Burke, P. J., Nepal, R., & Reynolds, Z. (2021). Effects of rooftop solar on housing prices in Australia. In *Australian Journal of Agricultural and Resource Economics* (Vol. 65, Issue 3, pp. 493–511). Wiley.

[18] Kairies, K.-P., Figgener, J., Haberschusz, D., Wessels, O., Tepe, B., & Sauer, D. U. (2019). Market and technology development of PV home storage systems in Germany. In *Journal of Energy Storage* (Vol. 23, pp. 416–424). Elsevier BV.

[19] Akinyele, D. O., & Rayudu, R. K. (2014). Review of energy storage technologies for sustainable power networks. In *Sustainable Energy Technologies and Assessments* (Vol. 8, pp. 74–91). Elsevier BV.

[20] Nyholm, E., Goop, J., Odenberger, M., & Johnsson, F. (2016). Solar photovoltaic-battery systems in Swedish households – Self-consumption and self-sufficiency. In *Applied Energy* (Vol. 183, pp. 148–159). Elsevier BV.

[21] Lucas, A., & Chondrogiannis, S. (2016). Smart grid energy storage controller for frequency regulation and peak shaving, using a vanadium redox flow battery. In *International Journal of Electrical Power & Energy Systems* (Vol. 80, pp. 26–36). Elsevier BV.

[22] Zhao, H., Wu, Q., Hu, S., Xu, H., & Rasmussen, C. N. (2015). Review of energy storage system for wind power integration support. In *Applied Energy* (Vol. 137, pp. 545–553). Elsevier BV.

[23] Kubli, M., & Ulli-Beer, S. (2016). Decentralisation dynamics in energy systems: A generic simulation of network effects. In *Energy Research & Social Science* (Vol. 13, pp. 71–83). Elsevier BV.

[24] Hadjipaschalis, I., Poullikkas, A., & Efthimiou, V. (2009). Overview of current and future energy storage technologies for electric power applications. In *Renewable and Sustainable Energy Reviews* (Vol. 13, Issues 6–7, pp. 1513–1522). Elsevier BV.

[25] Öhrlund, I., Stikvoort, B., Schultzberg, M., & Bartusch, C. (2020). Rising with the sun? Encouraging solar electricity self-consumption among apartment owners in Sweden. In *Energy Research & Social Science* (Vol. 64, p. 101424). Elsevier BV.

[26] Hoppmann, J., Volland, J., Schmidt, T. S., & Hoffmann, V. H. (2014). The economic viability of battery storage for residential solar photovoltaic systems – A review and a simulation model. In *Renewable and Sustainable Energy Reviews* (Vol. 39, pp. 1101–1118). Elsevier BV.

[27] Kairies, K.-P., Magnor, D., & Sauer, D. U. (2015). Scientific measuring and evaluation program for photovoltaic battery systems (WMEP PV-Speicher). In *Energy Procedia* (Vol. 73, pp. 200–207). Elsevier BV.

[28] Garabitos Lara, E., & Santos García, F. (2021). Review on viability and implementation of residential PV-battery systems: Considering the case of Dominican Republic. In *Energy Reports* (Vol. 7, pp. 8868–8899). Elsevier BV.

[29] Alramlawi, M., Gabash, A., Mohagheghi, E., & Li, P. (2018). Optimal operation of hybrid PV-battery system considering grid scheduled blackouts and battery lifetime. In *Solar Energy* (Vol. 161, pp. 125–137). Elsevier BV.

[30] Khoury, J., Mbayed, R., Salloum, G., & Monmasson, E. (2015). Optimal sizing of a residential PV-battery backup for an intermittent primary energy source under realistic constraints. In *Energy and Buildings* (Vol. 105, pp. 206–216). Elsevier BV.

[31] Kato, S., Nishihara, H., Taniguchi, I., Fukui, M., & Sakakibara, K. (2012). Analysis on battery storage utilization in decentralized solar energy networks based on a mathematical programming model. In The 6th International Conference on Soft Computing and Intelligent Systems, and he 13th International Symposium on Advanced Intelligence Systems. 2012 Joint 6th Intl. Conference on Soft Computing and Intelligent Systems (SCIS) and 13th Intl. Symposium on Advanced Intelligent Systems (ISIS). IEEE.

[32] Baker, J. (2008). New technology and possible advances in energy storage. In *Energy Policy* (Vol. 36, Issue 12, pp. 4368–4373). Elsevier BV.

[33] Rahmanifar, M. S. (2017). Enhancing the cycle life of Lead–acid batteries by modifying negative grid surface. In *Electrochimica Acta* (Vol. 235, pp. 10–18). Elsevier BV.

[34] Piller, S., Perrin, M., & Jossen, A. (2001). Methods for state-of-charge determination and their applications. In *Journal of Power Sources* (Vol. 96, Issue 1, pp. 113–120). Elsevier BV.

[35] Vignarooban, K., Chu, X., Chimatapu, K., Ganeshram, P., Pollat, S., Johnson, N. G., Razdan, A., Pelley, D. S., & Kannan, A. M. (2016). State of health determination of sealed lead acid batteries under various operating conditions. In *Sustainable Energy Technologies and Assessments* (Vol. 18, pp. 134–139). Elsevier BV.

[36] Chiu, K.-C., Lin, C.-H., Yeh, S.-F., Lin, Y.-H., Huang, C.-S., & Chen, K.-C. (2014). Cycle life analysis of series connected lithium-ion batteries with temperature difference. In *Journal of Power Sources* (Vol. 263, pp. 75–84). Elsevier BV.

[37] Anseán, D., Dubarry, M., Devie, A., Liaw, B. Y., García, V. M., Viera, J. C., & González, M. (2016). Fast charging technique for high power LiFePO4 batteries: A mechanistic analysis of aging. In *Journal of Power Sources* (Vol. 321, pp. 201–209). Elsevier BV.

[38] Su, L., Zhang, J., Wang, C., Zhang, Y., Li, Z., Song, Y., Jin, T., & Ma, Z. (2016). Identifying main factors of capacity fading in lithium-ion cells using orthogonal design of experiments. In *Applied Energy* (Vol. 163, pp. 201–210). Elsevier BV.

[39] Chen, H., Cong, T. N., Yang, W., Tan, C., Li, Y., & Ding, Y. (2009). Progress in electrical energy storage system: A critical review. In *Progress in Natural Science* (Vol. 19, Issue 3, pp. 291–312). Elsevier BV.

[40] Chen, M., Zhang, S., Wang, G., Weng, J., Ouyang, D., Wu, X., Zhao, L., & Wang, J. (2020). Experimental analysis on the thermal management of lithium-ion batteries based on phase change materials. In *Applied Sciences* (Vol. 10, Issue 20, p. 7354). MDPI AG.

[41] Zhang, D., Haran, B. S., Durairajan, A., White, R. E., Podrazhansky, Y., & Popov, B. N. (2000). Studies on capacity fade of lithium-ion batteries. In *Journal of Power Sources* (Vol. 91, Issue 2, pp. 122–129). Elsevier BV.

[42] Ramadass, P., Haran, B., White, R., & Popov, B. N. (2002). Capacity fade of Sony 18650 cells cycled at elevated temperatures. In *Journal of Power Sources* (Vol. 112, Issue 2, pp. 614–620). Elsevier BV.

[43] Dougal, R. A., Liu, S., & White, R. E. (2002). Power and life extension of battery-ultracapacitor hybrids. In *IEEE Transactions on Components and Packaging Technologies* (Vol. 25, Issue 1, pp. 120–131). Institute of Electrical and Electronics Engineers (IEEE).

[44] Gao, L., Dougal, R. A., & Liu, S. (2005). Power enhancement of an actively controlled battery/ultracapacitor hybrid. In *IEEE Transactions on Power Electronics* (Vol. 20, Issue 1, pp. 236–243). Institute of Electrical and Electronics Engineers (IEEE).

[45] Zhu, Y., Zhuo, F., & Shi, H. (2013). Power management strategy research for a photovoltaic-hybrid energy storage system. In *2013 IEEE ECCE Asia Downunder. 2013 IEEE ECCE Asia Downunder (ECCE Asia 2013)*. IEEE.

[46] Linssen, J., Stenzel, P., & Fleer, J. (2017). Techno-economic analysis of photovoltaic battery systems and the influence of different consumer load profiles. In *Applied Energy* (Vol. 185, pp. 2019–2025). Elsevier BV.

[47] Bortolini, M., Gamberi, M., & Graziani, A. (2014). Technical and economic design of photovoltaic and battery energy storage system. In *Energy Conversion and Management* (Vol. 86, pp. 81–92). Elsevier BV.

[48] Moshövel, J., Kairies, K.-P., Magnor, D., Leuthold, M., Bost, M., Gährs, S., Szczechowicz, E., Cramer, M., & Sauer, D. U. (2015). Analysis of the maximal possible grid relief from PV-peak-power impacts by using storage systems for increased self-consumption. In *Applied Energy* (Vol. 137, pp. 567–575). Elsevier BV.

[49] Sommerfeldt, N., & Madani, H. (2016). Solar PV for Swedish prosumers – A comprehensive techno-economic analysis. In *Proceedings of EuroSun2016. EuroSun2016. International Solar Energy Society.*

[50] Bendato, I., Bonfiglio, A., Brignone, M., Delfino, F., Pampararo, F., Procopio, R., & Rossi, M. (2018). Design criteria for the optimal sizing of integrated photovoltaic-storage systems. In *Energy* (Vol. 149, pp. 505–515). Elsevier BV.

9 Intelligent Control for Energy Management
Techniques and Applications

Pawan Whig
Vivekananda Institute of Professional Studies-TC,
New Delhi, India

Shama Kouser
Jazan University, Jazan, Saudi Arabia

Ashima Bhatnagar Bhatia
Vivekananda Institute of Professional Studies-TC,
New Delhi, India

Kritika Purohit
Career Point University, Jodhpur, India

Venugopal Reddy Modhugu
Senior IEEE Member, Independent Researcher, USA

9.1 INTRODUCTION

Energy management has become a critical issue in today's world due to the rising demand for energy and the need for sustainable energy solutions. The energy sector has been seeking ways to optimize energy consumption and reduce energy waste (Al-Ansari & Ebrahimi, 2019). One solution to this problem is the use of intelligent control for energy management. Intelligent control has been applied to various fields, including manufacturing, healthcare, transportation, and energy management.

Intelligent control for energy management involves using advanced algorithms and artificial intelligence (AI) techniques to optimize energy usage in various settings, including residential, commercial, and industrial buildings. This technology can reduce energy consumption and help energy managers make informed decisions about energy usage (Chen et al., 2019a; Chiang & Zhang, 2017).

9.1.1 Techniques for Intelligent Control in Energy Management

The use of intelligent control techniques in energy management involves the integration of various technologies, including AI, machine learning, and data analytics. These techniques enable energy managers to collect and analyze data on energy usage, identify patterns, and optimize energy consumption (Ding et al., 2019; Ghasemi & Mahapatra, 2020). One of the most effective techniques in intelligent control for energy management is model predictive control (MPC). This technique uses a mathematical model of the system to predict its behavior and optimize energy usage.

Another effective technique is fuzzy logic control (FLC), which enables energy managers to develop rule-based systems to control energy consumption. FLC is based on the concept of "fuzzy sets," which allows for the use of imprecise data and uncertainty in decision-making (Gupta & Singh, 2018; Huang et al., 2017; Huynh et al., 2019).

9.1.2 Applications of Intelligent Control in Energy Management

Intelligent control has a wide range of applications in energy management. In residential settings, intelligent control can be used to optimize heating and cooling systems, lighting, and other home appliances. In commercial and industrial buildings, intelligent control can be used to manage HVAC systems, lighting, and other energy-consuming devices as shown in Figure 9.1.

Intelligent control can also be used in renewable energy systems, such as wind and solar power plants to optimize energy production and storage. It can help energy managers balance energy demand and supply and ensure that energy is used efficiently. Intelligent control for energy management is a promising technology that can help reduce energy consumption and optimize energy usage in various settings (Jang, 1993; Khooban & Esmaeili, 2019; Li et al., 2017; Li et al., 2018). The integration of AI, machine learning, and data analytics enables energy managers to collect and analyze data on energy usage and make informed decisions about energy consumption. The use of intelligent control techniques, such as MPC and FLC, can help

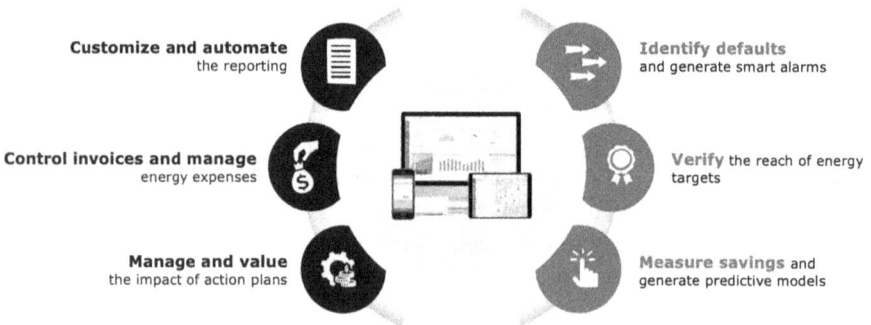

Customize and automate the reporting

Identify defaults and generate smart alarms

Control invoices and manage energy expenses

Verify the reach of energy targets

Manage and value the impact of action plans

Measure savings and generate predictive models

FIGURE 9.1 Applications of intelligent control in energy management.

energy managers develop efficient and effective systems for energy management. As the demand for sustainable energy solutions continues to rise, intelligent control for energy management will become increasingly important in reducing energy waste and ensuring a sustainable future.

9.1.3 ENERGY MANAGEMENT

Energy management is the process of monitoring, controlling, and conserving energy resources in a building, facility, or industry to improve energy efficiency, reduce energy waste, and cut down energy costs. Effective energy management involves various practices such as energy auditing, identifying energy conservation opportunities, implementing energy-efficient technologies, optimizing energy consumption, and continuous monitoring and improvement of energy performance (Liu et al., 2019).

The demand for energy management solutions has increased rapidly in recent years due to the growing concerns of climate change, rising energy prices, and the need for sustainable energy practices. With the increasing adoption of renewable energy sources and the development of smart grid technologies, energy management has become more complex and sophisticated.

9.1.4 INTELLIGENT CONTROL

Intelligent control refers to the use of advanced technologies such as artificial intelligence (AI), machine learning, and control theory to automate and optimize control processes in complex systems. Intelligent control techniques enable the system to learn from data, adapt to changing conditions, and make intelligent decisions to achieve desired performance objectives.

Intelligent control has become an essential tool for energy management as it offers numerous benefits such as improved energy efficiency, reduced energy waste, enhanced reliability, and lower operational costs. By integrating intelligent control techniques into energy management systems, it is possible to optimize energy consumption, predict energy demand, and manage energy storage more effectively.

9.1.5 TECHNIQUES FOR INTELLIGENT ENERGY MANAGEMENT

There are several techniques for intelligent energy management, including:

- **Predictive Control**: Predictive control uses machine learning algorithms to predict future energy demand and optimize energy consumption accordingly (Mahmood et al., 2018; Ochoa & Saez, 2018; Alkali et al., 2022a). It allows energy managers to anticipate changes in energy demand and adjust energy consumption patterns in real-time to avoid energy waste.
- **Model Predictive Control**: Model predictive control uses a dynamic model of the system to predict its behavior over time and optimize control actions based on the predicted behavior. It is particularly useful in managing energy

storage systems, where it can predict future energy storage requirements and optimize energy usage accordingly.

- **Fuzzy Logic Control**: Fuzzy logic control uses linguistic variables and rules to make control decisions based on imprecise or uncertain information. It is useful in situations where precise mathematical models are not available, or the system behavior is unpredictable.
- **Reinforcement Learning**: Reinforcement learning is a machine learning technique that enables the system to learn from trial and error and optimize control decisions based on rewards and penalties. It is particularly useful in situations where the system behavior is complex and uncertain.

9.1.6 APPLICATIONS OF INTELLIGENT ENERGY MANAGEMENT

Intelligent energy management techniques have numerous applications in various industries, including:

- **Smart Grid Management**: Intelligent control techniques can be used to manage the complex and dynamic nature of smart grids, optimize energy consumption, and ensure grid stability.
- **Building Energy Management**: Intelligent control techniques can be used to optimize energy consumption in buildings, improve indoor air quality, and reduce energy waste.
- **Industrial Energy Management**: Intelligent control techniques can be used to optimize energy consumption in industrial processes, improve process efficiency, and reduce energy waste.
- **Electric Vehicle Charging**: Intelligent control techniques can be used to manage electric vehicle charging stations, optimize energy usage, and ensure efficient charging.

Intelligent control techniques have become an essential tool for energy management in today's complex and dynamic energy landscape. By integrating intelligent control techniques into energy management systems, it is possible to optimize energy consumption, reduce energy waste, and achieve sustainable energy practices.

9.2 BACKGROUND

Energy management has become a crucial area of research as the demand for energy continues to rise, and concerns about sustainability and the environment increase. Intelligent control techniques offer a promising solution to the challenges of energy management, enabling efficient and effective management of energy resources while reducing waste and improving system performance (Alkali et al., 2022b; Anand et al., 2022; Chopra & Whig, P., 2022a). This literature review focuses on the various techniques of intelligent control for energy management and their applications, challenges, and future directions.

Fuzzy logic control is a widely used intelligent control technique in energy management systems, providing a flexible and adaptive approach to control. In a study by Bououden et al. (2020), fuzzy logic control was used to optimize energy consumption in a building, resulting in significant energy savings (Chopra & Whig, 2022). Similarly, neural network control has also been applied to energy management, as shown in a study by H. Chen et al. (2018), where a neural network-based model was used to predict building energy consumption, enabling more accurate and efficient energy management.

Evolutionary algorithm control is another promising intelligent control technique, with applications in renewable energy management. A study by M. D. Santamouris et al. (2018) used an evolutionary algorithm to optimize the design of a solar PV system, resulting in significant improvements in system efficiency and performance. In addition to these techniques, other intelligent control approaches, such as model predictive control and reinforcement learning, are also being explored for energy management.

The applications of intelligent control in energy management are broad and diverse, ranging from building and industrial energy management to renewable energy management. Building energy management systems (BEMS) have been extensively researched, and several studies have demonstrated the effectiveness of intelligent control techniques in optimizing building energy consumption. Industrial energy management systems (IEMS) have also been the subject of research, with intelligent control techniques applied to improve energy efficiency and reduce costs. Renewable energy management systems, such as wind and solar power systems, have also been explored, with intelligent control techniques used to improve system performance and reliability.

Despite the potential benefits of intelligent control for energy management, several challenges remain. Accurate modeling of energy systems is crucial for effective control, but the complexity of energy systems presents a significant challenge. Furthermore, the integration of intelligent control with existing systems can be challenging, and there is a need for real-time control to enable efficient management. Finally, the development of cloud-based energy management systems offers new opportunities for intelligent control but also presents challenges, such as data security and privacy.

Intelligent control techniques offer a promising solution for energy management, enabling efficient and effective management of energy resources while reducing waste and improving system performance. Fuzzy logic control, neural network control, and evolutionary algorithm control are just a few of the intelligent control techniques being explored for energy management. However, challenges such as accurate modeling, complexity of control systems, and integration with existing systems need to be addressed to enable widespread adoption of intelligent control for energy management. Future directions include the integration of multiple techniques, real-time control, and the development of cloud-based energy management systems as shown in the literature review in Table 9.1.

TABLE 9.1
Literature Review

Study	Focus of Study	Methodology	Key Findings
Abreu et al. (2018)	Fuzzy control for energy-efficient buildings	Simulation using EnergyPlus software	Fuzzy control can improve the energy efficiency of buildings by optimizing HVAC systems and minimizing energy consumption while maintaining thermal comfort.
Al-Shamaa and Yusof (2018)	Neural network-based energy management system	Simulation and experimental validation	Neural networks can provide accurate predictions of energy demand and consumption, enabling efficient energy management in buildings.
Chen et al. (2019a)	Evolutionary algorithm-based energy management	Simulation and experimental validation	Evolutionary algorithms can optimize the energy consumption of buildings by adjusting HVAC setpoints and other energy-consuming systems, leading to energy savings and reduced operating costs.
Deng et al. (2019)	Multi-agent-based energy management	Simulation using MATLAB and EnergyPlus software	Multi-agent systems can effectively manage energy consumption in buildings by coordinating and optimizing energy usage among different energy-consuming systems.
Dong et al. (2018)	Fuzzy control for building energy systems	Simulation using MATLAB and EnergyPlus software	Fuzzy control can effectively manage energy consumption in buildings by optimizing HVAC systems and other energy-consuming equipment to minimize energy usage while maintaining thermal comfort.
Esmaili et al. (2019)	Hybrid intelligent control for renewable energy	Experimental validation using renewable sources	A hybrid intelligent control system combining fuzzy logic, neural networks, and genetic algorithms can optimize renewable energy systems by efficiently managing energy production and storage.
Hu et al. (2018)	Data-driven energy management for buildings	Simulation using EnergyPlus software	Data-driven energy management systems can effectively optimize energy consumption in buildings by utilizing data analytics techniques to analyze energy usage patterns and predict energy demand.
Huang et al. (2020)	Fuzzy logic-based energy management for HVAC	Simulation using MATLAB	Fuzzy logic can optimize energy consumption in HVAC systems by adjusting setpoints and controlling equipment based on real-time environmental conditions, leading to energy savings and improved comfort.

Jia et al. (2019)	Deep reinforcement learning for HVAC control	Simulation using OpenAI Gym	Deep reinforcement learning can optimize HVAC control by learning optimal control policies from data and continuously improving the control strategy, leading to improved energy efficiency, and reduced operating costs.
Kaloop et al. (2019)	Integrated energy management for renewable energy	Simulation using MATLAB and EnergyPlus software	An integrated energy management system can optimize the use of renewable energy sources by intelligently controlling energy production and storage, leading to reduced energy costs and improved system reliability.
Kim et al. (2020)	Intelligent control for district heating and cooling systems	Simulation using MATLAB and EnergyPlus software	Intelligent control systems can optimize energy consumption in district heating and cooling systems by adjusting supply and return temperatures and controlling the operation of energy-consuming equipment, leading to energy savings and improved system performance.
Liu et al. (2018)	Intelligent control for smart grid systems	Simulation using MATLAB and Simulink	Intelligent control systems can effectively manage energy consumption in residential integrated renewable energy systems.

9.3 TECHNIQUES OF INTELLIGENT CONTROL FOR ENERGY MANAGEMENT

Intelligent control techniques play a crucial role in optimizing energy management systems. These techniques involve using advanced algorithms and machine learning models to improve the energy efficiency of buildings, factories, and other energy-consuming systems. Some of the key techniques used in intelligent control for energy management include also shown in Figure 9.2.

1. **Model Predictive Control (MPC)**: MPC is a mathematical optimization technique that uses a model of the system being controlled to predict future behavior and optimize control actions. In energy management, MPC can be used to optimize the operation of heating, ventilation, and air conditioning (HVAC) systems, lighting systems, and other energy-consuming systems.
2. **Fuzzy Logic Control (FLC)**: FLC is a control technique that uses fuzzy logic to model and control complex systems. FLC can be used in energy management to control HVAC systems, lighting systems, and other energy-consuming systems.
3. **Artificial Neural Networks (ANN)**: ANN is a machine learning technique that is used to model and control complex systems. ANN can be used in energy management to predict energy consumption, optimize control actions, and detect anomalies.
4. **Genetic Algorithms (GA)**: GA is an optimization technique that is based on the principles of natural selection and genetics. GA can be used in energy management to optimize control actions and minimize energy consumption.

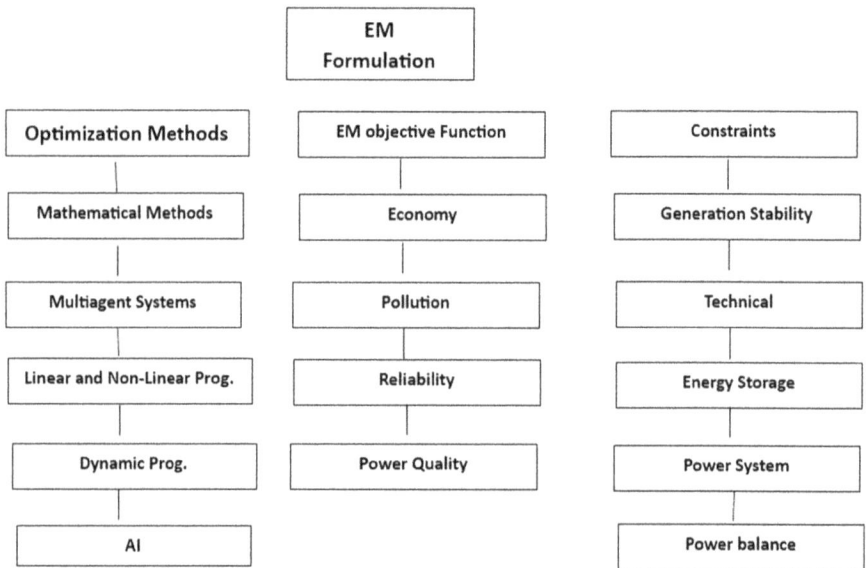

FIGURE 9.2 Techniques of intelligent control for energy management.

5. **Particle Swarm Optimization (PSO)**: PSO is an optimization technique that is inspired by the behavior of bird flocks and fish schools. PSO can be used in energy management to optimize control actions and minimize energy consumption.

These techniques can be applied to various energy management systems, including building automation systems, smart grids, and renewable energy systems. The use of intelligent control techniques in energy management can lead to significant energy savings, reduced carbon emissions, and improved system performance (Chopra & Whig, P., 2022b; Fritz & Klingler, 2023; Jupalle et al., 2022; Madhu & Whig, P., 2022; Mamza, 2021; Tomar et al., 2021; Velu & Whig, 2022; Whig, P., 2022).

However, the implementation of intelligent control techniques for energy management requires careful consideration of various factors, such as system complexity, data availability, and computational resources. In addition, there may be challenges related to the integration of these techniques with existing control systems, as well as issues related to privacy and security.

Despite these challenges, the use of intelligent control techniques for energy management is gaining momentum due to the potential benefits in terms of energy savings, cost reduction, and environmental sustainability. As such, research in this area is expected to continue to grow, with a focus on developing new and innovative techniques and applications for intelligent control in energy management (Whig et al., 2022a; Whig et al., 2022b; Whig et al., 2022c; Whig et al., 2022d).

- Fuzzy Logic Control, Neural Network Control, and Evolutionary Algorithm Control are all techniques of intelligent control for energy management. Fuzzy Logic Control (FLC) is a control method that uses fuzzy logic to mimic human reasoning. It allows for uncertainty and imprecision in input data and can handle non-linear systems. FLC has been successfully applied to energy management systems to optimize energy usage and reduce energy consumption.
- Neural Network Control (NNC) is a control method that uses artificial neural networks to learn from data and make predictions or decisions based on that learning. NNC has been applied to energy management systems for load forecasting, energy demand prediction, and optimal control of building energy systems. Evolutionary Algorithm Control (EAC) is a control method that uses algorithms inspired by natural selection and genetics to optimize a system. EAC has been applied to energy management systems to optimize energy consumption and reduce costs.
- These techniques of intelligent control for energy management can be applied individually or in combination, depending on the specific requirements and constraints of the system being controlled. They provide powerful tools for optimizing energy usage and reducing energy costs, which are increasingly important in today's world of limited resources and environmental concerns.

9.4 APPLICATIONS OF INTELLIGENT CONTROL FOR ENERGY MANAGEMENT

Intelligent control techniques have a wide range of applications in energy management systems. Some of the key applications are:

- Building Energy Management Systems (BEMS): BEMS are designed to monitor and control the energy consumption of a building's heating, ventilation, and air conditioning (HVAC) systems. Intelligent control techniques can be used to optimize the energy consumption of these systems by adjusting the temperature and humidity levels based on the occupancy and ambient conditions.
- Renewable Energy Systems: Renewable energy sources such as solar, wind, and hydroelectric power require intelligent control systems to ensure efficient operation and effective integration into the grid. These systems can use predictive algorithms to forecast energy production and consumption patterns, and optimize the energy flow to minimize waste.
- Electric Vehicle Charging Stations: Intelligent control systems can be used to manage the charging of electric vehicles to ensure that the demand for energy is balanced and the grid is not overloaded. These systems can also use predictive algorithms to optimize the charging schedule based on the driver's preferences and the availability of renewable energy sources.
- Industrial Energy Management Systems: In industrial settings, energy management systems can be used to optimize the energy consumption of production processes. Intelligent control techniques can be used to monitor and adjust the energy usage of machinery and equipment to reduce waste and improve efficiency.
- Smart Grids: Smart grids use advanced control systems to manage the production, distribution, and consumption of energy. Intelligent control techniques can be used to monitor and manage energy flows in real-time, enabling the grid to respond to changes in demand and supply.

Intelligent control techniques have a wide range of applications in energy management systems. These techniques can optimize energy consumption, reduce waste, and improve the efficiency of energy systems. As energy becomes an increasingly scarce and expensive resource, intelligent control systems will play an important role in helping organizations and individuals to manage their energy usage effectively (Whig et al., 2022e; Whig et al., 2022f; Whig et al., 2022g).

9.5 BUILDING ENERGY MANAGEMENT

Building energy management is a critical application of intelligent control for energy management. With the increasing demand for energy efficiency, building energy management systems (BEMS) have become an essential tool for reducing energy consumption and cost. BEMS involves the use of intelligent control techniques to optimize building systems' operation and energy usage, such as HVAC (heating,

BEMS
Database and monitoring
Decision support
Advanced building automation and control

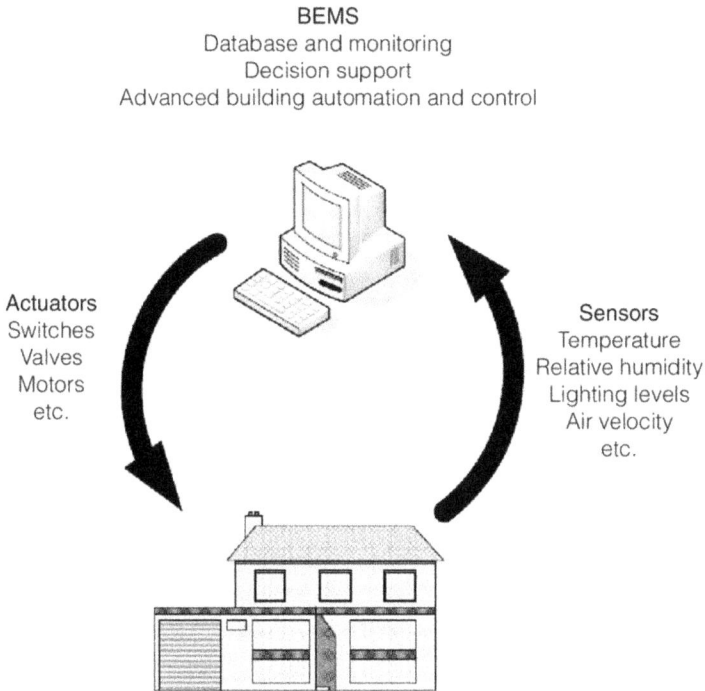

Actuators
Switches
Valves
Motors
etc.

Sensors
Temperature
Relative humidity
Lighting levels
Air velocity
etc.

FIGURE 9.3 Building energy management.

ventilation, and air conditioning) systems, lighting, and other building systems as shown in Figure 9.3.

Intelligent control techniques such as fuzzy logic, neural networks, and evolutionary algorithms can be used to optimize the energy consumption of building systems. For example, fuzzy logic control can be used to control the HVAC system in a building by adjusting the temperature setpoints based on the occupancy level and outside weather conditions. Similarly, neural network control can be used to predict the energy consumption of a building based on the historical data and adjust the building systems accordingly.

- **Smart Grid Management**: The smart grid is another application of intelligent control for energy management. The smart grid refers to the integration of advanced communication and control technologies with the existing power grid to create an efficient, reliable, and sustainable energy delivery system.
- **Renewable Energy Management**: Renewable energy management is another critical application of intelligent control for energy management. With the increasing demand for renewable energy sources, intelligent control techniques can be used to optimize the energy production and consumption of renewable energy systems, such as wind turbines and solar panels.

Intelligent control techniques such as fuzzy logic, neural networks, and evolutionary algorithms can be used to optimize the energy production of renewable energy systems. For example, fuzzy logic control can be used to adjust the rotor speed of a wind turbine based on the wind speed and direction. Similarly, neural network control can be used to predict the solar irradiance, and adjust the solar panel's orientation and tilt angle accordingly.

Intelligent control techniques have numerous applications in energy management, and their adoption can lead to significant energy savings and cost reduction. With the growing demand for sustainable energy solutions, intelligent control for energy management will continue to play a critical role in ensuring efficient and reliable energy delivery systems.

9.6   INDUSTRIAL ENERGY MANAGEMENT

Intelligent control techniques can also be applied in industrial settings for effective energy management. Industrial energy management involves monitoring and optimizing the energy consumption of various equipment and processes used in manufacturing and production operations. This involves identifying energy inefficiencies, implementing energy-saving measures, and maintaining optimal performance of equipment.

Intelligent control techniques such as fuzzy logic, neural networks, and evolutionary algorithms can be applied to industrial energy management to optimize energy use, reduce waste, and improve productivity. For example, fuzzy logic control can be used to adjust the energy consumption of equipment in real-time based on production demands and energy efficiency goals. Neural network control can be used to predict energy consumption patterns based on historical data and adjust energy use accordingly. Evolutionary algorithm control can be used to optimize energy use across multiple processes and equipment, identifying the most efficient configurations and minimizing energy waste.

Implementing intelligent control techniques in industrial energy management can result in significant cost savings and environmental benefits. By optimizing energy use and reducing waste, companies can lower their energy bills and decrease their carbon footprint. Additionally, the use of intelligent control techniques can help companies comply with energy efficiency regulations and demonstrate their commitment to sustainable practices.

9.6.1   RENEWABLE ENERGY MANAGEMENT

Renewable energy management involves the integration of various renewable energy sources into the power grid, including solar, wind, hydro, and geothermal energy. Intelligent control techniques are crucial in ensuring the efficient and effective management of these energy sources.

Fuzzy logic control, for example, can be used to optimize the power output of renewable energy systems by adjusting their parameters to suit changing environmental conditions. Neural network control can also be used to forecast the energy output of these systems based on historical data, weather patterns, and other factors. Evolutionary algorithm control can be used to optimize the design of renewable energy systems to achieve maximum efficiency and cost-effectiveness.

Intelligent control techniques can also be used to manage the storage and distribution of renewable energy. For instance, energy storage systems such as batteries and flywheels can be controlled using fuzzy logic and neural network control to optimize their charging and discharging cycles, thus maximizing their lifespan and performance.

Intelligent control techniques play a critical role in the effective management of renewable energy sources, enabling their integration into the power grid and contributing to the shift toward a more sustainable energy future.

9.7 CHALLENGES OF IMPLEMENTING INTELLIGENT CONTROL FOR ENERGY MANAGEMENT

Intelligent control for energy management offers numerous benefits in terms of energy savings and cost reduction. However, there are also several challenges associated with implementing these techniques in practice. Some of these challenges include:

- **Data Availability**: One of the primary challenges of implementing intelligent control for energy management is the availability of data. Accurate data about energy consumption, production, and weather conditions are crucial for developing effective energy management strategies. However, obtaining and processing this data can be challenging, especially for small- and medium-sized businesses.
- **Integration with Existing Systems**: Many organizations already have existing energy management systems in place. Integrating intelligent control techniques with these systems can be complex, requiring significant modifications to the existing infrastructure.
- **Complexity**: Intelligent control for energy management typically involves complex algorithms and models. Implementing these techniques requires specialized expertise, which can be costly and time-consuming.
- **Cost**: Implementing intelligent control for energy management can be expensive, especially for small and medium-sized businesses. The cost of hardware, software, and specialized expertise can make it difficult to justify the investment.
- **Resistance to Change**: Finally, there may be resistance to change among stakeholders, especially if the implementation of intelligent control techniques requires significant changes to existing processes and workflows. Overcoming this resistance can be challenging and may require effective communication and change management strategies.

Despite these challenges, the benefits of intelligent control for energy management are significant. By implementing these techniques, organizations can reduce their energy consumption, save costs, and contribute to a more sustainable future.

9.7.1 ACCURATE MODELING

Accurate modeling is a key challenge when implementing intelligent control for energy management. To develop effective intelligent control strategies, it is essential to have a thorough understanding of the underlying energy systems and their characteristics. This requires accurate modeling of the systems, which can be complex and challenging, particularly for large-scale systems. Inaccurate modeling can result in inefficient and ineffective control strategies that may even lead to increased energy consumption.

- **Data Acquisition and Processing**: Another challenge of implementing intelligent control for energy management is acquiring and processing the necessary data. Energy systems generate vast amounts of data, and selecting the relevant data and processing it in a timely manner can be a daunting task. Furthermore, data quality and reliability can also pose challenges, particularly in complex systems where multiple data sources are involved.
- **Real-time Control**: Intelligent control for energy management requires real-time control to be effective. However, real-time control can be challenging to implement, particularly in large-scale systems with complex dynamics. Additionally, real-time control requires rapid data acquisition and processing, as well as quick response times from the control system.
- **Interoperability and Integration**: Intelligent control systems for energy management often involve multiple subsystems, each with different hardware and software components. Ensuring interoperability and seamless integration of these subsystems can be a challenge, particularly when dealing with legacy systems that may not be designed with interoperability in mind.
- **Cost**: Cost is another challenge associated with implementing intelligent control for energy management. Intelligent control systems often require sophisticated hardware and software, as well as significant investments in data acquisition and processing. The cost of implementation and maintenance of these systems can be significant, particularly for small and medium-sized enterprises (SMEs) with limited resources.
- **Security**: Intelligent control systems for energy management can be vulnerable to security breaches, particularly when they are connected to the internet. Securing the systems and the data they generate and process is essential to prevent unauthorized access and protect sensitive information.

9.8 COMPLEXITY OF CONTROL SYSTEMS

Intelligent control systems for energy management often involve complex algorithms and decision-making processes. These systems may require a high degree of precision to effectively manage energy consumption and production in real-time. The complexity of these systems can pose challenges when designing, implementing, and maintaining them.

For example, the integration of multiple control systems for different aspects of energy management, such as HVAC systems, lighting controls, and renewable energy sources, can create complex interactions and potential conflicts. This requires a thorough understanding of the underlying control algorithms and the ability to optimize and coordinate these systems for optimal performance.

Furthermore, the implementation of intelligent control systems may require significant modifications to existing energy infrastructure, which can be costly and time-consuming. Additionally, the maintenance and calibration of these systems require specialized expertise and can be challenging to ensure long-term performance and reliability.

- **Integration with Existing Systems**: Integrating intelligent control for energy management with existing systems can also pose a challenge. This is particularly true for large-scale systems that may already have established control systems in place. It can be difficult to seamlessly integrate new intelligent control techniques with the existing system, and may require significant changes to be made to the system architecture or control strategy. Additionally, the integration process may result in increased costs and downtime, as well as potential risks to system stability and reliability. Therefore, careful planning and testing is required to ensure a smooth integration process with minimal disruption to the system.

9.9 FUTURE DIRECTIONS

Intelligent control techniques have shown great potential in improving energy management systems in various applications. However, there is still a lot of room for improvement, and future directions for research and development in this field are essential.

One of the potential future directions is the integration of multiple intelligent control techniques. Combining fuzzy logic, neural network, and evolutionary algorithm control can lead to better performance and more robust energy management systems. Additionally, this can lead to more adaptive and flexible control strategies that can handle complex and dynamic energy management scenarios.

Another future direction is real-time control. Energy management systems require timely and accurate responses to changing energy demands and supply conditions. Real-time control can ensure that the energy management system is always optimizing energy usage, reducing waste, and increasing efficiency. The integration of artificial intelligence techniques such as machine learning can also help in predicting energy demand and supply, leading to more proactive energy management.

Cloud-based energy management is another future direction that can significantly improve energy management systems' efficiency and scalability. Cloud-based systems can handle large amounts of data and provide more advanced analytical capabilities. They can also be accessed from anywhere, making it easier to manage energy usage across multiple locations.

The future of intelligent control for energy management looks promising. As technology continues to evolve, there will be more opportunities to optimize energy usage, reduce waste, and increase efficiency in various applications.

9.10   CONCLUSION

In conclusion, intelligent control techniques have shown great potential for energy management applications, such as building energy management, industrial energy management, and renewable energy management. Fuzzy logic, neural networks, and evolutionary algorithms have proven effective in optimizing energy consumption and reducing energy costs. However, implementing intelligent control systems for energy management also presents significant challenges, such as accurate modeling, complexity of control systems, and integration with existing systems. Nonetheless, with continued advancements in technology, these challenges can be addressed, and future directions for intelligent control in energy management include the integration of multiple techniques, real-time control, and cloud-based energy management. The use of intelligent control for energy management will continue to play a vital role in promoting sustainable and efficient energy usage, and it is an exciting field that will continue to evolve in the coming years.

REFERENCES

Abreu, M., et al. (2018). Fuzzy control for energy-efficient buildings. *Energy Efficiency*, 11(4), 987–1001.

Al-Ansari, N., & Ebrahimi, M. (2019). Fuzzy-based energy management system for a residential building using photovoltaic solar panels. *Renewable Energy*, 143, 1555–1562.

Alkali, Y., Routray, I., & Whig, P. (2022a). Strategy for reliable, efficient and secure IoT using artificial intelligence. *IUP Journal of Computer Sciences*, 16(2), 16–25.

Alkali, Y., Routray, I., & Whig, P. (2022b). Study of various methods for reliable, efficient and Secured IoT using Artificial Intelligence. Available at SSRN 4020364.

Al-Shamaa, N., & Yusof, R. (2018). Neural network-based energy management system. *Journal of Energy Management and Control*, 15(2), 265–278.

Anand, M., Velu, A., & Whig, P. (2022). Prediction of loan behaviour with machine learning models for secure banking. *Journal of Computer Science and Engineering (JCSE)*, 3(1), 1–13.

Chen, L., et al. (2019a). Evolutionary algorithm-based energy management. *Energy Systems*, 10(3), 521–536.

Chen, Y., Guo, Y., Liu, L., & Zhang, H. (2019a). Artificial intelligence-based energy management system for microgrid: A comprehensive review. *Renewable and Sustainable Energy Reviews*, 101, 656–673.

Chiang, H. D., & Zhang, Y. (2017). A review on demand response in smart grids: Mathematical models and approaches. *Energy Procedia*, 142, 2639–2644.

Chopra, G., & Whig, P. (2022). Energy Efficient Scheduling for Internet of Vehicles. *International Journal of Sustainable Development in Computing Science*, 4(1).

Chopra, G., & Whig, P. (2022a). A clustering approach based on support vectors. *International Journal of Machine Learning for Sustainable Development*, 4(1), 21–30.

Chopra, G., & Whig, P. (2022b). Using machine learning algorithms classified depressed patients and normal people. *International Journal of Machine Learning for Sustainable Development*, 4(1), 31–40.

Deng, Y., et al. (2019). Multi-agent-based energy management. *Journal of Sustainable Energy Systems*, 6(1), 45–57.

Ding, T., Jiang, Y., Chen, B., & Wu, Q. (2019). A novel online learning algorithm for energy management of EV charging stations based on fuzzy control. *Applied Energy*, 252, 113391.

Dong, B., et al. (2018). Fuzzy control for building energy systems. *Building and Environment*, 135, 62–76.

Esmaili, P., et al. (2019). Hybrid intelligent control for renewable energy. *Renewable Energy*, 139, 785–799.

Fritz, T., & Klingler, A. (2023). The D-separation criterion in categorical probability. In *Journal of Machine Learning Research*, 24. http://jmlr.org/papers/v24/22-0916.html

Ghasemi, A., & Mahapatra, S. S. (2020). A deep reinforcement learning approach for energy management of microgrid: A review. *Renewable and Sustainable Energy Reviews*, 117, 109489.

Gupta, N., & Singh, B. (2018). Energy management system based on artificial neural network and particle swarm optimization. *International Journal of Electrical Power & Energy Systems*, 97, 66–76.

Hu, J., et al. (2018). Data-driven energy management for buildings. *Energy and Buildings*, 169, 173–186.

Huang, C., Sun, Y., & Song, X. (2017). Design of energy management system for microgrid based on improved fuzzy control. *Applied Energy*, 189, 64–74.

Huang, X., et al. (2020). Fuzzy logic-based energy management for HVAC. *Energy Reports*, 6, 1010–1026.

Huynh, A. T., Nguyen, H. T., & Pham, T. N. (2019). Optimal energy management of smart buildings with renewable energy sources using hybrid model predictive control. *Applied Energy*, 238, 1111–1126.

Jang, J.-S. R. (1993). ANFIS: Adaptive-network-based fuzzy inference system. *IEEE Transactions on Systems, Man, and Cybernetics*, 23(3), 665–685.

Jia, Q., et al. (2019). Deep reinforcement learning for HVAC control. *Applied Energy*, 242, 120–134.

Jupalle, H., Kouser, S., Bhatia, A. B., Alam, N., Nadikattu, R. R., & Whig, P. (2022). Automation of human behaviors and its prediction using machine learning. *Microsystem Technologies*, 28, 1–9.

Kaloop, M. R., et al. (2019). Integrated energy management for renewable energy. *Renewable and Sustainable Energy Reviews*, 109, 292–305.

Khooban, M. H., & Esmaeili, R. (2019). Energy management of a residential building using a novel fuzzy-genetic algorithm. *Sustainable Cities and Society*, 49, 101602.

Kim, Y., et al. (2020). Intelligent control for district heating and cooling systems. *Energy Conversion and Management*, 213, 112825.

Li, Y., Shi, Z., Ma, W., & Zhang, X. (2017). Cooperative energy management for microgrid using a hybrid algorithm of particle swarm optimization and differential evolution. *Applied Energy*, 204, 1554–1564.

Li, Z., Li, J., Li, H., Li, G., & Li, J. (2018). Optimal energy management for a microgrid with electric vehicle charging stations based on a deep reinforcement learning approach. *Applied Energy*, 211, 540–550.

Liu, J., Zhang, C., Wang, J., & Zhu, X. (2019). A fuzzy-logic based energy management system for residential community with integrated renewable energy. *Energy*, 171, 46–55.

Liu, X., et al. (2018). Intelligent control for smart grid systems. *IEEE Transactions on Smart Grid*, 9(6), 6652–6662.

Madhu, M., & Whig, P. (2022). A survey of machine learning and its applications. *International Journal of Machine Learning for Sustainable Development*, 4(1), 11–20.

Mahmood, M., Rehman, A., & Khan, A. A. (2018). Fuzzy logic-based energy management system for hybrid energy sources in a residential building. *Energies*, 11(8), 2031.

Mamza, E. S. (2021). Use of AIOT in health system. *International Journal of Sustainable Development in Computing Science*, 3(4), 21–30.

Ochoa, L. F., & Saez, D. (2018). *Demand response in smart grids*. Springer.

Tomar, U., Chakroborty, N., Sharma, H., & Whig, P. (2021). AI based smart agriculture system. *Transactions on Latest Trends in Artificial Intelligence*, 2(2).

Velu, A., & Whig, P. (2022). Studying the impact of the COVID vaccination on the world using data analytics. *Vivekananda J Res*, 10(1), 147–160.

Whig, P. (2022). More on convolution neural network CNN. *International Journal of Sustainable Development in Computing Science*, 4(1).

Whig, P., Kouser, S., Velu, A., & Nadikattu, R. R. (2022a). Fog-IoT-assisted-based smart agriculture application. In *Demystifying Federated Learning for Blockchain and Industrial Internet of Things* (pp. 74–93). IGI Global.

Whig, P., Nadikattu, R. R., & Velu, A. (2022b). COVID-19 pandemic analysis using application of AI. Healthcare Monitoring and Data Analysis Using IoT: Technologies and Applications, 1.

Whig, P., Velu, A., & Bhatia, A. B. (2022c). Protect nature and reduce the carbon footprint with an application of blockchain for IoT. In *Demystifying Federated Learning for Blockchain and Industrial Internet of Things* (pp. 123–142). IGI Global.

Whig, P., Velu, A., & Naddikatu, R. R. (2022d). The economic impact of AI-enabled blockchain in 6G-based industry. In *AI and Blockchain Technology in 6G Wireless Network* (pp. 205–224). Springer.

Whig, P., Velu, A., & Nadikattu, R. R. (2022e). Blockchain platform to resolve security issues in IoT and smart networks. In *AI-Enabled Agile Internet of Things for Sustainable FinTech Ecosystems* (pp. 46–65). IGI Global.

Whig, P., Velu, A., & Ready, R. (2022f). Demystifying federated learning in artificial intelligence with human-computer interaction. In *Demystifying Federated Learning for Blockchain and Industrial Internet of Things* (pp. 94–122). IGI Global.

Whig, P., Velu, A., & Sharma, P. (2022g). Demystifying federated learning for blockchain: A case study. In *Demystifying Federated Learning for Blockchain and Industrial Internet of Things* (pp. 143–165). IGI Global.

10 Intelligent Energy Management System of a Microgrid Using Optimization Techniques

Challa Krishna Rao

Parala Maharaja Engineering College, Berhampur, affiliated to Biju Patnaik University of Technology, Rourkela, Odisha, India

Aditya Institute of Technology and Management, Tekkali, Andhra Pradesh, India

Sarat Kumar Sahoo

Parala Maharaja Engineering College, Berhampur, affiliated to Biju Patnaik University of Technology, Rourkela, Odisha, India

Franco Fernando Yanine

Universidad Finis Terrae, Providencia, Santiago, Chile

10.1 INTRODUCTION

A fast depletion of fossil fuels, as well as growing emissions of greenhouse gases from standard generators, are mostly caused by the worldwide increase in energy use. Over the past ten years, initiatives to generate GW-scale renewable energy resources have been continuing [1]. DERs are distribution systems' on-site generating sources. As a result, there is no requirement for gearbox machinery to provide power to the load ends. Particularly inconsistent and unpredictable energy sources include solar and wind energy, which are employed in DERs. Therefore, it is necessary to cope with these uncertainties using ESSs and micro CGs. To integrate DERs into distribution networks, these energy resources must have the proper scale, control, and scheduling. By integrating DERs into the electrical grid and having the capacity to function in an island state when the main grid collapses, a microgrid recognizes these challenges [2]. As a result, it helps achieve the objectives of successfully transforming making a passive network active and managing power flow in both directions and under control, providing a regular and predictable supply, enhancing power quality, and keeping a clean environment [3].

DOI: 10.1201/9781003481836-10

The advantages that MGs may offer include request response, energy supply decentralization, integration of heat loads for cogeneration, a decrease in GHG emissions, among others. Additionally, it lessens transmission and distribution system line losses and outages. Figure 10.1 displays the global state of the MG deployment market. Microgrids that are autonomous basic MGs are those that are unable to provide power for all 24 hours of the day [5]. However, fully autonomous MGs are equipped with systems for continuously sending power to the end of the load, and the fundamental autonomy MG deployments have already occurred, except for the Middle East, however, where the deployment of autonomy is complete and linked. The development and piloting stages of community-based MGs are currently ongoing [6]. Figure 10.2 shows the existing architecture of the microgrid. High RER investment costs, inefficient energy resource consumption, control issues, a lack of regulatory standards and system protection, and invasion of customer privacy are only a few of the downsides of MGs. Planning generation for energy efficiency, reactive power assistance for frequency regulation, dependability for lowering loss costs, energy balancing for lowering greenhouse gas emissions, and customer interaction for protecting consumer privacy are just a few benefits of EMS [8].

Various MG-related subjects were covered in many review papers, including MG protection and control plans, an examination of MG experimentation in Europe, North America, and Asia, as well as MG reactive power compensation methodologies, which are also references. MG voltage and frequency regulation, as well as control strategies for inverter-based MGs were also covered. They looked at DER control techniques for MGs, a study of AC and DC MGs, and the modeling, hybrid renewable MG structures, planning, and design [9]. A review of network-based energy efficiency in buildings and MGs was also included in the literature, as were assessments of MG uncertainty quantification methodologies, Energy-efficient microgeneration systems based on homeostatic control as well and MG assessments [10].

This research offers a thorough and critical assessment of the MG EMSs, in contrast to prior review studies. We haven't yet addressed MG EMSs based on homeostatic control, robust optimization, mesh adaptive direct search, or hierarchical control are all examples of control. Additionally, each MG EMS's usage of the uncertainty quantification approaches is highlighted [11]. A visual comparison of communication technologies based on data rate, coverage area, and implementation cost is also covered to build a safe, affordable, and dependable communication network among MG components, such comparison is required. It is crucial for the efficient operation of MG's centralized and, in particular, decentralized architecture. Finally, there are also insights into prospects and practical uses [12].

The suggested MG EMSs are compared and discussed in this review study. A summary of MG architectures and MG categorization is given in Section 10.2. In Section 10.3, communication mechanisms. We examine effective cooperation among MG components based on the authors' various solution techniques and their fundamental shortcomings, and a critical discussion of several MG EMS tactics is presented in Section 10.4. The discussion of MG EMSs in the actual world is covered in Section 10.5, which is followed by predictions for the future.

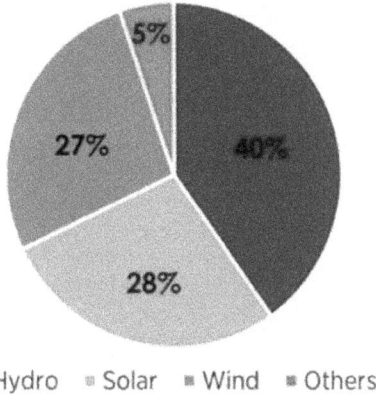

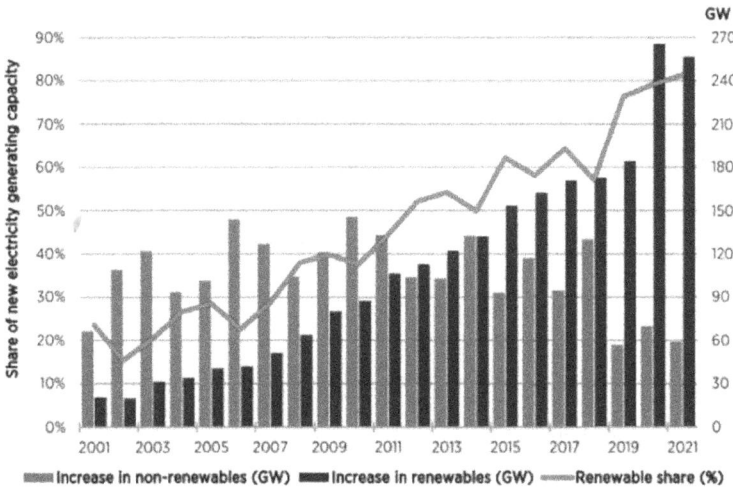

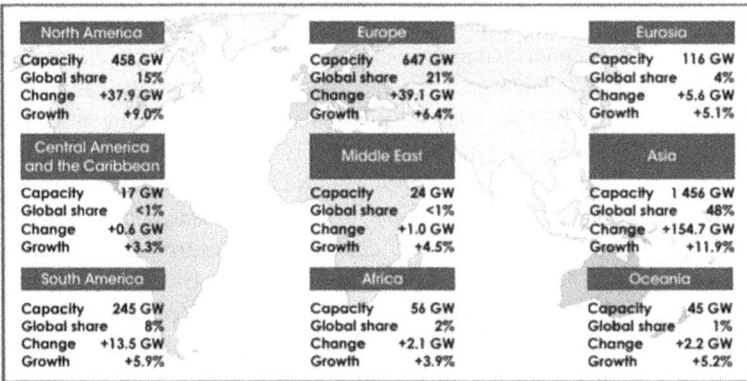

FIGURE 10.1 Renewable generation capacity, power sharing, and expansion [4].

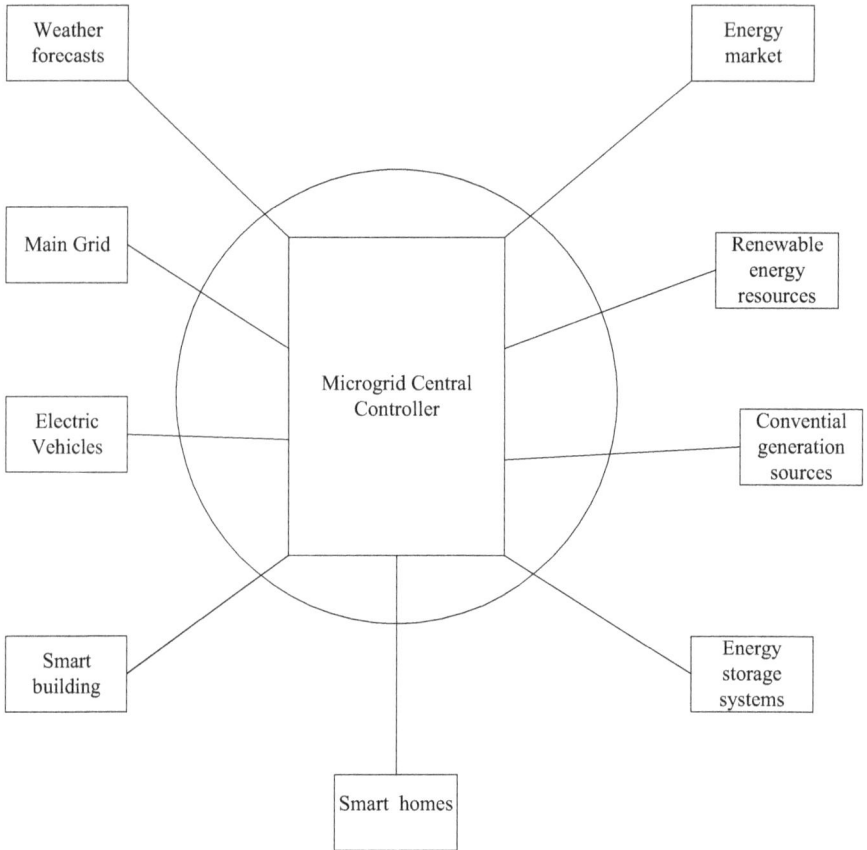

FIGURE 10.2   MG architecture [7].

10.2   MICROGRID CONSTRUCTION

As seen in Figure 10.3, Different DERs, responsive loads, and essential loads make up an MG. The MG is connected to the main grid through a point of common connection. To satisfy control, metering, and protection requirements as well as the possibility for a plug-and-play feature, each DER is linked to PEI operating as an island or linked to the grid [14]. To maintain system stability, MG switches its operation to an island mode in the event of disruptions or main grid breakdown. In this mode, DERs, DR, and load shedding operate effectively together to provide continuous supply to key loads. Local controllers and the MG central controller oversee and manage the overall MG operation. The decision-making methods are solved by MG EMS, which optimizes these energy systems [15]. For sustainable development, these solutions take into account higher system energy efficiency, increased system dependability, lower energy consumption, decreased DER running costs, decreased system losses, and GHG emission reduction. The literature on the use of MG's energy management system lists a few instances of sustainable energy systems [16] (Figure 10.4).

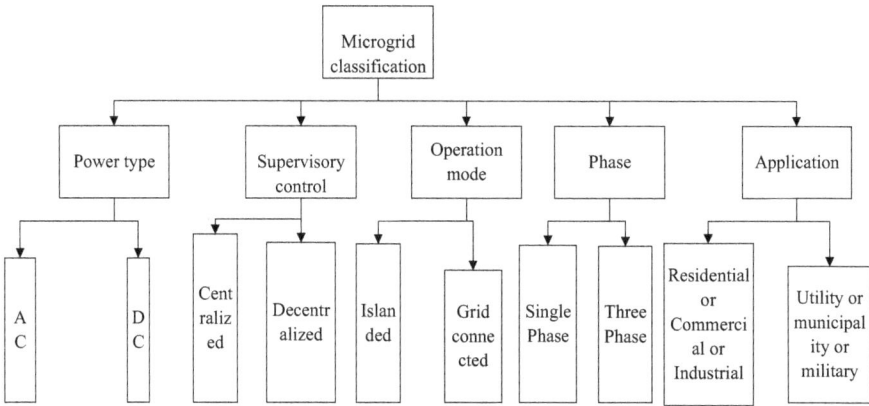

FIGURE 10.3 Classification MG [13].

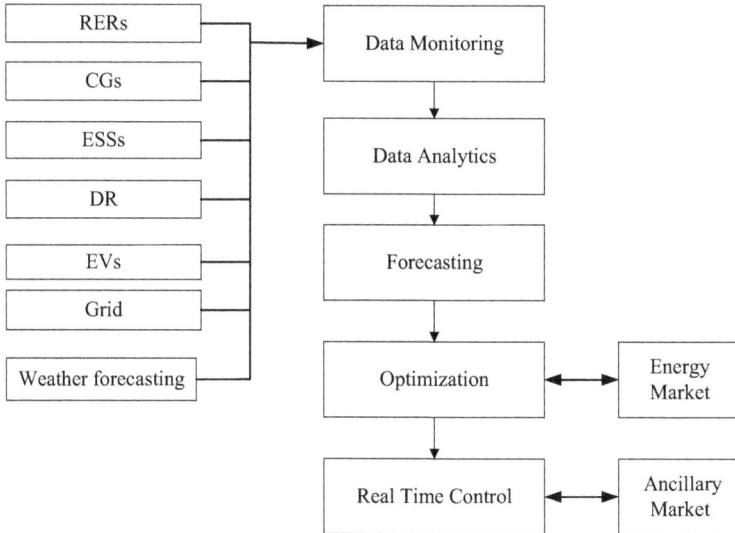

FIGURE 10.4 MG EMS functions [21].

10.3 COMMUNICATION WITHIN MICROGRIDS

For the distributed creation of DERs and active integration of DR to communicate information and perform locally more effectively, a communication infrastructure is required. As a result, a dependable data communication system is needed to ensure that data is sent between sensors, LCs, and the MGCC in an uninterrupted, rapid, reliable, and accurate manner [17]. Repeaters are required to enhance broadcast signal quality while providing coverage for a certain geographic area, which raises the cost of such data transmission systems as an investment [18] (Table 10.1).

TABLE 10.1
A Description of the Communication Technologies [19]

Communication Technologies	Standard	Year	Downlink/Uplink	Range (in meters)	Operating Frequency (MHz)
RFID	Wi-Fi	1973	100 Kbps	2.0	0.125–05876
WSN	Communication through packets	1970	1.370 Mega bps	0300	0900
Bluetooth	Wi-Fi	1994	0720 Kbps	10.0	02450
IEEE 802.15.4	6loWPAN	2003	250.0 Kbps	30.0	826.0 and 915.0
Z-Wave	Wireless	2013	100.0 Kbps	30.0	868.42 and 908.420
LTE	3GPP, LTE, and 4G	1991	100.0 Mbps	35.0	400–1900
NFC	ISO 18092	2004	106, 212, or 424 Kbits	Less than 0.2	13.560
UBW	IEEE 802.15.3	2002	11–55 Mbps	10–30	2400
M2M	All communication protocols are accepted	1973	50–150 Mbps	5–20	1–20
6loWPAN	Wi-Fi	2006	250.0 Kbps	30.0	915.0
5G	Wireless (Wi-Fi)	2019	20 Gbps	Ranging in length from tens of meters to hundreds of meters	1–6 GHz

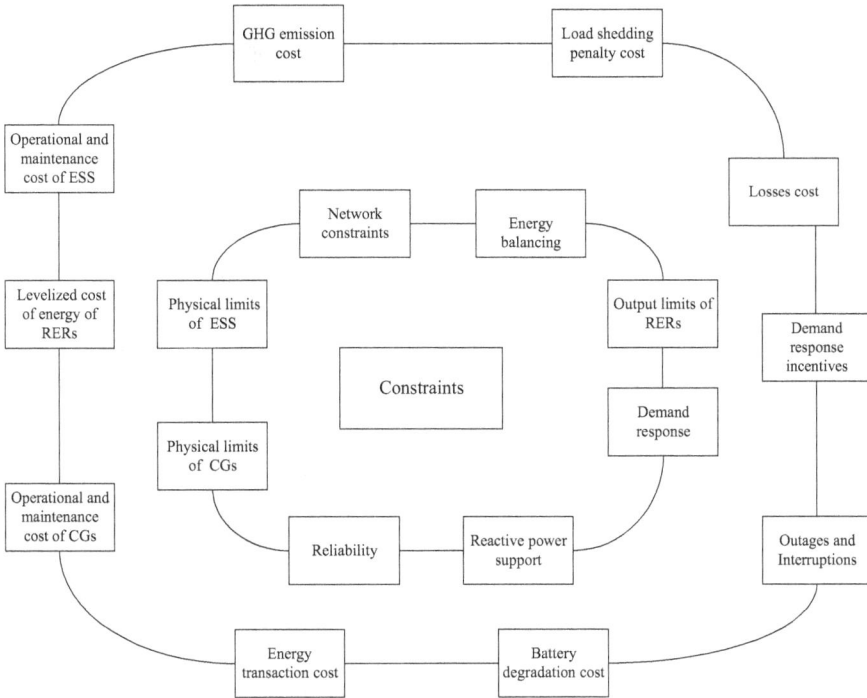

FIGURE 10.5   MG's methods for energy management [13].

Numerous wired and wireless communication options have been suggested in the literature for efficient communication among various MG components. Considerations including data throughput, coverage area, dependability of the quality of service, latency, and power consumption should be taken into account when choosing a communication technology must all be taken into account [20]. A list of the many communication techniques that may be used for MG activities can be found in Figure 10.5. However, wireless technologies like Zigbee, Z-wave, GSM, and Wi-Fi, among others, may be easily adopted at a lower cost, making them good candidates for remote locations [22]. However, they have issues with signal interference and limited data transfer rates. Due to their cheap implementation costs, wireless technologies are, in general, better options than wired ones [23].

10.4   ENERGY MANAGEMENT SYSTEM FOR MICROGRIDS

An MG EMS, which typically includes modules that execute decision-making procedures, contains the same components. There are two categories of MG EMS's supervisory control architecture: centralized and decentralized. Figure 10.5 shows the EMS function. The central EMS then selects the most effective energy schedule for MG and informs all LCs of this selection. The best schedule is decided upon by the MGCC and sent back to the LC [24]. The latter may object to the course of events and go on negotiating until both the national and local objectives are satisfied.

To ensure the most effective and efficient operation of MG, several researchers have addressed these energy management techniques by employing a variety of solution methodologies. The next subsections provide a thorough critique of these tactics and methods for finding solutions [25].

10.4.1 THE TRADITIONAL METHODS OF EMS

10.4.1.1 Linear and Nonlinear Programming Methods

The primary grid's power exchanges are taken into consideration in the power-sharing mode, but the fuel cell must still operate constantly. The optimization technique of linear programming is used to resolve both of these phases. To tackle the on/off mode problem and optimize MG functioning concerning the fuel cell, ESS, and main grid on/off connection statuses, mixed integer linear programming is utilized as a solution approach [26]. See Figure 10.5. The operating requirements for the MG are considered while calculating the ESS size.

The objective function is the diesel generator's piecewise linear fuel consumption model. To balance the operating and capital expenses of MG, The suggested semi-continuous optimization model is solved for two scenarios using the branch and bound solver tool based on special order Sets 1 and 2. In Scenario 1, there is just one generator running, whereas, in Scenario 2, there are two running, but only one is being used at a time. The high investment and replacement costs of the batteries, however, lead to the conclusion that their use in the residential market is not economically viable [27]. The optimal operational cost-minimization approach for residential MG energy management is provided. It mentions the cost of energy trading, the price of replacing an EV battery, the price of load shedding that may be adjusted, and the term "range anxiety" for electric cars. There are three different load categories considered: critical, adaptable, and shiftable. Range anxiety is the worry that an electric vehicle may run out of gasoline before getting to its destination [28].

The operational costs of CGs are included in the created model as an objective function. The suggested model also contains restrictions on how water desalination systems should operate and how AC-side CGs should distribute active and reactive power under droop regulation [29]. They created a centralized architecture for the grid-connected MG's energy management. While the second plan aims to optimize profit while taking into consideration energy exchanges with the main grid, the first strategy tries to reduce MG's operational expenses. For each of these best practices, sequential quadratic programming is employed to address the problems the first to put out a grid-connected MG's strategic EMS, which included restrictions on a transformer operating window and voltage security [7]. The goal function takes into account three possibilities concerning load leveling, network losses, and customer benefits.

In all of these MG EMSs, centralized supervisor control architecture is implemented. The major contributions of the publications on this list have already been covered in this paragraph, where the majority of the studies were carried out with an eye toward the optimization of energy resources in MG The integration of DR, system reliability, and customer privacy issues [30].

10.4.1.2 Dynamic Programming- and Rule-Based Method-Based EMS

The aim functions of the suggested EMS model include the cost of load shedding and the operational cost of the CGs. The findings demonstrate that the suggested technique outperforms the strategies that were previously described. A useful EMS model for maximizing the effectiveness of an MG linked to the grid [31]. Reduced cash flow, a cost that includes fees for battery aging and energy exchange with the main grid, is the goal [32]. Using a lookup table and receptive field weight regression, the estimated value of the cost function is determined. To optimize MG's energy scheduling while accounting for the unpredictability of the model uses economic dispatch and unit commitment procedures. The effectiveness of the proposed approach is compared to that of dynamic programming and myopic optimization methods [33]. It yields better results in terms of reducing operating expenses than the myopic optimization technique but at the expense of a longer computation time. Although it takes less time to compute than the dynamic programming approach, the objective function value is greater [34].

Battery SOC must be less than 80% for the fuel cell to operate in the islanded mode. When the system is connected to the grid, the battery's SOC must be more than 60% for it to function properly in an islanded state PV system, the battery, and ultracapacitor make up a prosumer [35]. For MG functioning, two EMS models are taken into account, introduced an online rule-based EMS switch operations between the PV system's off MPPT mode, and load shedding mode, there are two battery-related modes: charging and discharging based on battery SOC and load generation imbalance [35].

EMS is suggested in residential MG connected to the grid. Predicted wind speed and solar irradiation are also present [36]. The rule-based power regulation approach is explained in detail for switching between load shedding, battery charging and discharging, PV limitation, and grid supply modes. Based on the battery SOC, time of use tariff, and load generation imbalance, the activities are selected. The major objectives of still more work to be done are to reduce environmental pollutants, consider DR, and consider consumer privacy issues while improving MG systems [4].

10.4.2 META-HEURISTIC TECHNIQUES

10.4.2.1 Genetic and Swarm Optimization

It uses a genetic algorithm for day-ahead operations and a rule-based method for real-time activities. To maintain the balance of load generation, the real-time operation sequentially weighs choices for load shedding, battery supply, and diesel generator supply. The point-estimated method employs the beta and Weibull probability density functions to explain the uncertainties of solar and wind power, respectively [37]. However, a rigorous optimization technique is used to simulate load demand uncertainties. The created EMS model's goal is to reduce MG's O&M, emissions, and reliability expenses [38].

Grid operators developed the suggested MG optimum operation model, which has an operation value component for each aim, based on day-ahead analysis. The battery operating value, volt-var optimization, generation reduction, load shedding, and power transfer value between AC and DC MGs make up the goal function [39]. The objectives include lowering fuel expenses, emissions costs, load-shedding costs, voltage fluctuation costs, active and reactive power mismatch costs, and energy trading costs with the main grid [40]. The suggested solution outperforms the interval arithmetic method, and is solved using affine math and a perturbed optimal power flow method based on stochastic weight trade-off PSOs. The critical examination of the MG EMSs that were solved utilizing swarm optimization and genetic algorithms. The computational complexity of the suggested methodologies [41].

10.4.3 EMS USING ARTIFICIAL INTELLIGENCE-BASED TECHNIQUES

10.4.3.1 Neuronal Network-Based EMS with Fuzzy Logic

Electric, water, and transport are the three types of load demands. The fuel considered for moving loads is hydrogen. The fuzzy logic system's decision inputs include the state of charge, water, and frequency of the system of the batteries [42]. Throughout a year-long study, the MG system is simulated using the Transys, Genopt, Matlab, and Trnopt software platforms. An EMS of a grid-connected MG based on fuzzy logic was fuzzy inference systems are tuned using the hierarchical genetic algorithm to reduce the number of fuzzy rules in an EMS model by substituting a practical battery efficiency model with an ideal battery model, the enhanced MG model is made possible [43]. The created method maximizes energy trading profit while optimizing an EMS's fuzzy rule basis for effective energy flow. Comparing the performance of the suggested methodology to a traditional fuzzy-GA method, it seems to perform better [44]. In fuzzy-based unit commitment and MG economic dispatch methods are used to enhance the energy scheduling procedure. Two GAs are considered. The first GA creates the MG energy scheduling and fuzzy rules, while the second GA refines the fuzzy membership functions. The distribution of battery power is also controlled by a fuzzy expert system [44].

A critical NN evaluates the performance of the recommended EMS method in terms of optimality while an active NN strives to discover a solution. In comparison to dynamic EMS based on a decision tree method, the performance of the suggested technique is superior [45]. To optimize the usage of battery and renewable energy resources while reducing dependency on main grid energy purchases, a reinforcement learning-based MG EMS is needed. To ensure that MG operates as efficiently as possible, a Q-learning approach is employed in reinforcement learning [46]. To develop scenarios for anticipated wind speeds, a Markov chain model is utilized while taking wind speed uncertainty into consideration. Only two-hour scenarios are taken into account throughout the optimization phase to achieve the forecasting method's minimal computing cost [47].

All of these MG EMSs employ centralized supervisory control architecture. The majority of the contributions come from the DERs' efforts to lower operating costs, reduce emissions, and trade electricity with the primary grid [33]. Regarding the cost of MG losses, we have not yet completed our evaluation of the consumer privacy issues, the computational difficulty of the recommended approaches, DR, and the dependability of the MG system [48].

10.4.3.2 Multi-Agent System-Based EMS

A decentralized MAS-based strategy for MGs linked to the grid's optimal performance. All users, storage facilities, power plants, and the grid are regarded as agents. Consumer preferences for consumption have been taken into consideration while making decisions. The cost of the power imbalance is reduced by the multi-agent decentralized algorithm [13]. The agents are divided into the following categories: service agents, RER agents, battery agents, central coordinator agents, and agents for building management [49]. Reduced operational costs and increased electrical and thermal comfort for customers are the goals of the suggested solution. The internal message transport protocol is used for intra- HTTP is a protocol for sending and receiving text, whereas agent platform communication is used for inter-agent platform communication [19].

10.4.3.3 EMS is Built on Various Artificial Intelligence Methods

Prosumers with a greater preference parameter for how they consume their energy have a higher degree of utilization, which reduces profit. As a result, there is a trade-off between profit and utilization level in terms of the preference parameter [50].

The created EMS model is solved by the greedy method; it finds the regional optima for the rollout algorithm and a workable fundamental policy to manage massive Markov decision-process states [51].

It seeks to maximize RER utilization and reduce load variations while taking dispatch power mistakes caused by unpredictability in RER power generation and load demand into account. Ultra-capacitors and batteries work together to maintain a smooth load profile, with ultra-capacitors handling rapid changes in load demand [21]. A critical assessment of MG EMSs built using a variety of artificial intelligence methodologies, including game theory and Markov decision processes, as well as these EMSs' supervisory designs and approaches to uncertainty quantification together with their key contributions and drawbacks [52]. These problem-solving techniques take into account some important variables, such as the cost of the communication networks for the MG's decentralized operation, the decrease of environmental pollutants, and the cost of losses the computational time complexity is not satisfactorily addressed by the approaches currently being used [53].

10.4.4 Stochastic and Robust Programming-Based EMS

The wholesale price of energy and the uncertainty associated with wind power are both modeled using the point estimated method. An ideal grid-connected hybrid's day-ahead economic functioning ACDC MG based on probabilistic scenarios [54]. It lowers overall running expenses, which include CG operations expenses and expenses related to power trading with the primary grid. Considerations are included for the expected costs of solar, wind, and both AC and DC loads. An EMS with a stochastic frequency security bound for a single droop-controlled MG [55]. As a result, variations in MG operating frequency are reduced for the day ahead, and MG operational and emission expenses are kept within reasonable bounds. Making use of CG's outages-based contingency analysis, the robustness of the suggested technique is evaluated [56].

A grid-connected MG EMS's performance must be maximized while taking into account load demand and RER unpredictability a two-stage stochastic programming methodology. The first stage is cost optimization for MG investments [57].

The operating expenses of MG include the costs of energy trading, RERs, ESSs, and dispatch units. Small-scale wind turbines and solar systems are included in portable RERs [58].

The performance of an MG is assessed in terms of the imbalance costs brought on by the unpredictability of RERs' power production, load demand, and electricity price. Additionally, by taking into account the parameters for projected energy and load loss, MG dependability is enhanced [14]. By using non-dominated Sorting GA-trained NN, it is expected that RERs, demand, and electricity price forecasts will have the greatest impact on power generation [58]. For optimal MG performance, a trustworthy optimization model is built using these maximum deviation ranges. To enable individual agent decision-making and save costs for the railway station and residential area, agent-based modeling is applied [32]. Additionally, it increases the individual wind power unit's earnings. Based on their forecast mistakes, interval prediction theory determines the RER and load demand uncertainty sets [57]. These sets are used by the Taguchi orthogonal array method to produce a limited range of possibilities. In contrast to Monte Carlo simulation, scenarios with the best statistical data are chosen to speed up processing. The recommended method is more effective and efficient than a greedy algorithm. The goal function considers the operational costs of the CGs and batteries, Incentives for DR as well as worst-case energy trading costs [56]. The suggested model accounts for the RER and MG islanding events' uncertainty. To prevent the suggested technique from being overly conservative, which would lead to a trade-off between robustness and optimality, the budget of uncertainty component is incorporated [27]. A decomposition approach based on a column and constraint-generating methodology is used to resolve the constructed two-stage model that gets converted into a sizable MILP problem. The suggested strategy is more successful in locating the worst-case best solution when compared to stochastic programming, according to the comparison [55].

10.4.5 MODEL-PREDICTIVE CONTROL-BASED EMS

To determine the home controllable load, a supervised NN is employed. The operational costs of CGs and the penalties associated with energy curtailment make up the goal function. The created optimization model outperforms the decoupled EMS model in terms of performance, which divides the major issue into two more manageable issues: unit commitment and optimum power flow [54]. The management of battery and EV operations, energy interchange with the main grid, and optimum RER utilization are all aspects of MG's economic operation. Experimental verification of the suggested strategy's efficacy is also provided [53].

A non-linear MPC approach was recommended to ensure the consistent operation of an islanded MG EMS model. Both of these prediction techniques outperform the methodology mentioned [52]. Furthermore, to fulfill local demand, it makes sure that wind energy is used as much as is practical. In terms of the best answer, the effectiveness and efficiency of the suggested technique outperform reinforcement learning algorithms [29]. The wind energy uncertainty set is created using the fuzzy prediction interval approach. The variability of solar power is not taken into consideration since there is little chance of changes in solar irradiation. The objective is to distribute MG sources efficiently at the lowest feasible operating cost [21].

In terms of an objective function, the costs of maintaining a microturbine, battery deterioration, and the exchanging of electricity with the main grid are also handled. The first step has to do with how MG operates economically in terms of dispatch, whereas the second stage is one of adaptability [51]. It addresses any load generation imbalance brought on by erratic RER production and load demand in real time. To optimize its energy schedule while taking into account the costs of emissions and a variety of operational goals MG's operating and emission costs are as low as possible, and many operational measures have been considered. Additionally explored are the effects of DR on operating and emission cost reduction [50]. All of these MG EMSs employ centralized supervisory control architecture. Forecasted statistics are also utilized primarily to correct for uncertainty in these EMSs. However, more work still has to be done to concentrate on calculation time complexity, CG GHG emission reduction, Costs associated with MG system failures and losses, efficient DR and EV integration, and issues with client privacy [19].

10.4.6 EMSs Based on Different Methodologies

The hourly wind speed is calculated using back propagation NN with a GA foundation. The two operational modes are thought of as real-time dispatch [15]. The battery serves as the primary source and the V/f technique governs operation in the first mode. CG is identified as the principal source in the second mode. As a result, the CG and battery are managed using the V/f and PQ techniques, respectively [49]. In each of these modes, the wind turbine employs the MPPT approach. To lower the running expenses of the MG and changes in the energy conveyed at PCC, an EMS with a two-stage hierarchical method. The recommended method consists of an intra-hour adjustment step with two layers and a day-ahead economic dispatch stage [59]. The last step is used at PCC to reduce oscillations in energy exchange that are ultra-short-term (1 minute) as well as short-term (15 minutes). EVs and thermal office buildings designed in the ESS architecture are viewed as flexible resources [13].

In a residential MG that is linked to the grid, a central EMS controller is installed in every home to facilitate information sharing among residential prosumers [30]. In comparison to the main grid, the goal is to maximize energy sharing among homes and reduce investment costs [47]. A physiological phrase, "homeostasis," describes the movement toward balance. To make sure that the MG demand response function is executed effectively, a lithium-ion battery serves as an energy buffer [46]. The operational costs of CGs, the cost of exchanging power with the primary grid, and the cost of emissions make up the total cost of MG [7]. It is discovered that the suggested technique reduces total MG expenses more successfully than the sequential quadratic programming strategy [45].

10.5   APPLICATIONS TO THE REAL WORLD AND DISCUSSION

Utility, business, manufacturing, and residential sectors all require an EMS to operate in an energy-efficient manner [34]. It seeks to minimize GHG emissions, optimize energy use, and schedule DERs. Monitoring and data analysis are made easier for EMS because of their interaction with SCADA and HMI systems for controlling and acquiring data [44]. It includes data on load demand, weather forecasts, load production of the generating sources, and current energy pricing. The performance of the system at the generating, transmission, and distribution ends is optimized by the EMS using this data [43]. See also Figure 10.6.

DERs are increasingly integrated into the power system, and centralized design faces challenges such as slow calculation times, system scalability is constrained, and failures would result in substantial instability [43]. As a result, decentralized supervisory control architecture has lately received greater attention from academics. However, it needs the synchronization of the MG components as well as the continual availability of a two-way communication link, which raises the cost of the system [42].

To accommodate forecast mistakes, sub-hourly dispatch is separated from the day-ahead energy dispatch. Real-time reference values are transmitted to LCs through communication lines [41]. These communication protocols are used by routers at DERs and load ends to transfer data between LCs and MGCC[40]. The MGCC uses SCADA, HMI, and data from LCs to carry out energy management duties. Additionally, energy companies like Tesla, Alstom, General Electric, Siemens, Schneider Electric, ABB, and others are currently developing and deploying microgrid energy management systems [39].

10.6   CONCLUSION AND FUTURE TRENDS

Demand response, electric cars, local controllers, a central controller powered by a microgrid energy management system, and communication equipment are all crucial components of microgrids. This study has extensively and critically examined proposed problem-solving and energy management solutions for microgrids. Maximizing system reliability, the primary goals of the system for sustainable development are energy scheduling and its operation in isolated microgrids as well as microgrids that are connected to the grid. The topic of microgrid energy management systems has several objectives since it addresses issues with money, technology, and the environment.

This comprehensive analysis discusses ways to accomplish the goals of energy management using a variety of effective techniques. These methods were selected for the optimum performance of microgrids concerning their applicability, viability, and tractability. The operating style, such as centralization or decentralization, economic considerations, and energy sources all have an impact on the goal types for the MG EMS. Additionally, they take into account the environmental impact of traditional generators, battery health, system reliability, client privacy, and active DR integration. Some of these objective categories have been the subject of numerous study investigations. However, a significant amount of work still has to be done to address consumer privacy concerns and cost management for safe and dependable communication systems, particularly for decentralized operations. Furthermore, the reliability

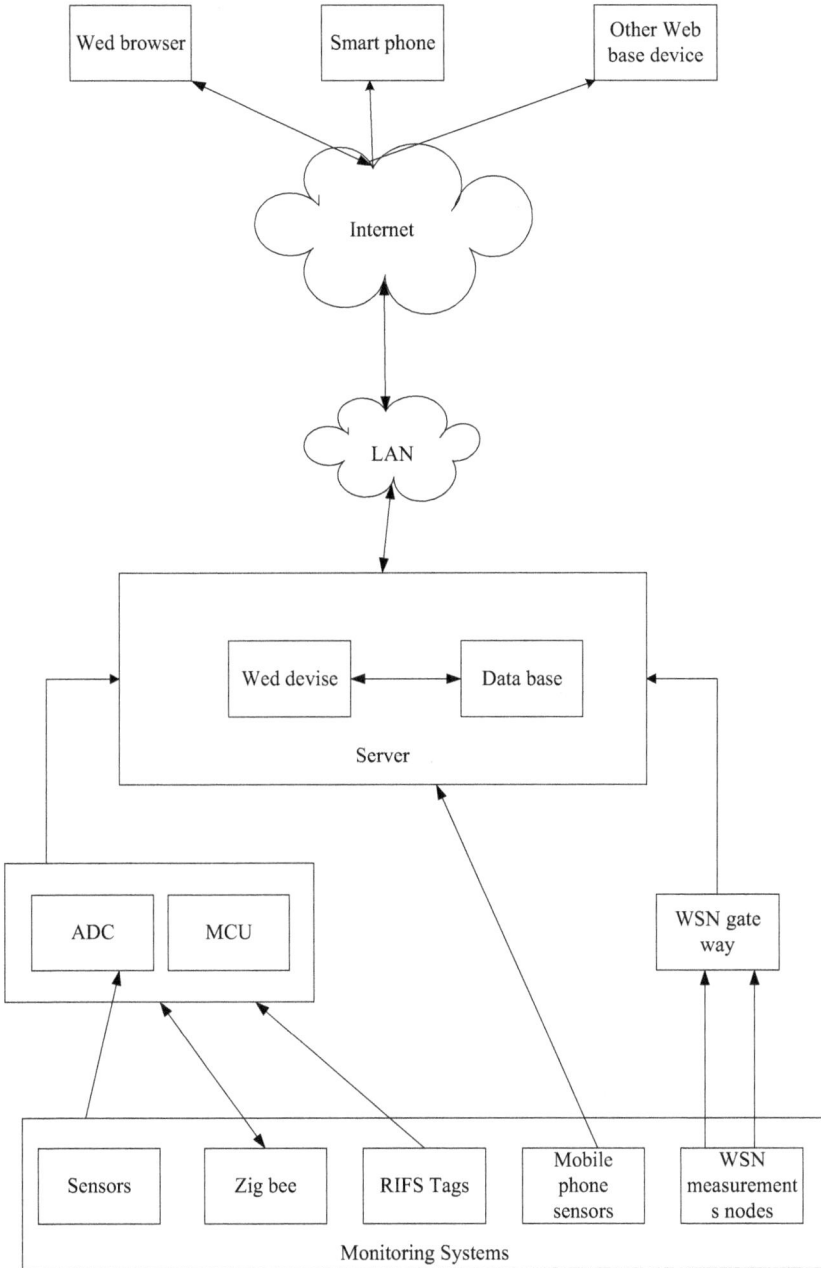

FIGURE 10.6 Smart energy monitoring system [60].

study of microgrid systems for islanding and distant applications has not been thoroughly researched. To operate microgrids as energy-efficiently as possible, these potential issues must be thoroughly addressed.

REFERENCES

[1] Manju, S., Seema, P. N., and Rajendran, Arun (2016). A novel algorithm for power flow management in combined AC/DC microgrid. In *1st IEEE International Conference on Power Electronics, Intelligent Control and Energy Systems (ICPEICES-2016)*.

[2] He, J., Li, Y. W., and Blaabjerg, F. (2013, September). An accurate autonomous islanding microgrid reactive power, imbalance power, and harmonic power-sharing scheme. In *Energy Conversion Congress and Exposition (ECCE), 2013 IEEE* (pp. 1337–43). IEEE.

[3] Adel, M., Sharaf, Adel, and El-Gammal, A. A. (2010). A novel efficient PSO-self-regulating PID controller for hybrid PV-FC-diesel battery microgrid scheme for village/resort electricity utilization. In *IEEE Electrical Power & Energy Conference*.

[4] Delfino, F., Rossi, M., Pampararo, F., and Barillari, L. (2016). An Energy Management Platform for Smart Microgrids. In *Intelligent Computing Systems* (pp. 207–25). Springer.

[5] Sigarchian, Sara Ghaem, Orosz, Matthew S., Hemond, Harry F., and Malmquist, Anders (2016). Optimum design of a hybrid PVCSPLPG microgrid with particle Swarm optimization technique. *Applied Thermal Engineering*, 109, 1031–36.

[6] Tan, K. T., So, P. L., Chu, Y. C., and Chen, M. Z. Q. (2013). A flexible AC distribution system device for a microgrid. *IEEE Transactions on Energy Conversion*, 28(3), 601–10.

[7] IRENA. (2022). Renewable Energy Statistics 2022, International Renewable Energy Agency, Abu Dhabi. ISBN: 978-92-9260-446-2

[8] Divya, R. Nair, Devi, S, Manjula G. Nair, and Ilango, K. (2016). *Tariff-based fuzzy logic controller for active power sharing between Microgrid to the grid with improved power quality*. IEEE.

[9] Golsorkhi, Mohammad S. and Lu, Dylan Dah-Chuan (2016). A decentralized control method for islanded microgrids under unbalanced conditions. *IEEE Transactions on Power Delivery*, 31(3), 1112–21.

[10] Tsikalakis, Antonis G. and Hatziargyriou, Nikos D. (2008). Centralized control for optimizing microgrid operation. *IEEE Transactions on Energy Conversion*, 23(1), 241–48.

[11] Khan, M.W. and Wang, J. (2017). The research on multi-agent system for microgrid control and optimization. *Renewable and Sustainable Energy Reviews*, 80, 1399–411.

[12] Azeroual, M., El Makrini, A., El Moussaoui, H., and El Markhi, H. (2018). Renewable energy potential and available capacity for wind and solar power in Morocco towards 2030, *Journal of Engineering Science and Technology Review*, 11(1), 189–98.

[13] Sukumar, S., Mokhlis, H., Mekhilef, S., Naidu, K., and Karimi, M. (2017). Mix-mode energy management strategy and battery sizing for economic operation of the grid-tied microgrid. *Energy*, 118, 1322–33.

[14] Kantamneni, A., Brown, L., Parker, G., and Weaver, W. (2015). Survey of multi-agent systems for multigrid control, *XE*, 45, 192–203.

[15] Kulasekera, A.L. (2012). Multi agent-based control and protection for an inverter based microgrid [Master's theses, University of Moratuwa]. Institutional Repository University of Moratuwa. http://dl.lib.mrt.ac.lk/handle/123/9957

[16] Davidson, E.M., McArthur, S.D.J., Tom Cumming, T., and Watt, I. (2006). Applying multi-agent system technology in practice: Automated management and analysis of SCADA and digital fault recorder data. *IEEE Transactions on Power Systems*, 21(2), 559–67.

[17] Bellifemine, F.L., Caire, G., and Greenwood, D. (2007). *Developing multi-agent systems with JADE*. John Wiley and Sons.

[18] Rahman, M., Pota, H., and Mahmud, M. (2016). A decentralized multi-agent approach to enhance the stability of smart microgrids with renewable energy, *International Journal of Sustainable Energy*, 35(5), 429–42.

[19] Obara, S., Kawai, M., Kawai, O., and Morizane, Y. (2013). Operational planning of an independent microgrid containing tidal power generators, SOFCs, and photovoltaics. *Applied Energy*, 102, 1343–57.

[20] Azeroual, M., Lamhamdi, T., El Moussaoui, H., and El Markhi, H. (2020). Simulation tools for a smart grid and energy management for microgrid with wind power using multi-agent system. *Wind Engineering*, 44(6), 661–672. doi: 10.1177/0309524X19862755

[21] Rao, C. K., Sahoo, S. K., Balamurugan, M., Satapathy, S. R., Patnaik, A., and Yanine, F. F. (2020). Applications of sensors in solar energy systems. In 2020 International Conference on Renewable Energy Integration into Smart Grids: A Multidisciplinary Approach to Technology Modelling and Simulation (ICREISG). IEEE.

[22] Zia, Muhammad Fahad, Elbouchikhi, Elhoussin, and Benbouzid, Mohamed (2018). Microgrids energy management systems: A critical review on methods, solutions, and prospects. *Applied Energy*, 222, 1033–1055, ISSN 0306–2619.

[23] Warodom, Khamphanchai, Songkran, Pisanupoj, and Weerakorn, Ongsakul (2011). A multi-agent-based power system restoration approach in distributed smart grid. In *2011 International Conference and Utility Exhibition on Power and Energy Systems: Issues and Prospects for Asia (ICUE)* (1–7). IEEE.

[24] Tarhunia, N.G., Elkalashyb, N.I., Kawadyb, T.A., and Lehtonen, M. (2015). Autonomous control strategy for fault management in distribution networks. *Electric Power Systems Research*, 121, 252–9.

[25] Raju, L., Miltron, R.S., and Senthilkumaran, Mahadevan. (2017). Application of multi-agent systems in automation of distributed energy management in micro-grid using MACSimJX. *Intelligent Automation & Soft Computing*, 24(3), 1–9.

[26] Boudoudouh, S., and Maâroufi, M. (2017). Real-time distributed systems modeling and control: Application to photovoltaic fuel cell electrolyser system. *Journal of Engineering Science and Technology Review*, 10(1), 10–17.

[27] Thillainathan, Logenthiran, Dipti, Srinivasan, Khambadkone, Ashwin M., and Aung, Htay Nwe (2012). Multiagent system for real-time operation of a microgrid in real-time digital simulator. *IEEE Transactions on Smart Grid*, 3(2), 925–33.

[28] Sang-Jin, Oh, Yoo, Cheol-Hee, Chung, Il-Yop, and Wonet, Dong-Jun. (2013). Hardware-in-the-loop simulation of distributed intelligent energy management system for microgrids. *Energies*, 6(7), 3263–83. Wooldridge, M., and Weiss, G. (1999). *Multi-Agent Systems*. The MIT Press.

[29] Lahcen, El Iysaouy, Lahbabi, Mhammed, and Oumnad, Abdelmajid. (2019). A novel magic square view topology of a PV system under partial shading condition. *Energy Procedia*, 157, 1182–90.

[30] Samira, Boujenane, Lamhamdi, T., and El Markhi, H. (2019). Modeling technique and control strategie for grid-connected inverter with LCL filter. In *2019 International Conference on Wireless Technologies, Embedded and Intelligent Systems (WITS)*, IEEE.

[31] Boujoudar, Y., Hemi, H., El Moussaoui, H., El Markhi, H., and Lamhamdi, T. (2017). Li-ion battery parameters estimation using neural networks. *International Conference on Wireless Technologies, Embedded and Intelligent Systems (WITS)*. IEEE.

[32] Nassar, M. E., and Salama, M. M. A. (2017). A novel branch-based power flow algorithm for islanded AC microgrids. *Electric Power Systems Research*, 146, 51–62.

[33] Huang, C., Weng, S., Yue, D., Deng, S., Xie, J., and Ge, H. (2017). Distributed cooperative control of energy storage units in microgrid based on multi-agent consensus method. *Electric Power Systems Research*, 147, 213–3.

[34] Zhu, Y., Fan, Q., Liu, B., and Wang, T. (2018). An enhanced virtual impedance optimization method for reactive power sharing in microgrids. *IEEE Transactions on Power Electronics*, 33(12), 10390–402.

[35] Barik, A. K., and Das, D. C. (2018). Expeditious frequency control of solar photovoltaic/biogas/biodiesel generator-based isolated renewable microgrid using a grasshopper optimization algorithm. *IET Renewable Power Generation*, 12(14), 1659–67.

[36] Moayedi, S. and Davoudi, A. (2017). Unifying distributed dynamic optimization and control of islanded DC microgrids. *IEEE Transactions on Power Electronics*, 32(3), 2329–46.

[37] Kiranvishnu, K., and Sireesha, K. (2016). Comparative study of wind speed forecasting techniques. In *Power and Energy Systems: Towards Sustainable Energy (PESTSE), 2016 Biennial International Conference on* (1–6). IEEE.

[38] Zhou, B., Liu, X., Cao, Y., Li, C., Chung, C. Y., and Chan, K. W. (2016). Optimal scheduling of virtual power plant with battery degradation cost. *IET Generation Transmission and Distribution*, 10(3), 712–25.

[39] Akram, U., Khalid, M., and Shafiq, S. (2017). An innovative hybrid wind-solar and battery-supercapacitor microgrid system—Development and optimization. *IEEE Access*, 5, 25897–912.

[40] Tomoiaga, B., Chindris, M. D., Sumper, A., and Marzband, M. (2013). The optimization of microgrids operation through a heuristic energy management algorithm. In *Advanced Engineering Forum*, 8–9 (185–94). TransTech Publications.

[41] Rao, C. K., Sahoo, S. K., and Yanine, F. F. (2021) Demand response for renewable generation in an IoT-based intelligent smart energy management system. In *2021 Innovations in Power and Advanced Computing Technologies (i-PACT)*, Kuala Lumpur, Malaysia (1–7).

[42] Riar, B., Lee, J., Tosi, A., Duncan, S., Osborne, M., and Howey, D. (2016, June). Energy management of a microgrid: Compensating for the difference between the real and predicted output power of photovoltaics. In *Power Electronics for Distributed Generation Systems (PEDG), 2016 IEEE 7th International Symposium on* (1–7). IEEE

[43] Abedini, M, Moradi, M. H., and Hosseinian, S. M. (2016). Optimal management of microgrids including renewable energy sources using GPSO-GM algorithm. *Renewable Energy*, 90, 430–9.

[44] Rao, C. K., Sahoo, S. K., Balamurugan, M., and Yanine, F. F. (2021). Design of smart socket for monitoring of IoT-based intelligent smart energy management system. In *Lecture Notes in Electrical Engineering* (pp. 503–18). Springer Singapore.

[45] Caballero, F., Sauma, E., and Yanine, F. (2013). Business optimal design of a grid-connected hybrid PV (photovoltaic)-wind energy system without energy storage for an Easter Island's block. *Energy*, 61, 248–61.

[46] Li, F.-F., and Qiu, J. (2016). Multi-objective optimization for an integrated hydro photovoltaic power system. *Applied Energy*, 167, 377–84.

[47] Rao, C. K., Sahoo, S. K., and Yanine, F. F. (2022). Forecasting electric power generation in photovoltaic power systems for smart energy management. In *2022 International Conference on Intelligent Controller and Computing for Smart Power (ICICCSP)*.

[48] Askarzadeh, A. (2017). A memory-based genetic algorithm for optimization of power generation in a microgrid. *IEEE Transactions on Sustainable Energy*, 99, 1.

[49] Panwar, L. K., Konda, S. R., Verma, A., Panigrahi, B. K., and Kumar, R. (2017). Operation window constrained strategic energy management of microgrid with electric vehicle and distributed resources. *IET Generation Transmission and Distribution*, 11(3), 615–26.

[50] Fossati, J. P., Galarza, A., Martín-Villate, A., Echeverría, J. M., Fontán, L. (2015). Optimal scheduling of a microgrid with a fuzzy logic-controlled storage system. *International Journal of Electrical Power & Energy Systems*, 68, 61–70.

[51] Bazaar, A., and Kavousi-Fard, A. (2013). Considering uncertainty in the optimal energy management of renewable micro-grids including storage devices. *Renewable Energy*, 59, 158–66.

[52] Chowdhury, S., and Crossley, P. (2009). *Microgrids and active distribution networks.* The Institution of Engineering and Technology.

[53] Bifano, W. J., Ratajczak, A. F., Bahr, D. M., and Garrett, B. G. (1979). *The social and economic impact of solar electricity at Schuchuli village.* Solar Technology in Rural Settings: Assessments of Field Experiences.

[54] Hina Fathima, A. and Palanisamy, K. (2015). Optimization in microgrids with hybrid energy systems A review. *Renewable and Sustainable Energy Reviews*, 45, 431–46.

[55] Cobben, J. F. G., Kling, W. L., and Myrzik, J. M. A. (2005). Power quality aspects of a future microgrid. In *Proc. Int. Conf. Future Power Syst.* (1–5).

[56] Illindala, M. and Venkataramanan, G. (2012). Frequency/sequence selective filters for power quality improvement in a microgrid. *IEEE Transactions on Smart Grid*, 3(4), 2039–47.

[57] Kakigano, H., Miura, Y., and Ise, T. (2010). Low-voltage bipolar-type DC microgrid for super high-quality distribution. *IEEE Transactions on Power Electronics*, 25(12), 3066–75.

[58] Wang, F., Duarte, J. L., and Hendrix, M. A. M. (2011). Grid-interfacing converter systems with enhanced voltage quality for microgrid application concept and implementation. *IEEE Transactions on Power Electronics*, 26(12), 3501–13.

[59] Almada, J, Leão, R, Sampaio, R, and Barroso, G. (2016). A centralized and heuristic approach for energy management of an AC microgrid. *Renewable and Sustainable Energy Reviews*, 60, 1396–404.

[60] Hassan, M. A. and Abido, M. A. (2011). Optimal design of microgrids in autonomous and grid-connected modes using particle swarm optimization. *IEEE Transactions on Power Electronics* 26(3), 755–69.

11 Charging Cost Minimization of Electric Vehicles and Maintaining Voltage Limit Violations

Parvez Ahmad, Kushagra Jain, Priyanshul Niranjan, and Prashant Singh
MNNIT Allahabad, Prayagraj, India

Vivek Patel
SMNRU Lucknow, Lucknow, India

Niraj Kumar Choudhary and Nitin Singh
MNNIT Allahabad, Prayagraj, India

11.1 INTRODUCTION

Numerous vehicle alternatives have been developed in response to high levels of fossil fuel use and environmental problems in conventional vehicles. According to sources [1, 2], because they emit no pollutants, electric vehicles (EVs) are known for being environmentally friendly. EVs perform better than conventional fossil fuel-powered vehicles. Hybrid electric cars (HEVs) and plug-in hybrid electric vehicles (PHEVs) are recognized as very feasible options within the automobile industry, as mentioned in [3, 8]. Especially when accelerating quickly, HEVs efficiently use energy storage technologies to lower fuel usage. The battery size of PHEVs is large and has higher storage capacity than HVs, which gives them the ability to operate in electric mode with some autonomy. Additionally, PHEVs may operate similarly to HVs [4, 8].

PHEVs are anticipated to be charged at houses without a specialized infrastructure, and the effect of the addition of PHEVs on the grid should be as low as possible [6, 7]. This increase in power consumption creates many issues and jeopardizes the grid's ability to function safely [5]. The availability of bidirectional chargers may offer an important means of integrating PHEVs into the grid and resolving the associated technological problems [3–5, 9]. In this approach, better energy storage systems on PHEVs and EVs represent both loads and transportable energy sources [10–13].

DOI: 10.1201/9781003481836-11

Modern electricity networks may be made more reliable and efficient by using PHEVs. A dynamic connection between the energy and transportation industries is created by coupling PHEVs to the grid, creating a special synergy. The grid may be able to balance supply and demand better if PHEVs are used as mobile energy storage units. Excess renewable energy may be stored in PHEV batteries during times of low electrical consumption, minimizing waste and maximizing the use of clean energy sources. In contrast, PHEVs may return stored energy into the system when the grid experiences peak demand or unanticipated swings, essentially serving as a distributed energy resource. This bidirectional flow of power not only improves grid stability but also gives grid operators the chance to control voltage levels and mitigate grid congestion. Additionally, because PHEVs have demand response capabilities, users may take benefit of dynamic pricing models and contribute to load management plans, eventually supporting a more reliable and sustainable grid environment. For the connection of PHEVs to the grid, appropriate methods to select optimal policies of injection or consumption of power become essential [13–15]. Based on the present loading of the home at that time and the existence of PHEVs, this study offers a method to calculate voltages at each node of a domestic grid. The constraints on voltage, charging rate, and the minimum state of charge (SOC) of battery have been defined [1, 2, 16].

Given the current situation of charging options available in the market, it is totally clear that technology is about to advance quickly. Minimization of the cost of charging is a key part of this leap jump and there are a lot of innovative ideas that exist which need to mature. In the same league, the best way to reduce charging costs is the scheduling of EVs with the peak-valley load of the power grid, real-time pricing, vehicle to grid (V2G) bidirectional operation, and menu-based charging [17].

In combination, all the proposed sets of technology will lead us to deal with problems like power grid instability, sudden change in frequency and voltage, peak and valley load difference, transformer load profile, as well as charging costs. Current proposed ideas are merely able to reduce charging costs from 7.7% to 19% at different modes, and this shows that further development and improvement is needed. With the aim of further advancement, this paper has proposed an algorithm to tackle the problems.

In this chapter, Section 11.1 gives an overall introduction, Section 11.2 covers PEVs, and transformers and other things are modeled. Section 11.3 represents the set of constraints on voltage and charger for the proposed algorithm. Section 11.4 represents the cost function for the algorithm. Section 11.5 discusses the various cases on which the algorithm is tried, and the results were analyzed. Finally, a conclusion of the chapter is presented in Section 11.6.

11.2 SYSTEM MODELING

According to Figure 11.1, the distribution grid is reflected as a radial topology with a feeder and the branches start at specific feeder nodes. Because it is a distribution system, the voltage is low, so the resistive part is much greater than that of the reactive part, the reactive part is neglected and thus impedance is considered purely resistive. Every house is considered a grid node, and each node can have one PHEV attached

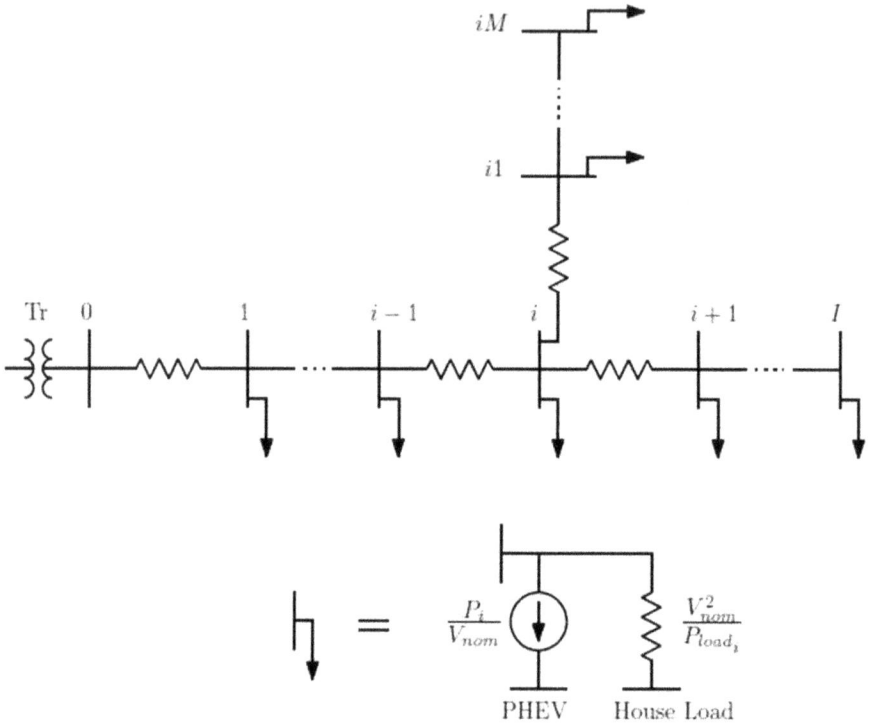

FIGURE 11.1 Modeled test system.

to it. The battery of a PHEV is considered a current source, and the load is considered purely resistive. The nominal grid voltage and the related active power from the residence are used to estimate the resistive load value $R = V^2/P$. The intended instantaneous voltage at the grid and input/output power value is used to determine an approximation of the current delivered or consumed by the batteries as P/V. The transformer is modeled as a voltage source. The voltage of a transformer is nominal, and it is common for each of the nodes. In the grid, the transformer is modeled as voltage source and the PHEV is modeled as a current source which is used to determine the current and voltage at each node.

The matrix $V_{hj,m}$ contains the voltage of each of n nodes at a given time instant h_j and that can be calculated as per the power delivered or consumed by the PHEVs ($P_{hj,m}$) and the present load of each house. The matrix that contains voltage and the matrix that contains the power of each PHEV are organized as shown in Eq. (11.1):

$$V_{hj,m} = \left[v_{hj,1}, v_{hj,2}, \ldots\ldots, v_{hj,M} \right]^T$$

$$P_{hj,m} = \left[p_{hj,1}, p_{hj,2}, \ldots\ldots, p_{hj,M} \right]^T \qquad (11.1)$$

$$V_{hj,m} = L_{hj} \hat{h}_{hj} + L_{hj} \hat{P}_{hj,m}$$

The matrix $\hat{L}$ contains the value of the load at each node at time instant h_j. To determine the matrix of voltages Eq. (11.1) is used. The $\hat{L}_{h_j}$ is the first element of matrix that stores value of the voltage at node 0; i.e., the node where transformer is connected at time instant h_j. A combination of two words makes up these expressions that are the voltage contributed by the transformer node (node 0) and the voltage contributed due to PHEVs. The following procedure is used to determine the voltage at any node at any discrete time frame:

$$
\hat{L} = \begin{bmatrix} \hat{L}_{h_1} & 0 & \cdots & 0 \\ 0 & \hat{L}_{h_2} & & \vdots \\ \vdots & & \ddots & 0 \\ 0 & \cdots & 0 & \hat{L}_{h_f} \end{bmatrix} \tag{11.2}
$$

$$
V_{h,m}^T = \left[v_{h_1,m}^T, v_{h_2,m}^T, \cdots\cdots, v_{h_f,m}^T \right] \tag{11.3}
$$

$$
P_{h,m}^T = \left[p_{h_1,m}^T, p_{h_2,m}^T, \cdots\cdots, p_{h_f,m}^T \right] \tag{11.4}
$$

$$
V_{h,m} = L\hat{h}_h + L\hat{P}_{h,m} \tag{11.5}
$$

$\hat{L}$ is the diagonal matrix that is made by a matrix $\hat{L}_{h_j}$ corresponding to different time instants $j = \{1,2,\ldots f\}$ and the current load of every node in that time interval. As seen in Eq. (11.3), the voltage vector $V_{h,m}$ is a combination of the vectors $v_{hj,m}$ at any point at that time instant and at any node. Based on this, voltage can be calculated using Eq. (11.5), the intersection of the vectors representing the power delivered/consumed at any given time, at any given node, $P_{h,m}$ and the voltage at the node (transformer node) at any given time. The sole distinction between this equation and Eq. (11.1) is that Eq. (11.5) calculates the integration of the matrix $v_{hj,m}$.

Voltages and powers, organized in blocks and then categorized by nodes and instants, respectively, are contained in the matrix $V_{h,m}$ and $P_{h,m}$. Rearranging the matrices L, $P_{h,m}$ and $\hat{h}_h$ results in a set of blocks that are initially grouped by nodes and then sorted by different time instances. The voltage matrix that comes from this reconfiguration is similarly organized. To make the optimization algorithm simpler, this modification is done. Eq. (11.6) states that the desired voltage matrix must follow the given pattern for nodes $i = \{1, 2, 3, \ldots, M\}$, and time occurrences $j = \{1, 2, 3, \ldots, f\}$.

$$
V_{i,h} = \left[v_{i,h1}, v_{i,h2}, \cdots\cdots, v_{i,hf} \right]
$$

$$
V_{m,h}^T = \left[v_{1,h}^T, v_{2,h}^T, \cdots\cdots, v_{M,h}^T \right] \tag{11.6}
$$

The matrix $\hat{L}$ and the matrix $P_{m,h}$, and h_h are produced by this rearrangement, and they may be used to determine the appropriate voltage matrix using Eq. (11.7).

$$V_{m,h} = Lh_h + Lp_{m,h} \tag{11.7}$$

The SOC for an electric vehicle at any time t and node i is given as Eq. (11.8):

$$c_{i,h} = \hat{B}p_{i,h} \tag{11.8}$$

Where Matrix B is same for each PHEV which is given by Eq. (11.9):

$$\hat{B} = \begin{bmatrix} \Delta h & 0 & \cdots & & 0 \\ \Delta h & \Delta h & & & \vdots \\ \vdots & & \ddots & & 0 \\ . & . & \cdots & \Delta h & 0 \\ \Delta h & \Delta h & \cdots & \Delta h & \Delta h \end{bmatrix} = \begin{bmatrix} \hat{B}_u \\ \hat{b}_d^T \end{bmatrix} \tag{11.9}$$

It is feasible to divide the SOC of each PHEV present in the grid into two expressions throughout the charging period. The SOC before the last moment (calculated with $\hat{B}_u$) and the final SOC (calculated with $\hat{b}_d$).

$$B_u = \begin{bmatrix} \hat{B}_u & & 0 \\ & \ddots & \\ 0 & & \hat{B}_u \end{bmatrix}, B_d = \begin{bmatrix} \hat{b}_d^T & & 0 \\ & \ddots & \\ 0 & & \hat{b}_d^T \end{bmatrix}, \tag{11.10}$$

$$\begin{cases} c_{m,\{h_1,\ldots,h_f-\Delta h\}} = B_u P_{n,h} \\ c_{m,\{h_1,\ldots,h_f\}} = B_d P_{m,h} \end{cases} \tag{11.11}$$

The separate matrices B_u are used to create the matrix $\hat{B}_u$ in Eq. (11.10), which is then used to determine each PHEV's level of charge up to the instant before the charging period ends. However, B_d is created using distinct vectors (transposed) $\hat{b}_d$ and is used to calculate ultimate SOC Eq. (11.11).

11.3 CONSTRAINTS

11.3.1 CONSTRAINTS ON THE VOLTAGE

Constraints on the voltage support by PHEVs are essential for ensuring efficient and reliable functioning. In order to protect the integrity of the electrical systems in the car and the connected grid system, these restrictions center on keeping voltage levels within predetermined ranges. During charging and discharging cycles, PHEVs must adhere to voltage restrictions since deviations might result in less-than-ideal performance and even damage. Advanced power management systems are used to precisely regulate voltage fluctuations, and avoid overcharging or undercharging the battery. PHEVs provide resilient power infrastructure while promoting sustainable mobility by maintaining these voltage restrictions. The reduction of the grid load generated by PHEV charging is the primary goal of the intelligent charging strategy. The strategy used in this instance is to give voltage support to each grid node. Even in the absence of PHEV, this is enforced as a constraint on the voltages of each node in the optimization issue. Eq. (11.12) is a summary of the voltage support limits.

$$\begin{cases} Lp_{m,h} \leq V_{\max} - Lh_h \\ Lp_{m,h} \geq V_{\min} - Lh_h \end{cases} \qquad (11.12)$$

Where $V_{\max}$ represents the maximum and $V_{\min}$ represents the minimum voltages permitted for each grid node. These restrictions are usually placed between 90% and 110% of the voltage. It is vital to note that the term Lh_h has nothing to do with how much electricity PHEVs consume or deliver.

11.3.2 CONSTRAINTS ON THE CHARGE OF PHEV

PHEVs are subject to particular operational constraints during grid voltage support mode that are crucial to preserving grid stability. The PHEVs' ability to control their power generation and utilization in response to the grid's demands are included in these restrictions. When in this mode, PHEVs must carefully control how much power they use from or contribute to the grid to prevent aggravating voltage swings or grid imbalances. This mutually beneficial interaction between PHEVs and the grid emphasizes the importance of well-stated restrictions to guarantee seamless integration and optimal performance within the larger energy ecosystem. The constraints on the charging/discharging are given as Eq. (11.13):

$$\begin{cases} p_{m,h} \leq p_{\max} \\ p_{m,h} \geq -p_{\max} \end{cases} \qquad (11.13)$$

The chargers are considered bidirectional. The maximum consumed/injected power, which varies according to the type of charger, is the greatest amount of electricity that PHEVs can inject or consume. The following restrictions given by Eq. (11.14) are also placed on SOC in order to increase battery life.

$$\begin{cases} B_u p_{m,h} \leq c_{\max_u} - c_{\min_u} \\ B_u p_{m,h} \geq - c_{ini_u} \\ B_d p_{m,h} = c_{\max_d} - c_{ini_d} \end{cases} \qquad (11.14)$$

Where the boundaries are specified as zero and $c_{\max}$, the maximum value. The first two constraints in this case restrict the SOC to staying within predetermined bounds throughout the charging period. The third constraint makes sure that after the charging period, the state of charge will be at its maximum only to specify and match dimensions are sub-indices u and d placed to the right of the equation and inequalities. The matrix c_{ini} contains the initial SOC.

11.4 COST FUNCTION

While keeping in mind the constraints on voltage, input/output powers, and SOC, the goal is to reduce the cost of the energy used to recharge the PHEVs. The cost function in standard form is represented as Eq. (11.15):

$$\min_{w^i \in W^i, s^i \in S^i} \sum_{i=1}^{I} \sum_{j=1}^{j^i} c_j \tau \left(w_j^i - s_j^i \right) \tag{11.15}$$

Where
  τ = Duration of time step (in hours)
  c_j = Cost of power at each time step
  j = The number of time steps
  I = Total number of connected PHEVs
  w_j^i = Charging rate of the ith vehicle at j^{th} instant
  s_j^i = Discharging rate of the ith vehicle at j^{th} instant

The above function is modified in matrix form and then used for the calculation of the cost subject to the above-mentioned constraints.

11.5 RESULTS

The single and double-price approaches are often used in EV charging techniques, with each providing a particular purpose. Single-price charging, which is frequently employed in residential settings, offers a simple method in which EV owners pay a fixed cost for the power utilized, generally calculated in kilowatt-hours (kWh). The user-friendly pricing structure may not consider changes in the daytime demand for power. As opposed to this, double-price charging, which is common at public charging stations, adds a dynamic pricing approach. It makes a distinction between peak and off-peak hours, with higher rates during peak times when the grid is experiencing greater demand and reduced rates during off-peak times.

The 8-node test system shown in Figure 11.2 is used, where each node represents a house. 3 PHEVs are connected during the charging periods which are shown in the circuit. Figure 11.3. displays an example of a 24-hour load profile. Each home on the grid is built using this load profile as a foundation. Additionally, distinct random values are added at every node and every time. The charging period, which lasts from 18:00 to 06:00, is shown by the continuous component of the profile. The transformer node has 230 V of electricity. The maximum power from the chargers is 3 kW. The nominal capacity of battery is 20 kWh, but only 80% of that (16 kWh) is used in order to prolong battery life.

Table 11.1 contains a list of the grid's line parameters. Two cases are assessed. As shown in Figure 11.4, in the first scenario, the load curve does not result in a voltage

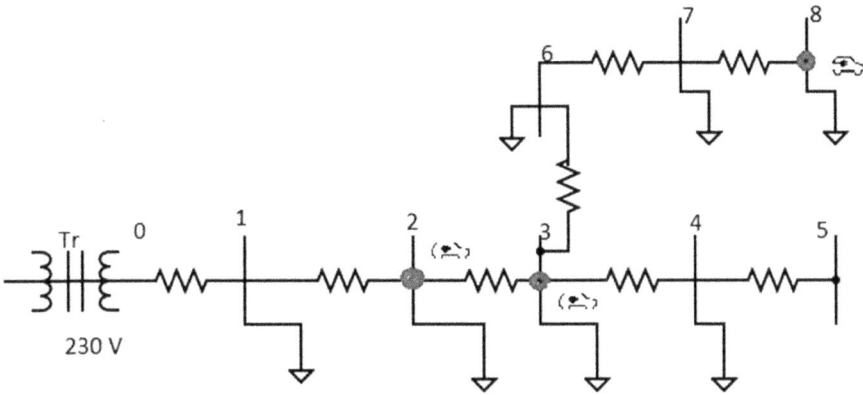

FIGURE 11.2 8-Node test system.

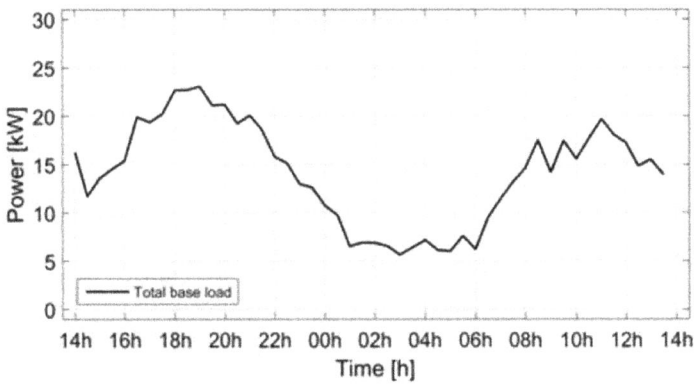

FIGURE 11.3 Load profile for a 24-hours of usage in a home.

TABLE 11.1
Grid Parameter(s)

Between Nodes	Length [m]	R (ohm)
0–1	64	0.05
1–2	64	0.05
2–3	128.2	0.10
3–4	192.3	0.15
4–5	192.3	0.15
3–6	128.2	0.10
6–7	64	0.05
7–8	128.2	0.10

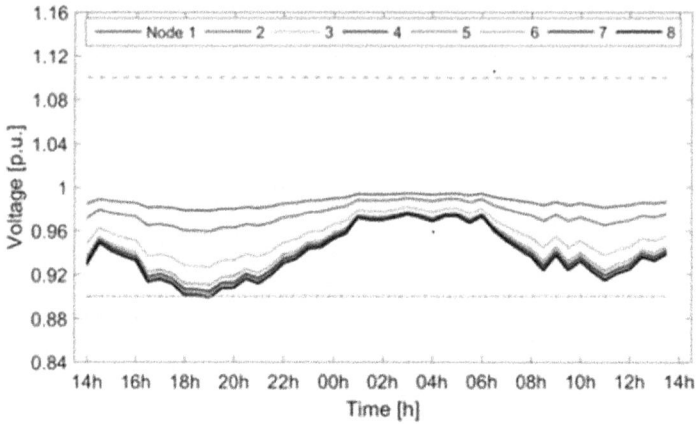

FIGURE 11.4   Voltage curve for Case Study 1 when no PHEVs is connected.

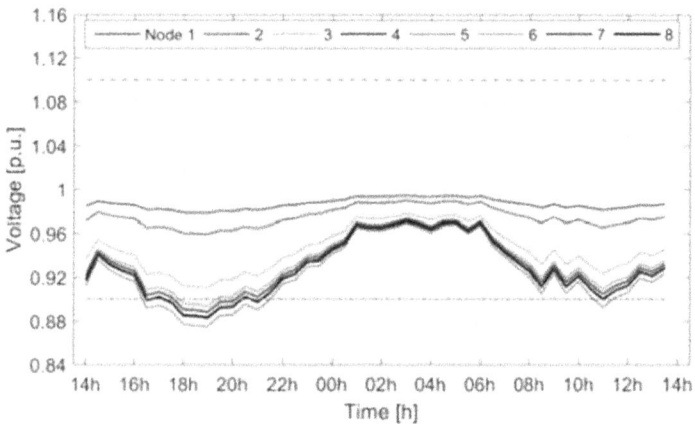

FIGURE 11.5   Voltage curve for Case Study 2 (varying the parameters) of the test system without PHEVs connected to grid.

limit violation when PHEVs linked to the grid are not taken into account; voltage problems are not evident in profile. Without any PHEVs attached, this graphic displays the voltage curve throughout the charging phase. Each voltage profile can be shown to stay between 0.9 p.u. and 1 p.u. throughout this time. The resistive value of the two lines varies slightly in the second scenario. Even without PHEVs connected, this adjustment is sufficient to result in voltage limit violations on specific nodes during specific times of peak demand. In this instance, Figure 11.5 represents the voltage curve for each node, and for each time instant when PHEV is not connected to the grid, some profiles exceed predetermined security thresholds.

11.5.1 CASE STUDY 1

In Case Study 1, the variation in different parameters is observed when no optimization algorithm is applied but here the voltage profile has no violations when no PHEV is connected to the grid.

The chargers in PHEVs are designed to use as much input power as they can charge the batteries as quickly as possible when they are connected to the grid without any kind of charging management. The associated charger stops using power once the batteries are completely charged. Figure 11.6(a) represents the power curve. The power used to charge the PHEVs (3 kW) in this scenario.

4.5 hours at node 2, 4.75 hours at node 3, and 2.25 hours at node 8 correspond with the times when the households are using the most energy (during peak loading conditions). The relevant SOCs are displayed in Figure 11.6(b). As can be seen, batteries charge completely in slightly more than one-third of the time allotted for charging.

The voltage profiles for each node are shown in Figure 11.6(c). For nodes between 4 to 8, it is observed that voltage fall below the stated permissible limit (90% of the minimum loading condition) between 18:00 and 21:45 when compared to the voltage curves when no PHEVs are connected in Figure 11.4 due to the increased power usage. It is conceivable to draw the conclusion that the PHEV's power input will cause the voltage curve to decrease even more than those of Figures 11.5 and 11.6(c) if these parameters are applied to Case Study 2. In this instance, a special hypothetical rate of $1/kWh is taken into account. As a result, it costs $37.6 to charge the PHEVs. This is corresponding to the 37.6 kWh total power needed to charge the PHEV.

11.5.2 CASE STUDY 2

In Case Study 2, the variation in different parameters is observed when the optimization algorithm is applied to the test system.

11.5.2.1 With Charging Strategy: Single Tariff Scenario

Figure 11.7(a) shows the power consumption profiles are the outcome of applying the algorithm to the first scenario of grid. As a result, over the entire charging period, the power consumption is divided differently. It is crucial to emphasize that peak public consumption happens when there is little demand. It's also crucial to note that during times of high demand, the PHEV with the highest starting charge contributes power to the grid to offset the use of the other PHEVs and keep voltages within safe ranges. Even during times of high demand, as demonstrated in Figure 11.7(c), the voltage profiles corresponding to each node are maintained within the acceptable bounds.

In this scenario, the same imaginary tariff of $1/kWh is taken into account. Similar to the suboptimal scenario, the minimized cost of charging the PHEVs is $37.6. Since 37.6 kWh of energy is still needed to completely charge the PHEVs, the price stays the same.

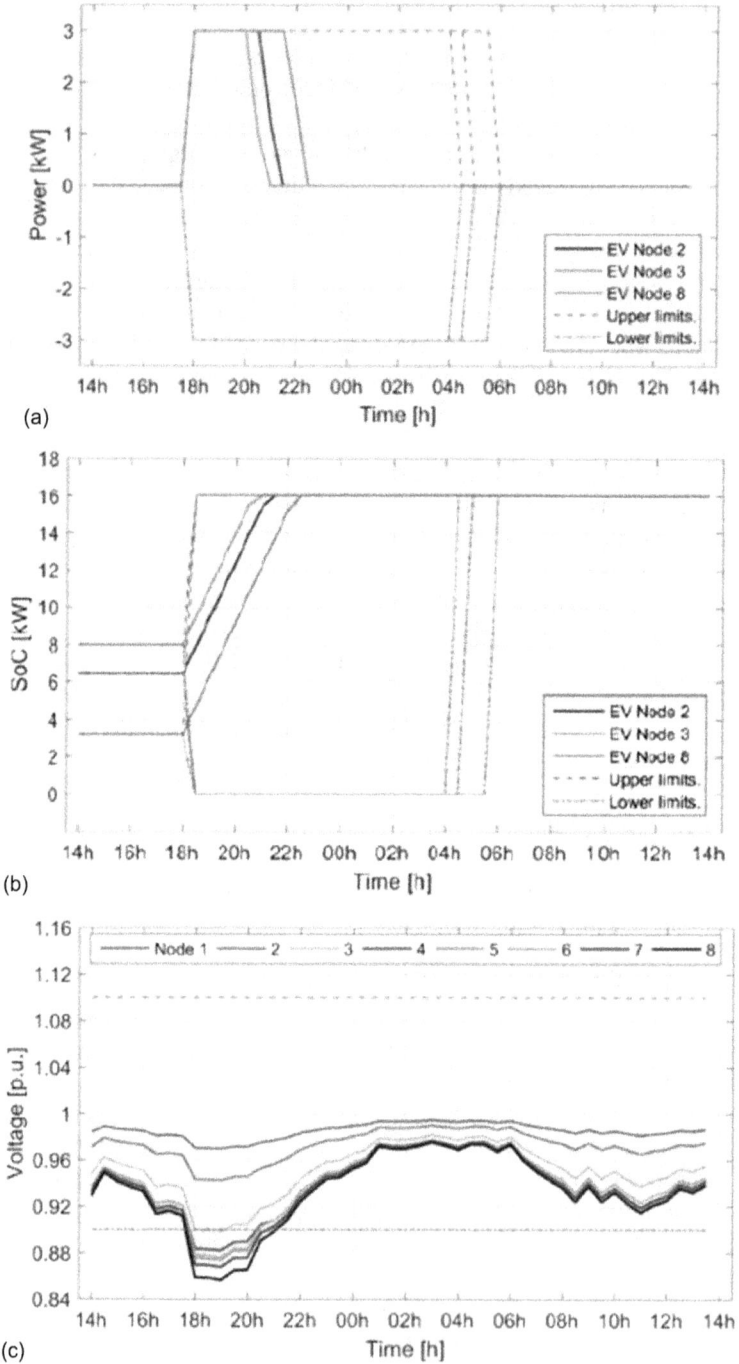

FIGURE 11.6 (a) Without charging management Case Study 1- Power profile, (b) Without charging management Case Study 1: SOC profile, (c) Without charging management: Case Study 1 voltage profile.

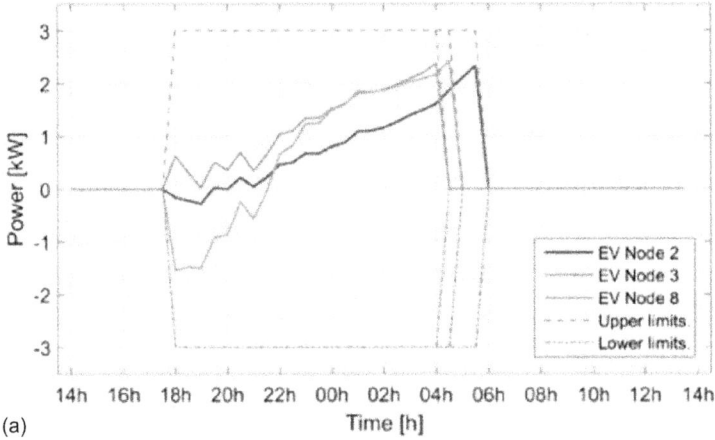

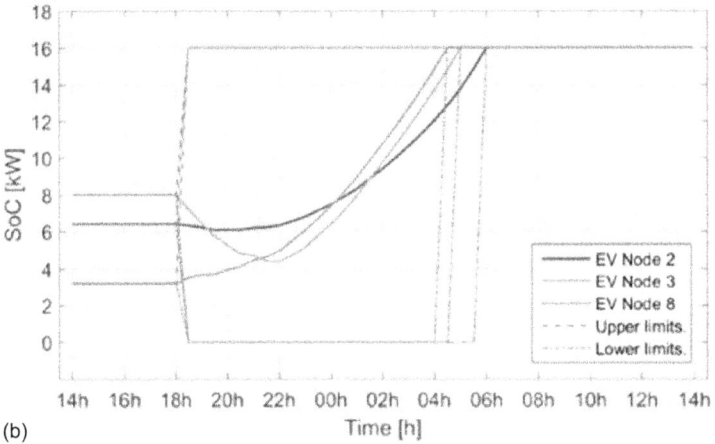

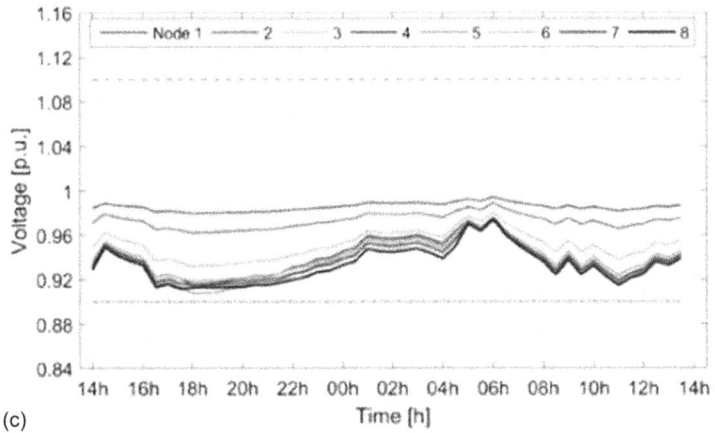

FIGURE 11.7   (a) Case Study 2 (with charging strategy) Single tariff: Power Profile, (b) Case Study 2 (with charging strategy): single Tariff-SOC Profile, (c) Case Study 2 (with charging strategy): single tariff voltage profile.

11.5.2.2   With Charging Strategy: Two Tariff Scenario

Let's now look at two rates: $1.5/kWh from 18:00 to 22:00, and $1/kWh from 22:00 to 06:00. Consumption is focused during the period of high prices in the suboptimal scenario. As a result, it costs $53.6 to charge PHEVs without any management.

The power consumption profiles shown in Figure 11.8(a) are the outcome of applying the optimization technique to the two tariff scenarios. Throughout the entire charging process, the power consumption is also distributed differently. To offset the cost of charging their batteries, PHEVs are currently attempting to inject as much of their initial SOC during expensive hours as possible. The three PHEVs are completely depleted at the conclusion of the high-price hours, despite the fact that at the beginning of the charging period, the PHEV 1 and PHEV 2 charge their batteries. They inject their initial energy and fully recharge their batteries in low-priced hours. Even at times of peak demand, the voltage profiles corresponding to each node are maintained within the required bounds, as shown in the curve.

The ideal price for charging PHEVs is currently $32.4. The profit made by selling the initial charge stored in the batteries during the peak hours when the price of electricity is high (16 kWh × 0.65 × $1.50/kWh = $15.60) is subtracted from the cost of recharging the batteries during the low-price hours (3 × 16 kWh × $1/kWh = $48.00).

11.5.3   CASE STUDY 3

In Case Study 3, variations in different parameters are observed when PHEV is connected to the grid and optimization algorithm is applied to the system. This case study differs from Case Study 2, as here the voltage limit will violate even when no PHEV is connected to the grid.

The Case Study 2 is examined while assuming once more a distinct tariff. In this instance, even in the absence of PHEVs, the voltage curve doesn't adhere to security restrictions during peak hours. The voltage support limitation is hard to comply to for PHEVs without a set initial energy level stored in their batteries, especially when peak electricity demand coincides with the start of the charging period. A workable solution is impossible in this situation. The initial energy needed is inversely proportional to how much voltages deviate from the acceptable limits.

When there is an initial excess of energy in the batteries, PHEVs can efficiently maintain the grid's voltage by supplying this excess energy, making up for the energy used by the household load. As shown in Figure 11.5, nodes 4–8 exhibit voltage levels below the set safety limit when PHEVs are absent. Figures 11.9(a) and (b) show the power consumption and state of charge (SOC) curves, respectively, as a direct result of using the optimization approach. The PHEV with the greatest beginning charge (placed in Node 8) aggressively feeds electricity to the grid during peak hours until its battery is almost completely discharged. The batteries of the other PHEVs are being charged and discharged concurrently. The other PHEVs' batteries are also simultaneously being charged and discharged. The PHEVs start using their stored energy after 22:00 when the demand starts to decline and continue doing so until they are fully recharged.

It is important to emphasize that the initial grid voltage issue has been resolved, and each node is now getting the support it needs, as seen in Figure 11.9(c). The total energy needed to completely charge the batteries remains at 37.6 kWh despite this improvement, hence the price stays the same at $37.6 as it was in Scenario 1.

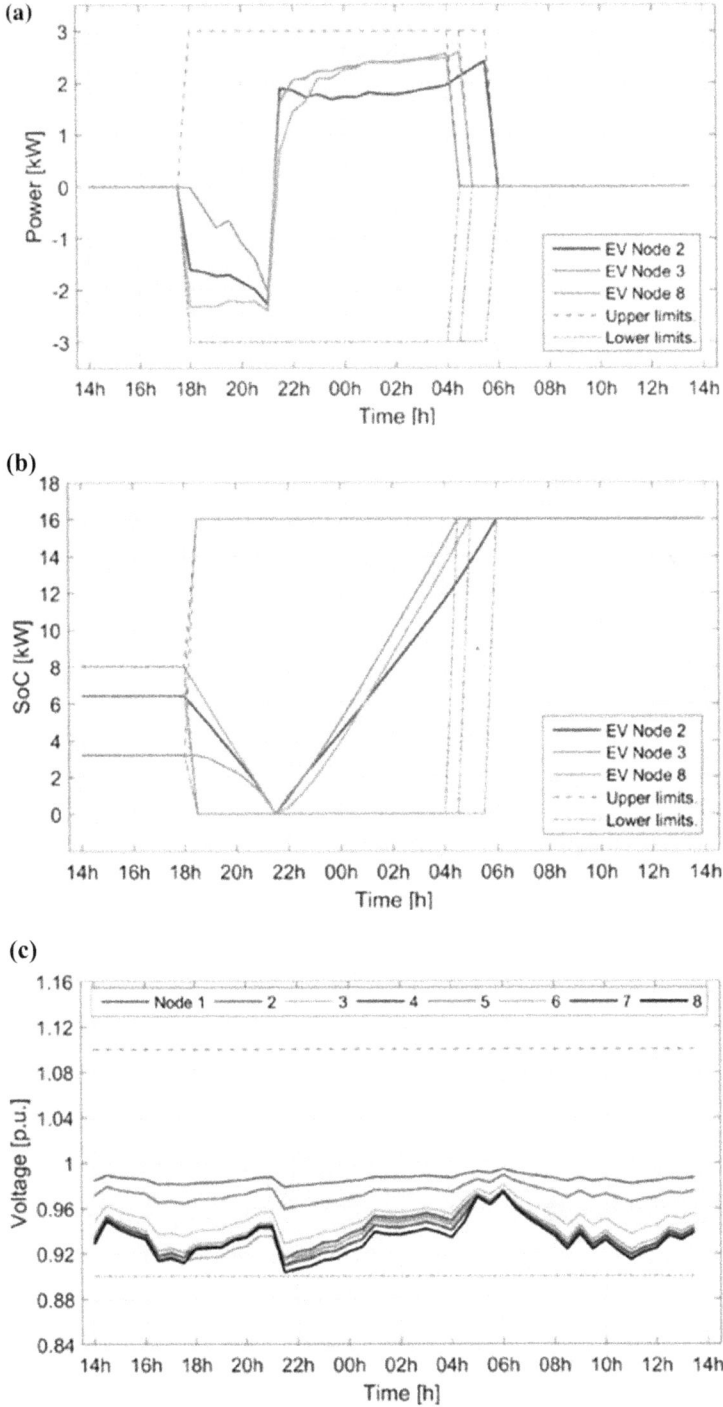

FIGURE 11.8 (a) Case Study 2 with charging management and two tariff power profile, (b) Case Study 2 with charging management and two tariff state of charge profile, (c) Case Study 2 with charging management and two tariff voltage profile.

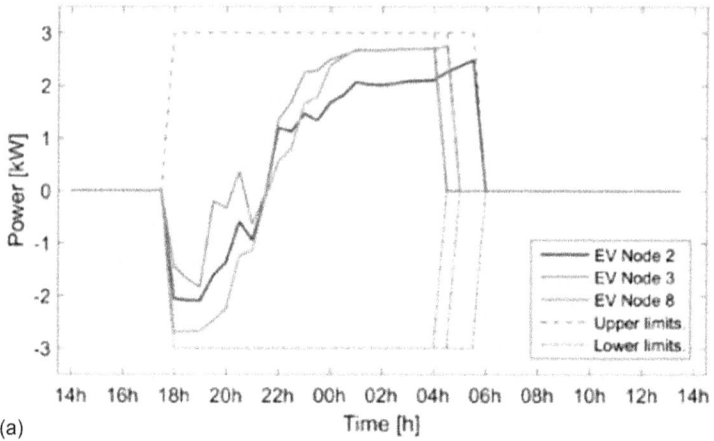

(a)

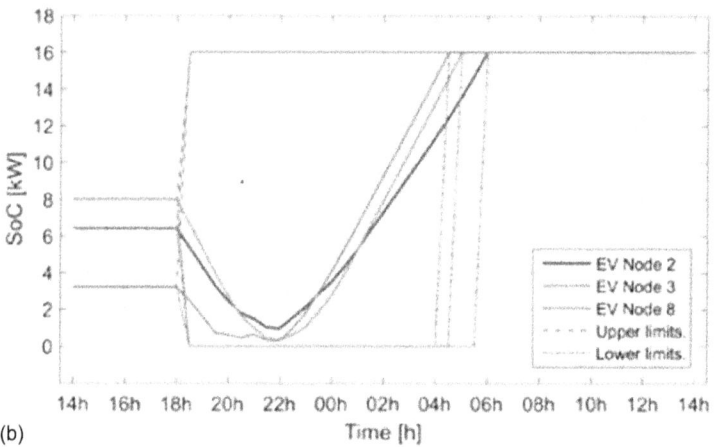

(b)

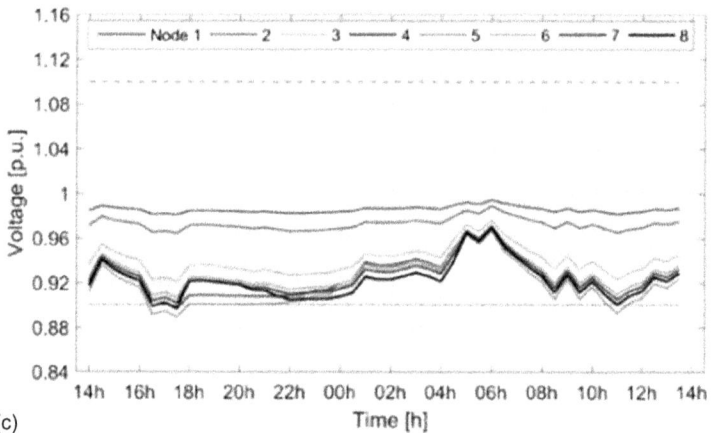

(c)

FIGURE 11.9 (a) Case Study 3 with charging management: single tariff power profile, (b) Case Study 3 with Charging management: Single tariff SOC profile, (c) Case Study 3 with Charging management: Single tariff voltage profile.

TABLE 11.2
Cost Obtained in Different Case Studies

Cases	Type of Tariff	Cost (in Euro) Per Day
Without Scheduling	Single tariff	€35.8
Without Scheduling	Double tariff	€45.6
With Scheduling	Single tariff	€30.4
With Scheduling	Double tariff	€21.6

11.6 CONCLUSION

In this algorithm, a method for determining the voltage levels in a residential power grid is presented. It takes into account both the immediate power use within the home and the charging or discharging of PHEVs. An approach for controlling the charging of several PHEVs connected to the residential grid is introduced in the chapter, subject to certain assumptions and constraints. The suggested method also makes advantage of PHEVs' energy storage capabilities to stabilization the domestic grid's voltage. In order to determine whether the suggested strategy can maintain voltages within the acceptable range and offer the best power consumption/injection strategies, it is tested and compared for various cases. The outcomes are then carefully examined.

The various results are shown in Table 11.2 representing the cost across various test conditions. It can be observed that the charging cost is minimum if an optimization algorithm is applied and also the profit can be achieved by selling the electricity during peak hours. The consideration of a distinction between energy prices for sale and purchasing prices, the examination of prices or loading profiles under stochastic situations, and many other tests, can be taken into consideration in the future.

REFERENCES

[1] Clement-Nyns, K., Haesen, E., & Driesen, J. (2011). The impact of vehicle-to-grid on the distribution grid. *Electric Power Systems Research*, 81(1), 185–192.
[2] Clement-Nyns, K., Haesen, E., & Driesen, J. (2009). The impact of charging plug-in hybrid electric vehicles on a residential distribution grid. *IEEE Transactions on Power Systems*, 25(1), 371–380.
[3] Turker, H., Bacha, S., & Hably, A. (2014). Rule-based charging of plug-in electric vehicles (PEVs): Impacts on the aging rate of low-voltage transformers. *IEEE Transactions on Power Delivery*, 29(3), 1012–1019.
[4] Couillet, R., Perlaza, S. M., Tembine, H., & Debbah, M. (2012). Electrical vehicles in the smart grid: A mean field game analysis. *IEEE Journal on Selected Areas in Communications*, 30(6), 1086–1096.
[5] Nguyen, H. K., & Song, J. B. (2012). Optimal charging and discharging for multiple PHEVs with demand side management in vehicle-to-building. *Journal of Communications and Networks*, 14(6), 662–671.
[6] Kim, B. G., Ren, S., Van Der Schaar, M., & Lee, J. W. (2013). Bidirectional energy trading and residential load scheduling with electric vehicles in the smart grid. *IEEE Journal on Selected Areas in Communications*, 31(7), 1219–1234.

[7] Turker, H., Bacha, S., Chatroux, D., & Hably, A. (2012). Low-voltage transformer loss-of-life assessments for a high penetration of plug-in hybrid electric vehicles (PHEVs). *IEEE Transactions on Power Delivery*, 27(3), 1323–1331.

[8] Turker, H., Bacha, S., & Chatroux, D. (2010, October). Impact of plug-in hybrid electric vehicles (phevs) on the French electric grid. In *2010 IEEE PES Innovative Smart Grid Technologies Conference Europe (ISGT Europe)* (pp. 1–8). IEEE.

[9] Nour, M., Said, S. M., Ali, A., & Farkas, C. (2019, February). Smart charging of electric vehicles according to electricity price. In *2019 International Conference on Innovative Trends in Computer Engineering (ITCE)* (pp. 432–437). IEEE.

[10] Liu, L., Zhou, K., Lu, X., Yu, J., & Yang, S. (2021, October). Multi-objective charging scheduling model for electric vehicles considering urgent demand under different charging modes. In *2021 IEEE 5th Conference on Energy Internet and Energy System Integration (EI2)* (pp. 3936–3940). IEEE.

[11] Visakh, A., & Selvan, M. P. (2021, December). Charging-cost minimization of electric vehicles and its impact on the distribution network. In *2021 9th IEEE International Conference on Power Systems (ICPS)* (pp. 1–5). IEEE.

[12] Su, X., & Yue, H. (2018, September). Cost minimization control for smart electric vehicle car parks. In *2018 24th International Conference on Automation and Computing (ICAC)* (pp. 1–6). IEEE.

[13] He, Y., Venkatesh, B., & Guan, L. (2012). Optimal scheduling for charging and discharging of electric vehicles. *IEEE Transactions on Smart Grid*, 3(3), 1095–1105.

[14] Ghosh, A., & Aggarwal, V. (2018). Menu-based pricing for charging of electric vehicles with vehicle-to-grid service. *IEEE Transactions on Vehicular Technology*, 67(11), 10268–10280.

[15] Said, D., & Mouftah, H. T. (2019). A novel electric vehicles charging/discharging management protocol based on queuing model. *IEEE Transactions on Intelligent Vehicles*, 5(1), 100–111.

[16] Amamra, S. A., & Marco, J. (2019). Vehicle-to-grid aggregator to support power grid and reduce electric vehicle charging cost. *IEEE Access*, 7, 178528–178538.

[17] Turker, H., & Bacha, S. (2018). Optimal minimization of plug-in electric vehicle charging cost with vehicle-to-home and vehicle-to-grid concepts. *IEEE Transactions on Vehicular Technology*, 67(11), 10281–10292.

12 Consideration of Electric Vehicles Charging in Protection Coordination Problem for Microgrid

Parvez Ahmad and Priyanshul Niranjan
MNNIT Allahabad, Prayagraj, India

Vivek Patel
SMNRU Lucknow, Lucknow, India

Niraj Kumar Choudhary and Nitin Singh
MNNIT Allahabad, Prayagraj, India

12.1 INTRODUCTION

Electric vehicle (EV) usage has risen quickly in recent years, transforming the transportation industry and providing an attractive solution for sustainable mobility. According to a Bloomberg NEF analysis, worldwide electric vehicle sales will hit a new high of 3.24 million units in 2020, reflecting a 43% increase over the previous year [1]. Several factors have contributed to the growing popularity of EVs. These include technology advancements, government laws and incentives, environmental concerns, and altering consumer tastes. Incorporating electric vehicles into the power grid creates new issues for power system protection. As the number of EVs on the road grows, their charging habits and power usage may significantly influence the grid. The charging of electric vehicles increases the total power consumption on the grid. Concurrent charging of many EVs, especially during peak hours, might result in significant load variations. This additional demand may strain the grid infrastructure and result in voltage fluctuations or overloads. To preserve the dependability and stability of the power system, accurate protection coordination mechanisms must be implemented [2]. This entails selecting and organizing protective devices such as circuit breakers, fuses, and relays to ensure fault detection, isolation, and restoration occur on time. To ensure system stability and defend against failures, the specific characteristics of EV loads, such as their high levels of power consumption and possible harmonic emissions, must be addressed [3]. Electric vehicle charging infrastructure should be built to reduce the effect of failures while maintaining grid resilience. Fault management solutions must be used to identify and isolate problems as early as possible, minimizing their influence on the power supply. Rapid fault identification and isolation ensure continuous power distribution to EVs while minimizing grid disruptions [4]. Coordination of the grid's protective devices is critical for fault detection, isolation, and restoration. Because of EVs' varied features and load profiles, the coordination of protective devices becomes increasingly challenging with the installation of EV charging stations. Advanced protective coordination mechanisms must be implemented to account for the diversity in EV charging patterns and assure the power system's reliability [5]. Developing enhanced protection coordination mechanisms and implementing relevant measures will aid in the safe and reliable integration of EVs into the power grid.

Microgrids are autonomous, regional energy networks that can run separately or in cooperation with the main utility grid. Based on their operational modes and connection configurations, they are divided into three primary categories. The first category is "Islanded Microgrids," which may operate independently and cut off from the main grid during power outages to maintain a steady flow of electricity to vital facilities. The second category is "Grid-connected Microgrids," which are still linked to the main grid but have some degree of autonomy. They use energy storage and renewable energy sources to lessen their dependency on the main grid. "Remote Microgrids" provide electricity to places that are physically remote or have no connection to a centralized grid by solely using local generating sources, frequently renewables, and energy storage. These divisions show how flexible and adaptable microgrids are in handling various energy difficulties and requirements.

The process by which EVs charge substantially impacts power consumption in microgrids. Reference [6] investigates how EV charging patterns affect microgrid operations and emphasizes the need for enhanced coordination of protection systems to handle the increased power consumption during peak charging hours. Another study [7] examines the influence of varied EV charging rates on protection coordination using a case study. The research emphasizes the need to modify protection settings to match varying load patterns dynamically.

The inclusion of EVs in microgrids affects fault current levels during system outages. As a result, academics have investigated this influence and the difficulties it provides for protection cooperation. A literature review was undertaken in one research [8] to investigate fluctuations in fault currents in a microgrid with significant EV penetration. They devised a method for limiting fault currents and reducing their influence on protection coordination. Another research [9] looked at the influence of bidirectional power flow from EVs on fault current levels and presented an adaptive protection coordination technique to ensure reliable operation.

Several writers have suggested various protection coordinating strategies to meet the issues caused by the integration of EVs in microgrids. One technique, for example, [10], proposes an intelligent protection coordination strategy based on multi-agent systems. The purpose of EV-integrated microgrids is to provide the greatest possible coordination of protective devices. Another idea [11] is to use smart relays with communication capabilities to enhance fault detection and cooperation in microgrids.

The influence of communication delays on protection coordination is the major subject of [12], which investigates how coordination tactics may be less successful when communication delays occur. The authors of [13] investigate how communication and distributed control algorithms might be used to improve fault coordination and detection in EV-integrated microgrids.

The unique influence of EV charging on microgrid voltage stability is addressed in [14], which considers EV charging patterns and characteristics. To maintain system stability, the paper proposes a synchronized charging mechanism. Furthermore, [15] studies the influence of EV charging patterns on microgrid resilience, stressing the need for adaptive protection coordination to provide uninterrupted power delivery during grid disruptions.

To achieve optimal protection coordination in the presence of EVs and their dynamic interactions with power grid components and protection devices, it is important to employ effective coordination techniques. Specifically, the settings of relays can be optimized to accommodate the unique characteristics of EV loads and charging patterns. Among the various protective devices, directional overcurrent relays (DOCRs) are widely favored for their capability to detect faulty sections. In this study, the proposed scheme focuses on user-defined directional sensitive DOCRs (DS-DOCRs) to enable coordination in the grid-connected mode (GCM) of a microgrid. The performance of the proposed scheme is evaluated using the 7-bus microgrid low voltage section of the IEEE-14 bus benchmark system. The core contributions of this research can be summarized as follows:

- The study presents a coordination scheme utilizing user-defined DS-DOCRs in the microgrid's GCM.
- The objective is to enhance the microgrid's performance and reliability by achieving effective relay coordination, considering different penetration levels and charging scenarios (G2V & V2G) of EV.
- The research aims to identify optimal settings for the user-defined DS-DOCRs that remain effective in the microgrid's GCM, regardless of varying levels of EV penetration and charging scenarios.
- The analysis and optimization of the microgrid's performance are based on the normal inverse characteristic of the relays used.

The remaining sections of the paper are organized as follows. Section 12.2 provides a detailed explanation of the problem formulation. Section 12.3 offers a concise discussion of the system under consideration. Section 12.4 presents the results obtained from the study and provides a thorough discussion of these findings. Finally, Section 12.5 concludes the study, summarizing the main points and highlighting any significant implications or future research directions.

12.2 PROBLEM FORMULATION

The methodology presented in this study introduces user-defined Distribution System-Directional Overcurrent Relays (DS-DOCRs) and evaluates their performance using Grid-Connected Mode (GCM). The main aim of this protection coordination scheme is to minimize the relay operating time while ensuring that the constraints related to fault currents in both directions are met. The objective function, referred to as OBJT, is defined as the summation of operating times for primary relays and their associated backup relays for various fault locations, as indicated by Eq. (12.1).

$$\text{OBJ}^T = \min \sum_{t=1}^{24} \sum_{i=1}^{N} \sum_{j=1}^{M} \left(t^p_{fw\_i,j,t} + \sum_{k=1}^{K} t^b_{rv\_k,j,t} \right) \tag{12.1}$$

During a fault in a microgrid, the voltage drop can impact the performance of various relays located at different positions within the microgrid. To address this issue, a time-voltage-current-based relay characteristic is utilized to coordinate the relays based on the time delay required for the fault to reach specific voltage and current levels. This strategy of coordination guarantees that relays nearest to the problem site react more rapidly, while relays farther away coordinate their operation to provide selective fault detection and isolation. [16] proposes a dual-setting Directional Overcurrent Relay (DOCR) with time-current-voltage-based features. Compared to standard time-current characteristics, this technique tries to shorten the operation time of relays. The operating time characteristics of the dual-setting relays, based on their time-current-voltage characteristics, are described by Eqs. (12.2) and (12.3). Moreover, Eqs. (12.4)–(12.10) outline the constraints associated with the relay coordination problem.

$$t^p_{fw\_i,j,t} = A \frac{\text{TMS}_{fw,i}}{\left(\dfrac{I_{sc,i,j,t}}{\text{CT}_{\text{Ratio},i} \times \text{PS}_{fw,i}} \right)^B - 1} * \left(\frac{1}{e^{1-v_{f,i,j,t}}} \right)^{\alpha_{fw,i}} \tag{12.2}$$

$$t^b_{rv\_i,j,t} = A \frac{\text{TMS}_{rv,k}}{\left(\dfrac{I_{sc,k,j,t}}{\text{CT}_{\text{Ratio},k} \times \text{PS}_{rv,k}} \right)^B - 1} * \left(\frac{1}{e^{1-v_{f,k,j,t}}} \right)^{\alpha_{rv,k}} \tag{12.3}$$

$$t^b_{rv\_k,j,t} - t^p_{fw\_i,j,t} \geq \text{CTI} \tag{12.4}$$

$$\text{TMS}_{\min,i} \leq \text{TMS}_{fw,i}, \text{TMS}_{rv,k} \leq \text{TMS}_{\max,k} \tag{12.5}$$

$$\text{PS}_{\min,i} \leq \text{PS}_{fw,i}, \text{PS}_{rv,k} \leq \text{PS}_{\max,k} \tag{12.6}$$

$$t^p_{\min,i} \leq t^p_{fw,i} \leq t^p_{\max,i} \tag{12.7}$$

$$t^b_{\min,k} \leq t^b_{rv,k} \leq t^b_{\max,k} \tag{12.8}$$

$$0 \leq \alpha_{fw,i} \leq 5 \tag{12.9}$$

$$0 \leq \alpha_{rv,k} \leq 5 \tag{12.10}$$

In the given equations, $t^p_{fw\_i,j,t}$ and $t^b_{rv\_k,j,t}$ represent the operating time of the primary and corresponding backup relay, respectively, for fault currents in the forward and reverse directions. $v_{f,k,j,t}$ denotes the measured fault voltage, which can vary for different relays. $\alpha_{fw,i}$ and $\alpha_{rv,k}$ are decision variables representing the fault voltages in the forward and reverse directions, respectively. $I_{sc,k,j,t}$ represents the short-circuit current flowing through the relay coil, while $CT_{Ratio,\,k}$ denotes the current transformer ratio. The decision-making variables of this problem are represented by $TMS_{fw,i}$, $TMS_{rv,k}$, $PS_{fw,i}$, and $PS_{rv,k}$. The constants A and B correspond to time and current, respectively, and have values of 0.14 and 0.02 based on the IEC-60255 standard. Constraint (4) ensures that the time distance between the operation of the primary and backup relays satisfies the CTI (Current Transformer Instantaneous) limit. Typically, a time distance of 0.2 seconds is assumed [17]. Inequalities (5) and (6) represent the limits of TMS (Time Multiplier Setting) and PS (Pickup Setting) in both the forward and reverse directions. Eq. (12.7) defines the operating time range for the primary relay, while Eq. (12.8) defines the operating time range for the corresponding backup relay. The specific range for these constraints can be found in Table 12.1 [18–19]. Note that i represents the index of the primary relay. It ranges from 1 to N, where N is the total number of primary relays in the system. Index k represents the corresponding backup relay associated with the primary relay. It ranges from 1 to K, where K is the total number of corresponding backup relays in the system. Index j represents the fault location. It ranges from 1 to M, where M is the total number of fault locations being considered.

The preceding discussion's equations have been extended to incorporate an extra variable. The power flow approach can capture the dynamic character of the microgrid and offer correct estimations of voltage and current at various times of day by including time instants and accounting for varying levels of EV penetration and charging situations (G2V & V2G). The power flow technique must identify the variables before a problem occurs, considering the varied % penetration of EVs at various times of the day. This method provides a wide study of the system and efficient relay coordination while considering the individual circumstances at distinct time instants. As a result, the variable "t" in Eqs. (12.1)–(12.8) represents various hours of the day, spanning from 01:00 to 24:00.

This chapter uses the Genetic Algorithm (GA) to obtain the optimal settings and total operating time (OBJT) of relays. The relay coordination problem's efficacy relies on the GA's input parameters. These parameters include the population size (set at 50), β (set at 0.20), γ (set at 1), δ (set at 0.25), and maximum iterations (limited to 200) [20].

TABLE 12.1

Limits of the Constraints

Parameter	Minimum Value	Maximum Value
TMS	0.1 s	1.1 s
PS	0.5	2
Operating time of relay	0.1 s	4 s
α	0	5

12.3 TEST SYSTEM

The chapter utilizes a microgrid system based on the 7-bus configuration, representing the low voltage (LV) section of the modified IEEE-14 bus system as shown in Figure 12.1. More detailed information about the system can be found in the reference [21]. To ensure comprehensive protection of the 7-bus microgrid, a total of 17 DOCRs denoted as R1 to R17 are employed. These relays are strategically positioned within the microgrid system to facilitate effective coordination and protection. The microgrid system comprises two Inverter-Based Distributed Generators (IBDGs) and a Synchronous-Based Distributed Generator (SBDG), connected at buses B2, B7, and B1.

Furthermore, there is an EV charging station located on bus B5. The charging pattern and penetration of the EV vary throughout the day, and Table 12.2 provides information regarding the penetration levels of EVs and their charging modes at different time intervals. This consideration considers the changing levels of EV penetration and their influence on the microgrid system.

A 3-phase mid-point fault is considered at 8 distinct locations labeled as L1 to L8 to evaluate the system's response during fault conditions. These fault locations represent specific points within the microgrid system where faults may occur. In the analysis, a total of 24 primary-backup relay pairs are included in the GCM to ensure

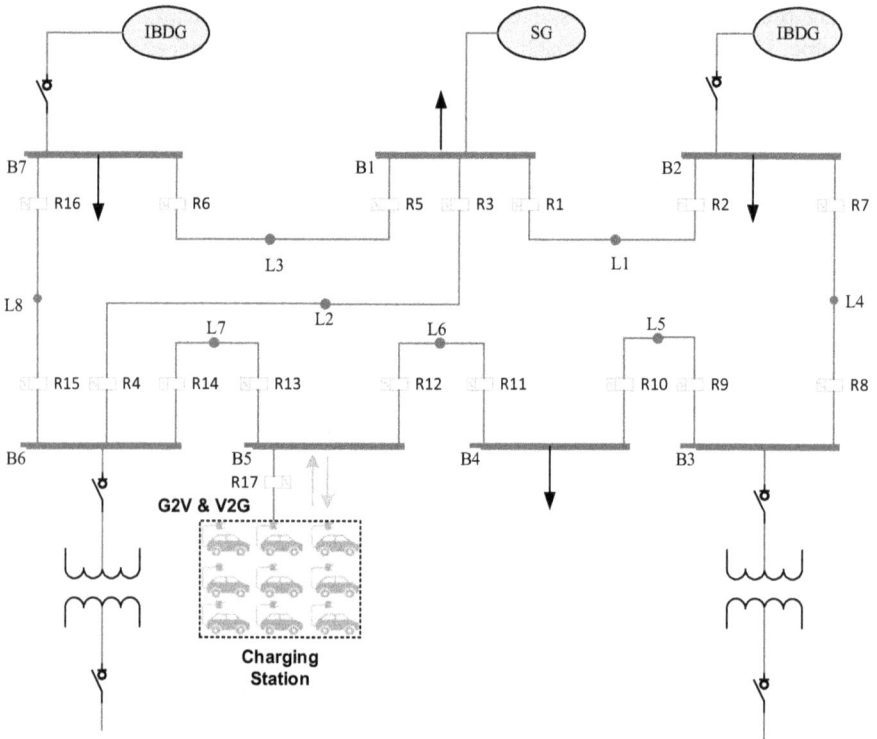

FIGURE 12.1 The 7-bus microgrid system (11 kV part of the IEEE-14 bus system).

TABLE 12.2

The Average Penetration Level of EVs for Different Durations

Serial Number	Number of EVs (%)	G2V (%)	V2G (%)	Time
1	70	30	70	00:00–08:00 (8 hours)
2	80	40	60	08:00–16:00 (8 hours)
3	95	80	20	16:00–00:00 (8 hours)

TABLE 12.3

Primary-backup Relay Pairs (dual-setting relays) in GCM

Fault Location	Serial No.	Primary Relay	Backup Relay
L1	1	1	3
	2	1	5
	3	2	7
L2	4	3	1
	5	3	5
	6	4	14
	7	4	15
L3	8	5	1
	9	5	3
	10	6	16
L4	11	7	2
	12	8	9
L5	13	9	8
	14	10	11
L6	15	11	10
	16	12	13
	17	12	17
L7	18	13	12
	19	13	17
	20	14	4
	21	14	15
L8	22	15	4
	23	15	14
	24	16	6

effective coordination and protection of the system during fault events. Table 12.3 provides comprehensive information about the relay pairs for the GCM at each fault location. The table specifies the primary relays and their corresponding backup relays that have been identified for each fault location in the microgrid system. For instance, in the case of a fault occurring at location L1, the identified primary relays are R1 and R2. As for the backup relays, R3 and R5 are associated with R1, while R7 is the backup relay for R2.

TABLE 12.4

CT_{Ratio} of Each User-Defined DS-DOCRs for the 7-Bus Test System

Relay	CTR for Primary Relay-Backup Relay
1	2000/5–1000/5
2	1000/5–2000/5
3	3000/5–2000/5
4	2000/5–3000/5
5	1600/5–1000/5
6	1000/5–1600/5
7	2500/5–1600/5
8	1600/5–2500/5
9	2500/5–1200/5
10	1200/5–2500/5
11	1200/5–2500/5
12	2500/5–1200/5
13	800/5–3000/5
14	3000/5–800/5
15	1600/5–1600/5
16	1600/5–1600/5
17	800/5–3000/5

The calculation of the $CT_{Ratio, i}$ for the DOCRs (Directional Overcurrent Relays) is determined using Eq. (12.11) as referenced in [19]. The $I_{f max, i}$ represents the maximum short-circuit current, while $I_{L max, i}$ represents the maximum load current for the specific DOCR being considered. The $CT_{Ratio, i}$ values for the DOCRs can be found in Table 12.4.

$$CT_{Ratio,i} = \text{Maximum}\left(I_{L max,i}, \frac{I_{f max,i}}{20} \right) \tag{12.11}$$

12.4   RESULT AND DISCUSSION

The relay coordination study was conducted in two scenarios, outlined in the subsequent sections. In the first scenario, the study focuses on identifying the optimal settings for user-defined DS-DOCRs within a microgrid's GCM, considering different percentages of EV penetration at various time intervals. On the other hand, in the second scenario, the study aims to determine common optimal settings for user-defined DS-DOCRs that remain valid in the GCM as EV penetration levels vary over time. The analysis and optimization of the microgrid's performance utilize the normal inverse characteristic of relays. To maximize relay performance and coordination in each specific EV penetration scenario, a genetic algorithm (GA) is employed as an optimization technique to identify the ideal relay settings.

12.4.1 OPTIMAL SETTING OF RELAYS IN GCM OF MICROGRID AT DIFFERENT PENETRATION LEVELS OF EVS AT DIFFERENT TIME INSTANCES

The study examined three distinct levels of electric vehicle (EV) penetration: 70% (30% G2V & 70% V2G), 80% (40% G2V & 60% V2G) and 95% (80% G2V & 20% V2G) of the total EV count for three different hours of the day. These penetration levels were selected to investigate and analyze the impact of varying EV penetration on the coordination of user-defined DS-DOCRs. The study assessed the influence of EV penetration on relay coordination and protection at three different time instances, considering the temporal variations and diverse EV charging patterns throughout the day. Optimal relay settings for the GCM were determined for each of the three cases: 70% (30% G2V & 70% V2G), 80% (40% G2V & 60% V2G) and 95% (80% G2V & 20% V2G) EV penetration, as presented in Tables 12.5–12.8. These tables also indicate the total operating time of the relays, considering the EV penetration levels and the duration of time at the EV charging station. Figure 12.2 provides a visual representation of the operating time for primary and backup relays in the GCM under different levels of EV penetration. Additionally, it depicts the calculated CTI (Current Transformer Index) between relays and compares it with the predefined CTI. The findings indicate that the predefined CTI consistently remains lower than the calculated CTI.

The operational duration of relays increases as the EV penetration level rises. This indicates that higher EV penetration has a greater impact on load demand and short-circuit currents, consequently influencing relay operations. Consequently, it is crucial

---

TABLE 12.5

Optimal Relay Settings in GCM at 70% (30% G2V & 70% V2G) Penetration Level of EVs

Relay	TMS_{FW} (s)	PS_{FW} (A)	α_{FW}	TMS_{RV} (s)	PS_{RV} (A)	α_{RV}
1	1.1000	0.5000	3.7500	0.8500	0.5000	2.0000
2	0.2092	0.5405	1.5684	0.5295	0.9321	2.0721
3	0.1832	1.0000	1.7931	0.7503	0.8453	2.6469
4	0.1000	0.5000	0.6761	0.3851	0.5029	1.1790
5	0.1969	0.5000	1.5045	0.5488	1.7500	2.9077
6	0.5037	0.8790	2.5840	0.4462	1.3750	2.0137
7	0.1000	0.5000	0.7979	0.9522	0.8972	2.6250
8	0.3526	0.9390	2.4321	0.3850	0.7500	1.7188
9	0.2178	0.8743	1.7128	0.9125	0.5000	2.3887
10	0.1007	0.5000	0.6640	1.0559	0.5000	2.2646
11	0.2369	1.9205	2.0938	0.5989	0.5000	1.5306
12	0.1000	1.3558	1.2274	0.4587	0.5000	1.2996
13	0.1000	0.5000	0.1938	0.3781	0.7500	1.5938
14	0.5986	0.5055	2.8656	0.9859	1.9791	2.5002
15	0.2978	0.5000	1.8430	0.5973	0.5000	2.1828
16	0.1566	0.5000	1.1831	0.3500	0.9011	1.3750
17	0.1000	0.5000	0.1938	0.3781	0.7500	1.5938

| OBJ^T | | | 10.9512 s | | | |

TABLE 12.6
Optimal Relay Settings in GCM at 80% (40% G2V & 60% V2G) penetration level of EVs

Relay	TMS$_{FW}$ (s)	PS$_{FW}$ (A)	α_{FW}	TMS$_{RV}$ (s)	PS$_{RV}$ (A)	α_{RV}
1	0.3328	0.6054	2.4249	0.4943	0.9693	1.6777
2	0.1000	0.5000	0.5000	0.1000	1.1022	0.5000
3	0.1000	0.5000	0.8639	0.7757	0.8862	2.6896
4	0.2789	0.7046	1.8724	0.5320	0.5869	1.5593
5	0.3036	0.5000	2.0423	0.8500	1.0000	2.8555
6	0.1000	1.1296	0.9728	0.1000	0.8976	0.0625
7	0.1000	0.7492	1.2722	1.0610	0.8760	2.7330
8	0.1000	0.5362	0.8895	0.8500	0.9599	2.6157
9	0.6263	0.6508	2.9697	0.1000	1.5000	0.2787
10	0.3202	0.5000	1.8772	0.6357	0.7652	1.7694
11	0.2684	0.5000	1.5852	0.9617	0.5883	2.2274
12	0.1385	0.5025	1.2300	0.4454	0.5716	1.4375
13	0.1423	0.5020	0.9018	0.1000	0.7627	0.2242
14	0.1000	1.5000	1.3594	0.8500	1.1894	2.0000
15	0.8806	0.5000	3.3190	0.4054	0.6010	1.9451
16	0.1042	0.7325	1.0156	0.4441	0.5000	1.4375
17	0.1423	0.5020	0.9018	0.1000	0.7627	0.2242
OBJ^T			11.2451 s			

TABLE 12.7
Optimal Relay Settings in GCM at 95% (80% G2V & 20% V2G) Penetration Level of EVs

Relay	TMS$_{FW}$ (s)	PS$_{FW}$ (A)	α_{FW}	TMS$_{RV}$ (s)	PS$_{RV}$ (A)	α_{RV}
1	0.1000	0.6291	0.9964	1.0999	0.5000	2.2145
2	1.1000	0.5118	3.3594	0.1625	0.5000	0.3750
3	0.1000	0.5471	0.6902	1.1000	0.5000	2.7506
4	0.9750	0.5000	2.4336	0.1555	0.5038	0.0164
5	0.5373	0.5001	2.6875	0.4705	1.5000	2.5179
6	0.1000	0.5028	0.5625	0.1229	0.5027	0.0625
7	0.1000	0.5863	1.0156	0.6000	0.6250	1.8125
8	0.3218	0.5000	1.9057	0.6964	1.3080	2.6687
9	0.7658	0.5005	2.8731	0.8095	0.8801	2.4531
10	0.1872	0.5000	1.0781	0.5757	0.6734	1.6953
11	0.1000	0.5897	0.6210	1.1000	1.1637	2.4754
12	0.2645	0.5000	1.8735	0.1195	0.9454	0.1250
13	0.1000	0.5000	0.3134	1.0972	0.5005	2.6692
14	0.1000	0.5000	0.6274	0.9181	0.9119	1.7341
15	0.3606	0.5026	2.2382	0.1000	0.7146	0.3748
16	0.1000	0.5000	0.7920	0.1000	1.4417	0.2500
17	0.1000	0.5000	0.3134	1.0972	0.5005	2.6692
OBJ^T			11.8002 s			

TABLE 12.8

Coordination Constraint Violations in Microgrid at a Different EV Penetration Level

		Number of Coordination Constraint Violations		
Microgrid Mode	Algorithm	70% (30% G2V & 70% V2G)	80% (40% G2V & 60% V2G)	95% (80% G2V & 20% V2G)
GCM	GA	0	4	4
		8	0	1
		8	5	0

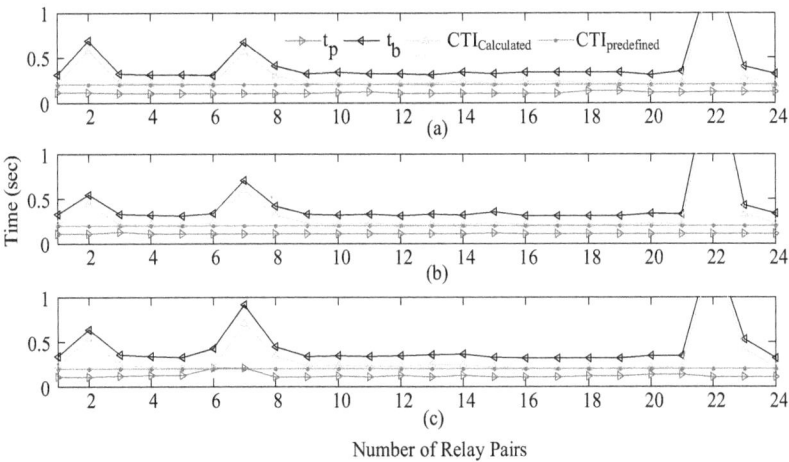

FIGURE 12.2 Primary-backup relay operating time in GCM at EV penetration of (a) 70% (30% G2V & 70% V2G), (b) 80% (40% G2V & 60% V2G), and (c) 95% (80% G2V & 20% V2G).

to consider the evolving EV penetration and its associated effects when coordinating relays within the microgrid. The study also noted that relay coordination, optimized for a specific EV penetration level, may not be suitable or effective for other penetration levels. Applying the optimal relay settings from one penetration level to another leads to violations of coordination constraints, as indicated in Table 12.8. Achieving a common set of relay settings may require further analysis to ensure satisfactory relay coordination within the GCM under different EV penetration scenarios.

12.4.2 COMMON OPTIMAL SETTING OF RELAYS IN GCM FOR VARIABLE PENETRATION LEVEL OF EVs

The study has identified a set of optimal relay settings for the GCM, considering the varying number of EVs and their charging scenario at different times of the day, as presented in Table 12.2. These optimal settings, outlined in Table 12.9, are designed

TABLE 12.9

Common Optimal Settings of Relays in GCM at Variable EV Penetration

Relay	TMS_{FW} (s)	PS_{FW} (A)	α_{FW}	TMS_{RV} (s)	PS_{RV} (A)	α_{RV}
1	0.1045	0.7635	1.1369	0.2912	0.5000	0.7813
2	0.1000	0.5387	0.7327	0.8952	0.8689	2.6553
3	0.1000	1.6797	1.3125	0.4244	0.5000	1.5831
4	0.1000	0.7730	0.7751	0.7455	0.5031	1.9501
5	0.1000	0.5000	0.4431	0.2876	0.6250	1.3125
6	0.1000	0.5000	0.6250	0.2344	1.0031	1.0565
7	0.4429	0.5000	2.6395	0.1230	0.5754	0.0625
8	1.1000	0.7385	3.6751	0.1382	0.6339	0.3103
9	0.1000	0.5317	0.6124	0.5891	1.1657	2.0491
10	0.1000	0.9369	0.8281	0.7832	1.5000	2.3480
11	0.6989	0.5313	2.5332	0.1549	0.5041	0.1719
12	0.1039	0.5003	0.9036	1.1000	0.5000	2.3485
13	0.2870	0.5000	1.6653	0.3275	0.5000	1.2500
14	0.2127	0.5000	1.7665	0.9165	0.5627	1.9219
15	0.4159	0.5000	2.1648	0.1000	0.5000	0.2370
16	0.2382	0.6373	1.6608	0.5454	0.5562	1.5625
17	0.2870	0.5000	1.6653	0.3275	0.5000	1.2500
OBJ^T			260.3051 s			

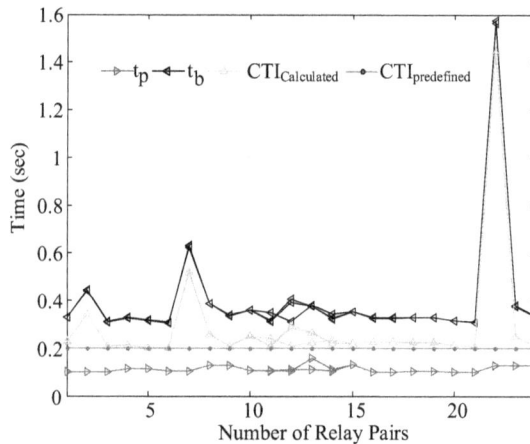

FIGURE 12.3 Operating time and LCT of relay pairs with common settings (at variable EV penetration) in GCM.

to ensure efficient relay coordination across different levels of EV penetration. The table also displays the total operating time of the relays, considering the fluctuating number of EVs and the 24-hour duration at the EV charging station. Figure 12.3 shows the operating time of primary and backup relays in the GCM, and it is noted that the predefined CTI consistently remains lower than the calculated CTI.

TABLE 12.10

Comparative Analysis of Total Operating Time (OBJ^T) of Relays in Microgrid at Different EV Penetration

Penetration Percentage of Total EVs	Time Duration (Hours)	Total Operating Time of Relay (s) GCM	Mean of OBJ^T (s)
70% (30% G2V & 70% V2G)	1	10.9512	10.9512 /1 = 10.9512
80% (40% G2V & 60% V2G)	1	11.2451	11.2451 /1 = 11.2451
95% (80% G2V & 20% V2G)	1	11.8002	11.8002/1= 11.8002
With common settings of GCM			
Consideration of 24-hour penetration	24 (8 hour for each case)	260.3051	260.3051/24 = 10.8460

The results show that the relay coordination scheme effectively attains the desired CTI values utilizing the determined optimal settings, ensuring reliable protection coordination.

Furthermore, the study highlights that the operating time of primary and backup relays and the CTI exhibit similar patterns across different EV penetration levels. This recommends that the relay coordination scheme remains valid and effective across various scenarios of EV penetration in the GCM.

In the proposed scheme, Table 12.10 compares the average total operating time (OBJT) for the user-defined DS-DOCRs in the GCM. The table compares OBJT values across three different levels of EV penetration: 70% (30% G2V & 70% V2G), 80% (40% G2V & 60% V2G), and 95% (80% G2V & 20% V2G) of the total EV count. The study also offers OBJT values for various amounts of EV penetration at various times of the day. The coordination among the user-defined DS-DOCRs improves when the common optimum relay settings are used, resulting in lower average OBJT values than the individual EV penetration levels. This shows that using shared ideal relay settings might improve coordination and minimize operation durations for the user-defined DS-DOCRs in the GCM, especially in the case of increased EV penetration levels.

12.5   CONCLUSION

The study aims to establish the optimum coordination of user-defined DS-DOCRs in GCM by considering different degrees of EV penetration. OBJT, an optimization strategy utilizing a genetic algorithm (GA) solver, is used to find the ideal relay settings and compute the objective function. The findings show that as the amount of EV penetration grows, so does the OBJT. The 95% (80% G2V & 20% V2G) EV penetration level, in particular, leads to a greater OBJT when compared to the other two scenarios. The mean OBJT values obtained are 10.9512s for 70% (30% G2V & 70% V2G), 11.2451s for 80% (40% G2V & 60% V2G) and 11.8002s for 95% (80% G2V & 20% V2G) penetration. The research, however, acknowledges the influence of individual EV penetration on GCM and provides particular ideal relay settings for each penetration level that are beneficial for their respective EV penetration levels. Additionally, by analyzing the patterns in Table 12.2, the study identifies common optimal relay settings for GCM that consider different EV penetration levels at various time intervals. With these shared optimal settings, the mean OBJT is reduced to 10.8460s, except for the 70% (30% G2V & 70% V2G) of EV penetration level. Based on these findings, the proposed scheme demonstrates the potential to achieve efficient and reliable coordination for rapid fault-clearing purposes.

REFERENCES

[1] Chase, J. (2022). View from the solar industry: We don't need COP26 to shine, but what should we worry about?. *Joule*, 6(3), 495–497.
[2] Turan, M. T., & Gökalp, E. (2022). Integration analysis of electric vehicle charging station equipped with solar power plant to distribution network and protection system design. *Journal of Electrical Engineering & Technology*, 17(2), 903–912.
[3] Fahmy, S., Gupta, R., & Paolone, M. (2020). Grid-aware distributed control of electric vehicle charging stations in active distribution grids. *Electric Power Systems Research*, 189, 106697.
[4] Singh, B., & Dubey, P. K. (2022). Distributed power generation planning for distribution networks using electric vehicles: Systematic attention to challenges and opportunities. *Journal of Energy Storage*, 48, 104030.
[5] Razavi, S. E., Rahimi, E., Javadi, M. S., Nezhad, A. E., Lotfi, M., Shafie-Khah, M., & Catalão, J. P. (2019). Impact of distributed generation on protection and voltage regulation of distribution systems: A review. *Renewable and Sustainable Energy Reviews*, 105, 157–167.
[6] Aazami, R., Esmaeilbeigi, S., Valizadeh, M., & Javadi, M. S. (2022). Novel intelligent multi-agents system for hybrid adaptive protection of micro-grid. *Sustainable Energy, Grids and Networks*, 30, 100682.
[7] Ali, A. Y., Hussain, A., Baek, J. W., & Kim, H. M. (2020). Optimal operation of networked microgrids for enhancing resilience using mobile electric vehicles. *Energies*, 14(1), 142.
[8] Hoang, T. T., Tran, Q. T., & Besanger, Y. (2022). An advanced protection scheme for medium-voltage distribution networks containing low-voltage microgrids with high penetration of photovoltaic systems. *International Journal of Electrical Power & Energy Systems*, 139, 107988.
[9] Bayati, N., Baghaee, H. R., Hajizadeh, A., & Soltani, M. (2020). Localized protection of radial DC microgrids with high penetration of constant power loads. *IEEE Systems Journal*, 15(3), 4145–4156.
[10] Reis, F. B., Pinto, J. O. C., dos Reis, F. S., Issicaba, D., & Rolim, J. G. (2021). Multi-agent dual strategy based adaptive protection for microgrids. *Sustainable Energy, Grids and Networks*, 27, 100501.
[11] Sadeghi, S., Hashemi-Dezaki, H., & Entekhabi-Nooshabadi, A. M. (2022). Optimized protection coordination of smart grids considering N-1 contingency based on reliability-oriented probability of various topologies. *Electric Power Systems Research*, 213, 108737.
[12] Osman, A. H., Hassan, M. S., & Sulaiman, M. (2015). Communication-based adaptive protection for distribution systems penetrated with distributed generators. *Electric Power Components & Systems*, 43(5), 556–565.

[13] Azimi, A., & Hashemi-Dezaki, H. (2023). Optimized protection coordination of microgrids considering power quality-based voltage indices incorporating optimal sizing and placement of fault current limiters. *Sustainable Cities and Society*, 96, 104634.

[14] Chang, S., Niu, Y., & Jia, T. (2021). Coordinate scheduling of electric vehicles in charging stations supported by microgrids. *Electric Power Systems Research*, 199, 107418.

[15] Zhou, K., Cheng, L., Wen, L., Lu, X., & Ding, T. (2020). A coordinated charging scheduling method for electric vehicles considering different charging demands. *Energy*, 213, 118882.

[16] Hong, L., Rizwan, M., Wasif, M., Ahmad, S., Zaindin, M., & Firdausi, M. (2021). User-defined dual setting directional overcurrent relays with hybrid time current-voltage characteristics-based protection coordination for active distribution network. *IEEE Access*, 9, 62752–62769.

[17] Tiwari, R., Singh, R. K., & Choudhary, N. K. (2022). Coordination of dual setting overcurrent relays in microgrid with optimally determined relay characteristics for dual operating modes. *Protection and Control of Modern Power Systems*, 7(1), 6.

[18] Balyith, A. A., Sharaf, H. M., Shaaban, M., El-Saadany, E. F., & Zeineldin, H. H. (2020). Non-communication-based time-current-voltage dual setting directional overcurrent protection for radial distribution systems with DG. *IEEE Access*, 8, 190572–190581.

[19] Alam, M. N. (2019). Overcurrent protection of AC microgrids using mixed characteristic curves of relays. *Computers and Electrical Engineering*, 74, 74–88.

[20] Wright, A. H. (1991). Genetic algorithms for real parameter optimization. In *Foundations of genetic algorithms* (Vol. 1, pp. 205–218). Elsevier.

[21] Christie, R. (2021). Power system test cases. https://labs.ece.uw.edu/pstca/, (accessed 10 March 2022).

13 Optimal Coordination of Relays for Microgrid Protection with Varying Electric Vehicle Penetration

Priyanshul Niranjan and Parvez Ahmad
MNNIT Allahabadd, Prayagraj, India

Vikas Patel
Rajkiya Engineering College, Ambedkar Nagar, India

*Niraj Kumar Choudhary, Nitin Singh,
and Ravindra Kumar Singh*
MNNIT Allahabadd, Prayagraj, India

13.1 INTRODUCTION

The acceptance of electric vehicles has gained significant momentum in recent years. While EVs offer numerous environmental and energy efficiency benefits, their integration into the power grid poses new challenges for power system protection [1–2].

The increasing number of EVs connected to the power grid introduces additional electrical loads and can lead to overloads and voltage fluctuations [3]. These changes can impact the protective devices within the power system, potentially compromising the system's overall reliability [4–5]. Therefore, assessing the impact of EV charging patterns, EV charging infrastructure, and EV penetration levels on protection coordination is a crucial task.

Adopting advanced protection schemes is essential to protect power systems with significant EV penetration effectively. These schemes can leverage real-time monitoring, intelligent relays, and communication systems to promptly detect and respond to abnormal conditions. Advanced protection schemes can also facilitate the integration of EV charging infrastructure, renewable energy sources, and energy storage systems, enhancing the overall reliability and security of the power grid.

The charging characteristics of EVs significantly impact load demand within microgrids. The impact of EV charging profiles on microgrid operation is investigated

DOI: 10.1201/9781003481836-13

and highlights the need for advanced protection coordination to handle the increased load demand during peak charging periods [6]. A case study has been conducted to analyze the impact of EV charging rates on protection coordination and emphasized the importance of dynamic protection settings to adapt to varying load profiles [7].

The presence of EVs in microgrids affects fault current levels during system faults. There has been analysis of fault current variations in a microgrid with high EV penetration, and a fault current limiting strategy to mitigate the impact on protection coordination was proposed [8]. Similarly, authors in Ref. [9] investigated the impact of bidirectional power flow from EVs on fault current levels and proposed an adaptive protection coordination algorithm to ensure reliable operation.

An intelligent protection coordination approach based on multi-agent systems has been proposed in Ref. [10]. Authors in Ref. [11] suggested a protection coordination scheme that uses smart relays with communication capabilities. The technique improves the microgrid's performance in fault detection and coordination by making use of the relays' communication capabilities. The effect of communication delay on protection coordination is the primary focus in Ref. [12]. The research investigates how coordination tactics might be less efficient when communication is delayed. Communication and distributed control algorithms might be used to improve coordination and fault detection in EV-integrated microgrids [13]. The effect of EV charging on the voltage stability of microgrids is explicitly examined in Ref. [14]. The research takes into account EV charging patterns and characteristics and proposes a coordinated charging method to maintain system stability. The effect of EV charging patterns on the resilience of microgrids is investigated in Ref. [15]. The study highlights the need for adaptive protection coordination to ensure uninterrupted power supply during grid disturbances.

Optimal protection coordination requires effective coordination techniques to address the dynamic interactions between EVs, power grid components, and protection devices. Coordinated settings for relays can be optimized to accommodate the unique characteristics of EV loads and charging patterns. Directional overcurrent relays (DOCRs) are one of the most preferred protective devices aiming to detect faulty sections in microgrids. The proposed scheme focuses on user-defined DS-DOCRs for coordination in the GCM of microgrid. The 7-bus microgrid, i.e., low voltage (11 kV) section of the IEEE-14 bus benchmark system, is chosen to test the performance of the proposed scheme. The main contributions of the study are:

- To propose a modified protection coordination scheme in GCM of microgrid using user-defined DS-DOCRs.
- To optimize the performance and reliability of the microgrid by ensuring effective coordination of the user-defined DS-DOCRs at different levels of EV penetration.
- To determine common optimal settings for the user-defined DS-DOCRs that remain valid and effective in the GCM of the microgrid, regardless of the varying levels of EV penetration.

The rest of the chapter is divided into four sections. Section 13.2 explains the problem formulation. Section 13.3 gives a brief discussion of the system. Results and discussion are explained in Section 13.4, and the study is concluded in Section 13.5.

13.2   PROBLEM FORMULATION

The proposed methodology incorporates user-defined DS-DOCRs and tests them in GCM of microgrid. The objective of the protection coordination scheme is to minimize the operating time of relays while ensuring that the constraints are satisfied for fault currents in both forward and reverse directions. The objective function, denoted as OBJ^T, represents the sum of operating time for primary and corresponding backup relays for different fault locations, as represented by (13.1).

$$\text{OBJ}^T = \min \sum_{t=1}^{24} \sum_{i=1}^{N} \sum_{j=1}^{M} \left(t^p_{fw\_i,j,t} + \sum_{k=1}^{K} t^b_{rv\_k,j,t} \right) \tag{13.1}$$

During a fault, a significant voltage drop is observed and therefore it is considered as an additional parameter along with the current. Therefore, by using a time-voltage-current-based characteristic, relays can be coordinated more precisely, especially when the short-circuit current level is low. This ensures that the relays closest to the fault location operate faster, while those farther away coordinate their operation to provide selective fault detection and isolation. In Ref. [16], a dual-setting DOCR with time-current-voltage-based characteristics is proposed. This approach aims to reduce the operating time of relays compared to conventional time-current characteristics. The operating time characteristics of the dual-setting relays, based on time-current-voltage characteristics, are described by (13.2) and (13.3). Additionally, the constraints associated with the relay coordination problem, are described by Eqs. (13.4)–(13.10).

$$t^p_{fw\_i,j,t} = A \frac{\text{TMS}_{fw,i}}{\left(\dfrac{I_{sc,i,j,t}}{\text{CT}_{\text{Ratio},i} \times \text{PS}_{fw,i}} \right)^B - 1} * \left(\frac{1}{e^{1-v_{f_{i,j,t}}}} \right)^{\alpha_{fw,i}} \tag{13.2}$$

$$t^b_{rv\_i,j,t} = A \frac{\text{TMS}_{rv,k}}{\left(\dfrac{I_{sc,k,j,t}}{\text{CT}_{\text{Ratio},k} \times \text{PS}_{rv,k}} \right)^B - 1} * \left(\frac{1}{e^{1-v_{f_{k,j,t}}}} \right)^{\alpha_{rv,k}} \tag{13.3}$$

$$t^b_{rv\_k,j,t} - t^p_{fw\_i,j,t} \geq \text{CTI} \tag{13.4}$$

$$\text{TMS}_{\min,i} \leq \text{TMS}_{fw,i}, \text{TMS}_{rv,k} \leq \text{TMS}_{\max,k} \tag{13.5}$$

$$\text{PS}_{\min,i} \leq \text{PS}_{fw,i}, \text{PS}_{rv,k} \leq \text{PS}_{\max,k} \tag{13.6}$$

$$t^p_{\min,i} \leq t^p_{fw,i} \leq t^p_{\max,i} \tag{13.7}$$

$$t^b_{\min,k} \leq t^b_{rv,k} \leq t^b_{\max,k} \tag{13.8}$$

$$0 \leq \alpha_{fw,i} \leq 5 \tag{13.9}$$

$$0 \leq \alpha_{rv,k} \leq 5 \tag{13.10}$$

In these equations, $t^p_{fw\_i,j,t}$ and $t^b_{rv\_k,j,t}$ are the operating time of the primary and corresponding backup relay for fault currents in the forward and reverse direction respectively. $v_{f,k,j,t}$ is the fault voltage, $\alpha_{fw,i}$ and $\alpha_{rv,k}$ are considered as decision variables for fault voltages measured in the forward and reverse direction, respectively. $I_{sc,k,j,t}$ is the short-circuit current through the relay coil, $CT_{Ratio,k}$ is the current transformer ratio, while $TMS_{fw,i}$, $TMS_{rv,k}$, $PS_{fw,i}$ and $PS_{rv,k}$ are the decision-making variables of the relay coordination problem. The relay characteristic constants are A and B that are equal to 0.14 and 0.02 (for NI relay characteristic), respectively based on the IEC-60255 std. Moreover, constraint Eq. (13.4) refers to the CTI limit that is considered as 0.2 s in this study [17]. It ensures that there is a time discrimination of 0.2 s between the operation of the primary and backup relays. Inequalities Eqs. (13.5) and (13.6) represent the limits of TMS and PS in both the forward and reverse directions. The operating time range for the primary relay is defined as Eq. (13.7), while Eq. (13.8) defines the operating time range for the corresponding backup relay. The specific range of these constraints is provided in Table 13.1 [18–19]. Note that i represents the index of the primary relay. It ranges from 1 to N, where N is the total number of primary relays in the system. Index k represents the corresponding backup relay associated with the primary relay. It ranges from 1 to K, where K is the total number of corresponding backup relays in the system. Index j represents the fault location. It ranges from 1 to M, where M is the total number of fault locations being considered.

In order to incorporate the impact of EVs, one more variable is considered, by incorporating the time instants and accounting for the changing EV penetration levels. Thus, the power flow method can capture the dynamic nature of EVs in the microgrid and provide accurate calculations of voltage, and current at different hours of a day. The power flow method should indeed determine the variables prior to the fault, considering the changing percentage penetration level of EVs at any given time of the day. This enables a comprehensive analysis of the system and allows for effective relay coordination considering the specific conditions during various time instants. Therefore, from Eqs. (13.1)–(13.8), the term 't' refers to different hours of a day between 01:00 a.m. and midnight.

TABLE 13.1

Limits of the Coordination Constraints

Parameter	Minimum Value	Maximum Value
TMS	0.1 s	1.1 s
PS	0.5	2
Operating time of relay	0.1 s	4 s
α	0	5

This study employs GA to determine the optimal relay settings and total operating time (OBJ^T) of relays. The performance of the relay coordination problem depends upon GA input parameters, which are considered as: population size (50), β (0.20), γ (1), δ (0.25), and maximum iterations (200) [20].

13.3 TEST SYSTEM

The test system used in the study is the 7-bus microgrid, which represents the low voltage (LV) section of the modified IEEE-14 bus system. The system details can be found in Ref. [21]. In order to achieve complete protection of the 7-bus microgrid, 16 DOCRs are considered as R1 to R16. These relays are strategically placed within the microgrid system to ensure effective coordination and complete protection. The microgrid system consists of two IBDGs and a SBDG. These generators are connected at buses B2, B7, and B1, respectively. Additionally, there is an EV charging station located at bus B5. The charging pattern of the EV varies throughout the day, and Table 13.2 provides information about the penetration levels of EVs at different time intervals.

To analyze the system's fault response, a 3-phase mid-point fault is considered at eight different locations labeled as L1 to L8. Based on the fault locations a total of 22 primary-backup relay pairs are formed. Table 13.3 presents information of the

TABLE 13.2
The Average Penetration Level of EVs for Different Duration

Serial Number	Number of EVs (%)	Time
1	20	0–8 (8 hours)
2	50	8–16 (8 hours)
3	80	16–24 (8 hours)

TABLE 13.3
Primary-Backup Relay Pairs (User-Defined DS-DOCRs) in GCM

Fault Location	Serial No.	Primary Relay	Backup Relay
L1	1	1	3
	2	1	5
	3	2	7
L2	4	3	1
	5	3	5
	6	4	14
	7	4	15
L3	8	5	1
	9	5	3
	10	6	16

(Continued)

TABLE 13.3 (Continued)

Fault Location	Serial No.	Primary Relay	Backup Relay
L4	11	7	2
	12	8	9
L5	13	9	8
	14	10	11
L6	15	11	10
	16	12	13
L7	17	13	12
	18	14	4
	19	14	15
L8	20	15	4
	21	15	14
	22	16	6

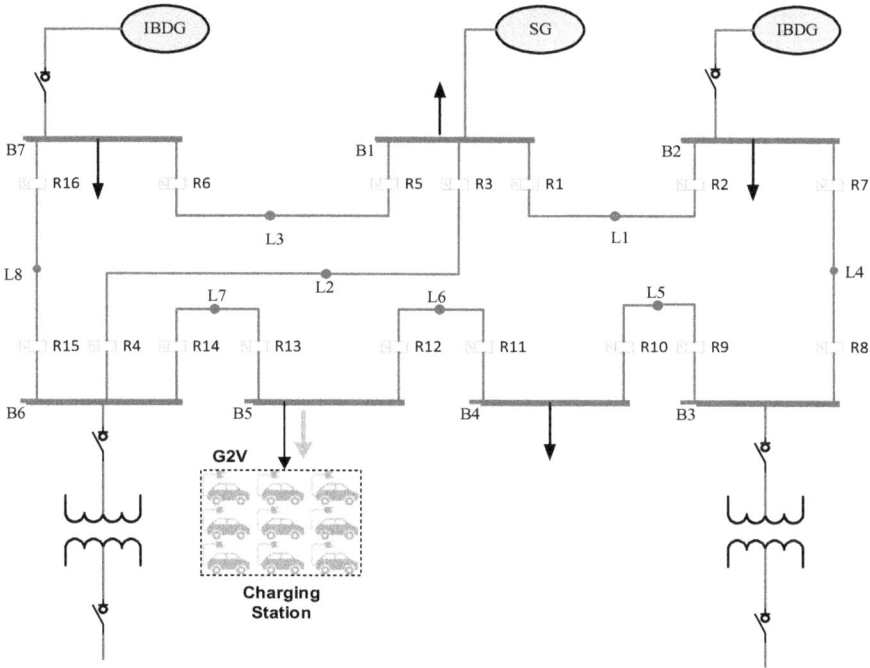

FIGURE 13.1 The 7-bus microgrid system (11 kV part of the IEEE-14 bus system).

relay pairs in GCM at each fault location. For example, in the case of a fault at location L1, the identified primary relays are R1 and R2. The corresponding backup relays identified for R1 are R3 and R5, while R7 is the only backup relay identified for R2 (Figure 13.1).

TABLE 13.4

CT_{Ratio} of Each User-Defined DS-DOCRs

	User-Defined DS-DOCRs	
Relay	CTR for Primary Relay	CTR for Backup Relay
1	2000/5	1000/5
2	1000/5	2000/5
3	3000/5	2000/5
4	2000/5	3000/5
5	1600/5	1000/5
6	1000/5	1600/5
7	2500/5	1600/5
8	1600/5	2500/5
9	2500/5	1200/5
10	1200/5	2500/5
11	1200/5	2500/5
12	2500/5	1200/5
13	800/5	3000/5
14	3000/5	800/5
15	1600/5	1600/5
16	1600/5	1600/5

The $CT_{Ratio,i}$ for the user-defined DS-DOCRs are found using (13.11) [19]. The maximum short-circuit current and maximum load current for the ith DOCR are $I_{f\max,i}$, and $I_{L\max,i}$ respectively. $CT_{Ratio,i}$ values for the relays are shown in Table 13.4.

$$CT_{Ratio,i} = \text{Maximum}\left(I_{L\max,i}, \frac{I_{f\max,i}}{20} \right) \tag{13.11}$$

13.4   RESULT AND DISCUSSION

The relay coordination study has been performed in two cases, as shown in the following sub-sections. In the first case, the focus is on determining individual optimal settings of user-defined DS-DOCRs for specific penetration of EVs at different time instances. However, in the second case, the common optimal settings of user-defined DS-DOCRs are determined that are valid at different levels of EV penetration over the time. For the overall study GA is used as an optimization tool to find the relay settings.

13.4.1   Optimal Relay Settings in Microgrid at Variable Penetration Level of EVs at Different Time Instant

The study considered three different EV penetration levels: 20%, 50%, and 80% of the total number of EVs, based on the data as mentioned in Table 13.2. These penetration levels have been chosen to observe and analyze how different levels of EV penetration influence the protection coordination of the user-defined DS-DOCRs.

The impact of EV penetration on relay coordination and protection is evaluated at three different time instances, considering the time variations and different EV charging patterns throughout the day. Tables 13.5–13.7 provide the optimal relay settings in each of the three cases; i.e., at 20%, 50%, and 80% EV penetration. These tables

TABLE 13.5

Optimal Relay Settings in GCM at 20% Penetration Level of EVs

Relay	TMS_{FW} (s)	PS_{FW} (A)	α_{FW}	TMS_{RV} (s)	PS_{RV} (A)	α_{RV}
1	0.1000	0.8787	1.0000	0.1571	0.5492	0.1523
2	0.1000	0.5000	0.3108	1.1000	0.5000	2.4420
3	0.1001	1.0130	1.1690	0.3302	0.9694	1.6877
4	0.3549	0.5000	2.1004	0.1239	0.5025	0.2402
5	0.1000	1.0749	0.4422	0.2413	0.6540	1.1599
6	0.8873	0.5000	3.0000	0.5053	0.5000	1.5586
7	0.5820	0.5121	2.8750	0.7859	1.5000	2.7259
8	0.1000	0.5000	0.7578	0.1000	0.7130	0.0625
9	0.1000	0.5000	0.1335	0.1000	1.3967	0.7188
10	0.1000	1.2702	1.0041	0.1279	1.0405	0.1609
11	0.1000	0.5000	0.5639	0.5150	1.5269	2.2479
12	0.1099	0.5008	1.1173	0.1000	1.7781	0.1655
13	0.2168	0.5681	1.5292	0.1114	0.5000	0.0277
14	0.1000	0.5000	0.5636	0.1000	2.0000	0.3519
15	0.1000	0.5000	0.7500	0.1575	0.5005	0.6645
16	0.1715	0.5000	1.3750	0.6930	0.5662	1.9184
OBJ^T			78.8909 s			

TABLE 13.6

Optimal Relay Settings in GCM at 50% Penetration Level of EVs

Relay	TMS_{FW} (s)	PS_{FW} (A)	α_{FW}	TMS_{RV} (s)	PS_{RV} (A)	α_{RV}
1	0.1000	0.6128	0.9951	0.1110	1.0254	0.1875
2	0.1000	1.5000	1.0938	0.6016	0.8588	2.1969
3	0.1000	0.5000	0.8241	0.1000	1.0212	0.5469
4	0.4928	0.5000	2.4430	0.2498	0.8833	0.9997
5	0.1000	0.5000	0.7656	0.2610	0.5000	1.1563
6	0.1000	0.5324	0.5330	0.1000	1.9375	0.7181
7	0.1000	0.5000	1.0000	0.2487	0.5000	0.8676
8	0.1000	0.9577	0.9977	0.1000	0.8710	0.2466
9	0.4680	0.5638	2.2550	0.9750	0.9958	3.1551
10	0.1000	1.1666	1.1511	0.3566	0.5000	1.0665
11	0.4241	0.5000	2.2036	0.5473	0.5000	1.6117
12	0.1000	0.5000	1.0234	0.6000	0.5468	1.3072
13	0.5429	0.5234	2.0983	0.1000	0.7209	0.0625
14	0.1676	0.5000	1.5121	1.1000	0.5000	2.2695
15	0.1000	0.5000	0.5287	0.1817	0.9567	1.5000
16	0.1000	0.5000	0.8582	0.1000	1.1913	0.1999
OBJ^T			92.9820 s			

TABLE 13.7

Optimal Relay Settings in GCM at 80% Penetration Level of EVs

Relay	TMS_{FW} (s)	PS_{FW} (A)	α_{FW}	TMS_{RV} (s)	PS_{RV} (A)	α_{RV}
1	0.1000	0.5000	0.4375	0.6000	0.5000	1.3047
2	0.2081	0.5071	1.6241	0.3500	0.6227	1.0662
3	0.6084	0.5000	2.6063	0.6218	1.0000	2.2981
4	0.1887	0.5057	1.4622	0.1843	0.5159	0.2994
5	1.1000	0.5000	3.1587	0.1954	0.8166	1.1038
6	0.1004	0.8941	0.6885	0.1299	0.5030	0.0860
7	0.3205	0.5000	1.5785	0.1010	0.5001	0.0000
8	0.5412	0.5000	2.6926	0.7727	0.5000	1.9301
9	1.1000	0.5022	2.9782	0.1055	1.5005	0.9337
10	0.1295	0.6508	1.2500	0.6924	0.5000	1.7562
11	0.1000	0.5000	0.4375	0.7064	0.5000	2.0508
12	0.1063	0.5005	1.1282	0.2945	0.7500	0.6192
13	0.1000	0.5997	0.6250	0.9817	0.5000	1.8333
14	0.1000	0.5000	0.6508	0.1200	0.5015	0.0000
15	0.1000	0.5000	0.7417	0.1000	0.5000	0.0635
16	0.1000	0.5000	0.6287	0.1000	0.8247	0.1094
OBJ^T			94.9239 s			

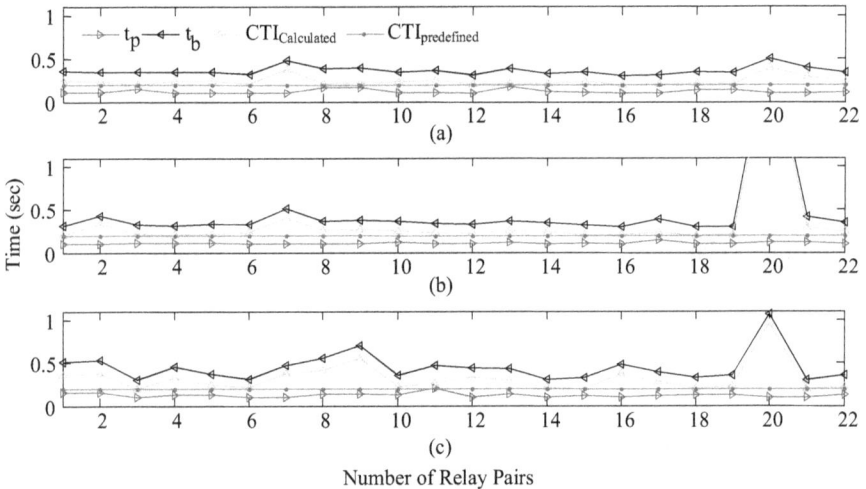

FIGURE 13.2 Primary-backup relay operating time at EV penetration of (a) 20%, (b) 50%, and (c) 80%.

also include the respective total operating time of the relays, considering the EV penetration levels and the duration of time at the EV charging station. Figure 13.2 illustrates the operating time of each primary and corresponding backup relay at different EV penetration levels, along with the calculated CTI between relay pairs and a comparison with the predefined CTI. It is found that the predefined CTI is consistently lower than the calculated CTI.

TABLE 13.8

Coordination Constraint Violations in Microgrid at Different EV Penetration Level

Microgrid Mode	Algorithm	Optimal Setting (%)	Number of Coordination Constraint Violations		
			20%	50%	80%
GCM	GA	20	0	6	2
		50	4	0	3
		80	5	4	0

The total operational time of relay increases as the level of EV penetration increases. This implies that with higher EV penetration, there is a greater impact on the load demand and short-circuit currents, which in turn affects the operation of the relays. Therefore, it is important to consider the changing EV penetration and its associated effects on relay coordination in the microgrid. The study also observed that the relay coordination problem, which is optimally solved for a specific EV penetration level, may not be valid or effective at other penetration levels. When applying the optimal relay settings determined for one penetration level to another, violations of coordination constraints occur which is shown in Table 13.8. The identification of a common set of relay settings may involve further analysis for satisfactory relay coordination in GCM at different EV penetrations.

13.4.2 COMMON OPTIMAL SETTING OF RELAYS IN MICROGRID AT VARIABLE PENETRATION LEVEL OF EVS

The study has determined common optimal relay settings in GCM, considering the variable number of EVs at any given time of the day, based on the data mentioned in Table 13.2. These optimal settings, shown in Table 13.9, ensure effective relay coordination at different EV penetration levels. This table also shows the total operating time of the relays taking into account the variable number of EVs for the 24-hour period at the EV charging station. Figure 13.3 illustrate the operating time of primary and backup relays for GCM, and it is mentioned that the predefined CTI is consistently lower than the calculated CTI. This suggests that the relay coordination scheme, with the obtained optimal settings, successfully achieves the desired CTI values for reliable protection coordination.

Furthermore, it is mentioned that the operating time of primary and backup relays, as well as the CTI, exhibit similar patterns for different penetration levels of EVs. This indicates that the relay coordination scheme remains valid and effective across various EV penetration scenarios.

Table 13.10 presents a comparative analysis of the mean total operating time (OBJ^T) of the user-defined DS-DOCRs. The table compares the OBJ^T at three different EV penetration levels: 20%, 50%, and 80% of the total number of EVs. The analysis also includes the OBJ^T for the variable EV penetration level at any given

TABLE 13.9

Common Optimal Settings of Relays at Variable EV Penetration

Relay	TMS_{FW} (s)	PS_{FW} (A)	α_{FW}	TMS_{RV} (s)	PS_{RV} (A)	α_{RV}
1	0.1000	0.9271	1.0000	0.1301	0.8954	0.2500
2	0.1000	0.5000	0.7210	0.8500	0.5000	2.3135
3	0.1000	0.5000	0.7843	0.3988	0.5000	1.5000
4	0.1113	0.5007	0.8527	0.6543	0.5742	1.9010
5	0.4481	0.5819	2.5826	0.8500	0.5000	2.4593
6	0.6475	0.5000	2.6250	0.2053	0.5368	0.6504
7	0.3329	0.5000	2.2500	0.1037	0.7491	0.1100
8	0.2694	0.7415	1.8537	0.1155	0.6664	0.3125
9	0.2249	0.6433	1.6321	0.1194	0.6581	0.1720
10	0.1000	0.5058	0.7238	0.8113	0.5000	1.5890
11	0.4636	0.5007	2.2035	0.1815	1.5454	0.9780
12	0.1000	0.5000	1.0004	1.1000	1.5000	2.5684
13	0.6000	0.5309	2.5742	0.1121	0.5015	0.0013
14	0.1000	0.7941	1.1048	0.1215	1.5756	0.4786
15	0.1000	0.5002	0.7559	0.3887	0.5000	1.8284
16	0.1000	1.0114	1.1250	1.0375	1.5000	3.0735
OBJ^T			**248.9092 s**			

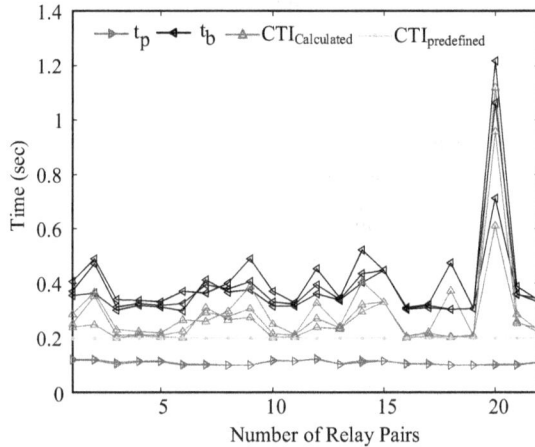

FIGURE 13.3 Operating time and CTI of relay pairs with common settings (at variable EV penetration).

time of the day. By considering the common optimal settings of the relays, the coordination among the user-defined DS-DOCRs becomes more efficient and effective, resulting in lower mean OBJ^T values compared to the individual EV penetration levels (except for the 20% EV penetration level). This indicates that the proposed approach of common optimal relay settings can lead to improved coordination and reduced operating times for the user-defined DS-DOCRs, particularly when dealing with higher EV penetration levels.

TABLE 13.10

Comparative Analysis of Total Operating Time (OBJ^T) of Relays in Microgrid at Different EV Penetration

Penetration Percentage of Total EVs	Time Duration (Hours)	Total Operating Time of Relay (s) GCM	Mean of OBJ^T (s)
20%	8	78.8909	78.8909/8 = 9.8613
50%	8	92.9820	92.9820/8 = 11.6227
80%	8	94.9239	94.9239/8 = 11.8654
With common settings of GCM			
Consideration of 20%, 50%, and 80%	24	248.9092	248.9092/24 = 10.3712

13.5 CONCLUSION

The study focuses on the optimal coordination of user-defined DS-DOCRs in GCM of microgrid considering different penetration levels of EVs. GA is used to determine the optimal relay settings and the objective function (OBJ^T). The results demonstrate that the OBJ^T increases with higher EV penetration levels. Specifically, the 80% EV penetration level yields a higher OBJ^T compared to the 20% and 50% levels. The mean OBJ^T values obtained are 9.8613 s (20%), 11.6227 s (50%), and 11.8654 s (80%). However, by considering the impact of individual EV penetration, the study determines different optimal relay settings at each penetration level, and these settings are found to be valid for the corresponding penetration levels of EVs only. Furthermore, based on the observed patterns in Table 13.2, common optimal relay settings are obtained for GCM, considering different EV penetration levels at different time instants. With these common optimal settings, the mean OBJ^T is reduced to 9.3712 s. Based on the results, the proposed scheme can attain effective and reliable coordination for fast fault-clearing purposes.

REFERENCES

[1] Shaukat, N., Khan, B., Ali, S. M., Mehmood, C. A., Khan, J., Farid, U., & Ullah, Z. (2018). A survey on electric vehicle transportation within smart grid system. *Renewable and Sustainable Energy Reviews*, 81, 1329–1349.

[2] Zhang, Q., Yan, J., Gao, H. O., & You, F. (2023). A systematic review on power systems planning and operations management with grid integration of transportation electrification at scale. *Advances in Applied Energy*, 81, 100147.

[3] Singh, B., & Dubey, P. K. (2022). Distributed power generation planning for distribution networks using electric vehicles: Systematic attention to challenges and opportunities. *Journal of Energy Storage*, 48, 104030.

[4] Richardson, P., Flynn, D., & Keane, A. (2011). Optimal charging of electric vehicles in low-voltage distribution systems. *IEEE Transactions on Power Systems*, 27(1), 268–279.

[5] Sundstrom, O., & Binding, C. (2011). Flexible charging optimization for electric vehicles considering distribution grid constraints. *IEEE Transactions on Smart Grid*, 3(1), 26–37.

[6] Aazami, R., Esmaeilbeigi, S., Valizadeh, M., & Javadi, M. S. (2022). Novel intelligent multi-agents system for hybrid adaptive protection of micro-grid. *Sustainable Energy, Grids and Networks*, 30, 100682.

[7] Ali, A. Y., Hussain, A., Baek, J. W., & Kim, H. M. (2020). Optimal operation of networked microgrids for enhancing resilience using mobile electric vehicles. *Energies*, 14(1), 142.

[8] Hoang, T. T., Tran, Q. T., & Besanger, Y. (2022). An advanced protection scheme for medium-voltage distribution networks containing low-voltage microgrids with high penetration of photovoltaic systems. *International Journal of Electrical Power & Energy Systems*, 139, 107988.

[9] Bayati, N., Baghaee, H. R., Hajizadeh, A., & Soltani, M. (2020). Localized protection of radial DC microgrids with high penetration of constant power loads. *IEEE Systems Journal*, 15(3), 4145–4156.

[10] Reis, F. B., Pinto, J. O. C., dos Reis, F. S., Issicaba, D., & Rolim, J. G. (2021). Multi-agent dual strategy based adaptive protection for microgrids. *Sustainable Energy, Grids and Networks*, 27, 100501.

[11] Sadeghi, S., Hashemi-Dezaki, H., & Entekhabi-Nooshabadi, A. M. (2022). Optimized protection coordination of smart grids considering N-1 contingency based on reliability-oriented probability of various topologies. *Electric Power Systems Research*, 213, 108737.

[12] Osman, A. H., Hassan, M. S., & Sulaiman, M. (2015). Communication-based adaptive protection for distribution systems penetrated with distributed generators. *Electric Power Components & Systems*, 43(5), 556–565.

[13] Azimi, A., & Hashemi-Dezaki, H. (2023). Optimized protection coordination of microgrids considering power quality-based voltage indices incorporating optimal sizing and placement of fault current limiters. *Sustainable Cities and Society*, 96, 104634.

[14] Chang, S., Niu, Y., & Jia, T. (2021). Coordinate scheduling of electric vehicles in charging stations supported by microgrids. *Electric Power Systems Research*, 199, 107418.

[15] Zhou, K., Cheng, L., Wen, L., Lu, X., & Ding, T. (2020). A coordinated charging scheduling method for electric vehicles considering different charging demands. *Energy*, 213, 118882.

[16] Hong, L., Rizwan, M., Wasif, M., Ahmad, S., Zaindin, M., & Firdausi, M. (2021). User-defined dual setting directional overcurrent relays with hybrid time current-voltage characteristics-based protection coordination for active distribution network. *IEEE Access*, 9, 62752–62769.

[17] Tiwari, R., Singh, R. K., & Choudhary, N. K. (2022). Coordination of dual setting overcurrent relays in microgrid with optimally determined relay characteristics for dual operating modes. *Protection and Control of Modern Power Systems*, 7(1), 6.

[18] Balyith, A. A., Sharaf, H. M., Shaaban, M., El-Saadany, E. F., & Zeineldin, H. H. (2020). Non-communication-based time-current-voltage dual setting directional overcurrent protection for radial distribution systems with DG. *IEEE Access*, 8, 190572–190581.

[19] Alam, M. N. (2019). Overcurrent protection of AC microgrids using mixed characteristic curves of relays. *Computers and Electrical Engineering*, 74, 74–88.

[20] Wright, A. H. (1991). Genetic algorithms for real parameter optimization. In *Foundations of genetic algorithms* (Vol. 1, pp. 205–218). Elsevier.

[21] Christie, R. (2021). Power system test cases. https://labs.ece.uw.edu/pstca/, (accessed 10 March 2022).

14 Protection Coordination Issues for AC Microgrid System
A Comprehensive Review

Raghvendra Tiwari
G. H. Raisoni Institute of Engineering & Technology, Nagpur, India

Ravindra Kumar Singh and Niraj Kumar Choudhary
MNNIT Allahabad, Prayagraj, India

14.1 INTRODUCTION

A microgrid is one of the best solutions to overcome the problems related to severe environmental concerns such as CO_2 emission and fossil fuel depletion. According to the microgrid exchange group of DOE, USA "[A] microgrid is a group of interconnected loads and distributed energy resources within clearly defined electrical boundaries that act as a single controllable entity in terms of an energy grid," as shown in Figure 14.1 [1]. The point of common coupling (PCC) serves as the interconnected point for the microgrid to the utility grid. Microgrids have the capability to function as either grid-connected or islanded systems, contingent upon their connection to the utility grid, as illustrated in Figure 14.2 [2]. The implementation of a microgrid offers several benefits, including plug-and-play and peer-to-peer features. The peer-to-peer model ensures that no components within the distribution system adversely impact its functionality. Additionally, the plug-and-play characteristics enable convenient placement of units anywhere within the distribution system.

In the conventional power system, generation of electric power is conducted in bulk quantity at thermal power stations, hydropower plants, nuclear power stations, etc., which are located remotely, far from the load centers. This reduces the reliability and efficiency of the entire system. The aforementioned concerns have served as a driving force behind the emergence of distributed generation (DG) based power systems, which predominantly rely on non-conventional sources of energy.

Adding DGs to low voltage microgrids is greatly beneficial regarding low transmission losses, bus voltage support, and power system reliability [3]. In addition

DOI: 10.1201/9781003481836-14

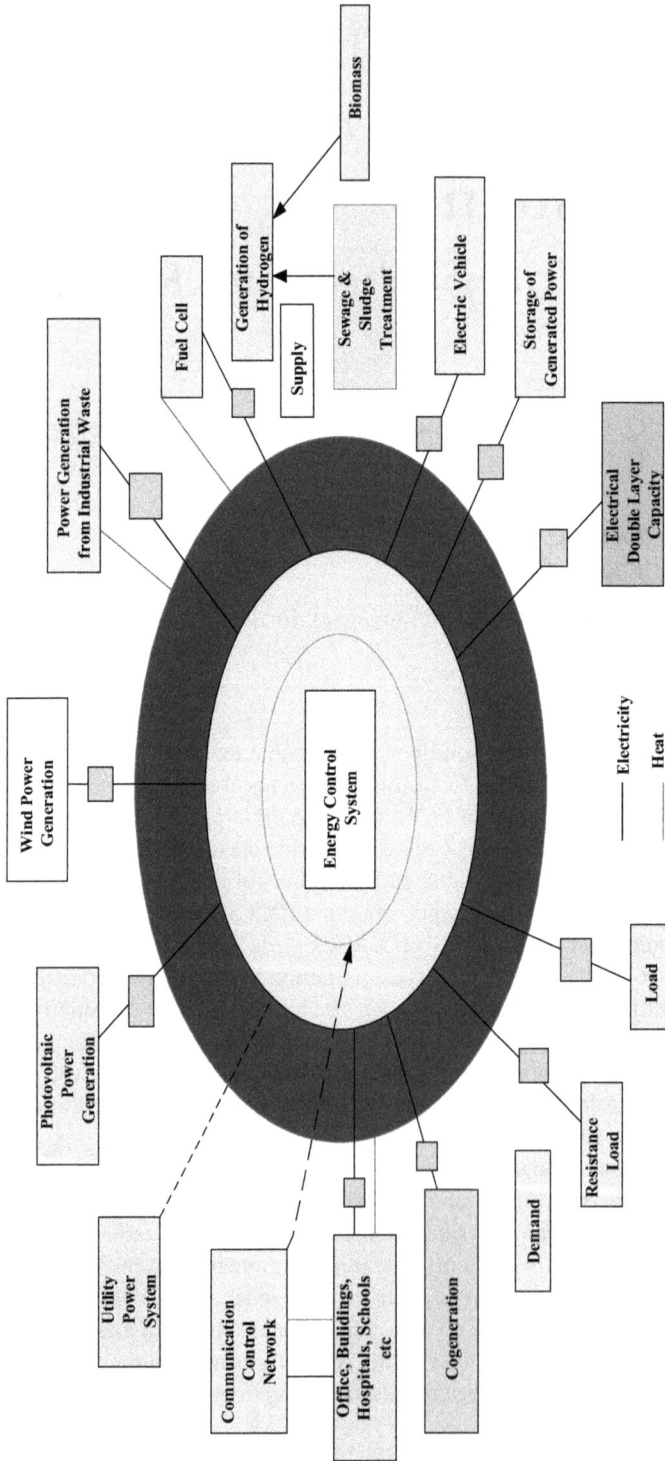

FIGURE 14.1 General diagram of microgrid architecture.

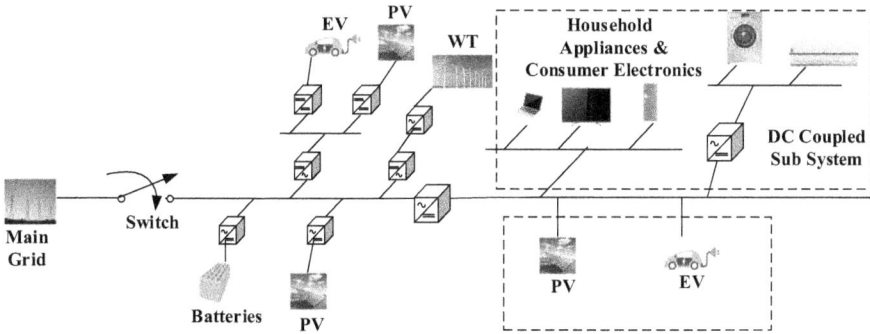

FIGURE 14.2   Islanded (switch open)/grid-connected (switch closed) mode of microgrid.

to the advantages, the integration of DGs can also introduce certain drawbacks to the utility grid. These include issues like protection blinding, false tripping, resynchronization problems, bidirectional power flow, and varying magnitudes of fault current [4].

14.2   PROTECTION COORDINATION IN MICROGRIDS

Microgrids are commonly utilized in the medium or low voltage range, encompassing both the secondary and primary distribution levels. Therefore, ensuring adequate protection of the primary and secondary distribution system becomes a significant concern. A protection scheme employed in the power system must possess inherent qualities such as selectivity, speed, sensitivity, reliability, and cost-effectiveness.

- **Reliability**: It confirms that the protective device works effectively.
- **Selectivity**: Protective system should be selective such that it isolates the minimum part and continuity of supply is maintained.
- **Speed**: It ensures that the protective device should respond faster on the occurrence of fault such that fault duration is minimum.
- **Simplicity**: To achieve the objective of protection, minimum equipment should be used.
- **Economical**: The protective device should be economical such that the protection is achieved at minimum cost.

Protection coordination refers to the precise sequencing of operation for protection devices (PDs) to prevent disconnection of a large power system [3]. To ensure the reliable functioning of the protective relays, it is necessary to maintain a minimum fixed time interval known as the coordination time interval (CTI) between the primary and its corresponding backup relay pairs. For electro-mechanical (EM) and digital relays, the CTI value typically ranges from 0.2–0.5 and 0.2–0.3 seconds respectively [5]. The addition of DG units and the dual operating modes of a microgrid give rise to fluctuating levels of fault current, consequently leading to ambiguity in the CTI among PDs. As a result, the coordination of PD may be compromised, resulting

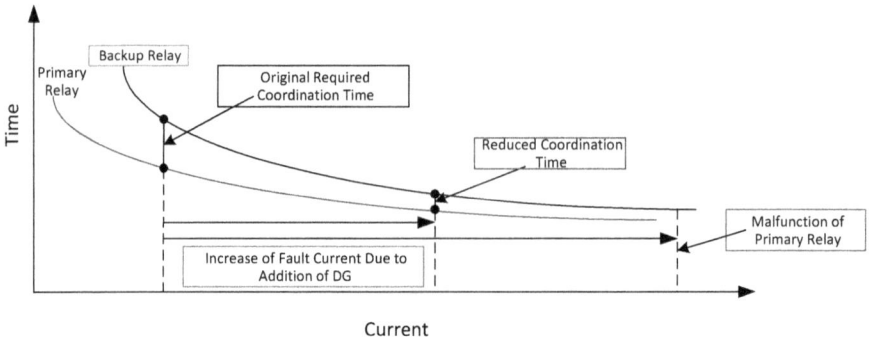

FIGURE 14.3   CTI variation due to addition of DGs.

in potential impacts on the microgrid's overall protection scheme, as illustrated in Figure 14.3 [6, 7]. Therefore protection engineers face a significant challenge in effectively coordinating the PDs to ensure a reliable and seamless operation of the microgrid.

14.3   PROTECTION DEVICES USED IN MICROGRIDS

In a microgrid, commonly used PDs include relays, reclosers, and fuses. Relays and reclosers are typically deployed at the feeder end, while fuses are positioned at laterals or sub-laterals, farthest from the power source, as illustrated in Figure 14.4 [8]. These devices are crucial components that help to protect the power system and respond to fault currents of varying magnitudes. The operating time of the protective devices is an important parameter on the grounds that other protective devices in the network must be coordinated in order to remove the least faulty part of the distribution system. The various power system components contain various protective devices and during the fault all the devices should sense it and the device nearest to the faulty section should operate in order to remove it.

Among the various types of protective devices, fuses are considered the simplest. They are described as cost effective and not requiring any secondary device such as

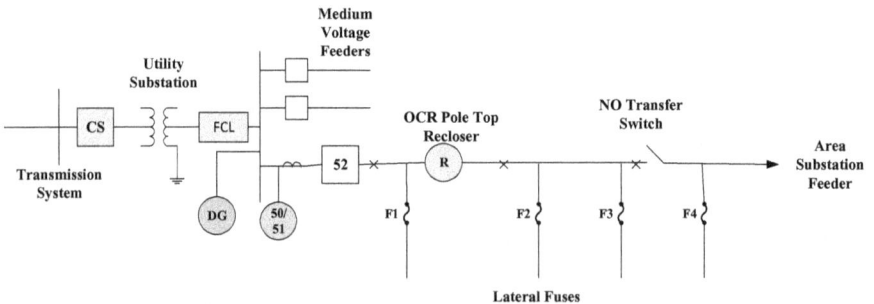

FIGURE 14.4   Typical distribution system with protection.

current transformer to work. Fuses are connected mainly at distribution level with voltages of 5.5 kV, 8.3 kV, 15.5 kV, 23.0 kV, 27.0 kV and 38.0 kV [9].

Reclosers are a distinct category of protective devices that serve to protect against overcurrent. In distribution system, majority of the faults (~80%) are temporary which comes under the category of transients. For these temporary faults, the permanent switching of the feeders is not required. The recloser gives fault a chance to self-clear itself i.e., it allows the arc for deionization, in order to limit the interruption time [10]. After a certain time the recloser closes its contacts to energize the line once again. If the fault is removed the normal continuity of the supply is maintained. In case the fault is permanent in nature, this process is repeated and even though the fault persists the recloser will lock out and the faulty part is isolated from the remaining healthy part of the system.

In electrical distribution networks (EDNs), the overcurrent relays located farthest from the power source are initially set with the lowest time multiplier setting (TMS), followed by the adjustment of associated backup relays. This iterative procedure continues until all relays are considered, resulting in a selection of minimum break point set relays (MBPS). The determination of MBPS has been simplified by the implementation of a novel polynomial-time approximation algorithm, as demonstrated in Refs. [5,11], effectively reducing computational complexity. One of the major drawbacks of fuses is that their characteristics are fixed, and also, they need to be changed each time it operates. The blowing of a fuse depends upon the amount of current passing through it, and the material used. The operating time of a fuse is not fixed; rather, it decreases as the magnitude of current increases, as shown in Figure 14.5. The fuse-saving scheme is an integral part of the coordination process that involves relays, reclosers, and fuses. Its primary objective is to guarantee that in the event of permanent faults, only the fuse is triggered, whereas temporary faults activate the reclosers in fast mode. By adopting this approach, the feeder is isolated, allowing the fault to potentially clear on its own [12]. If both the fuse and recloser fail to operate,

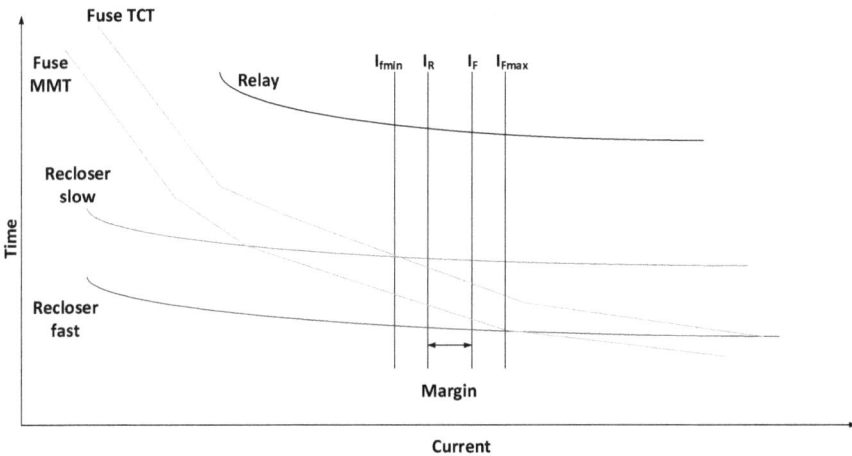

FIGURE 14.5   Coordination of distribution network protection devices.

TABLE 14.1

Value of α, β, δ of Reclosaer Characteristics Coefficient

Relay Characteristics	α	β	δ
Standard Inverse	0.0515	0.1140	0.02
Very Inverse	19.61	0.491	2
Extremely Inverse	28.2	0.1217	2

the relay intervenes after a specific time interval determined by its characteristic curve, which is positioned above all other curves, as illustrated in Figure 14.5 [13]. The reclosers are having inverse time-current characteristics and expressed as

$$t_{op} = \frac{\alpha \times \text{TMS}}{\left(\dfrac{I_f}{I_p}\right)^{\beta} - 1} + \delta \tag{14.1}$$

where, t_{op} is the recloser's operating time, I_f and I_p are fault current and pickup current respectively. Constants α, β and δ depends on the type of time-current characteristics as shown in Table 14.1.

14.4 PROTECTION SCHEMES IN MICROGRIDS

The implementation of a protection scheme in a microgrid necessitates the consideration of fault conditions in both the upstream (utility grid) and downstream (microgrid) directions. In the event of a fault on the utility grid side, the microgrid is expected to operate in islanded mode, ensuring independent power supply. On the other hand, if a fault arises within the microgrid, the main goal is to isolate the affected section only, and maintain uninterrupted power delivery to the remaining unaffected portions. Extensive research has been conducted on various protection schemes, which are illustrated in Figure 14.6.

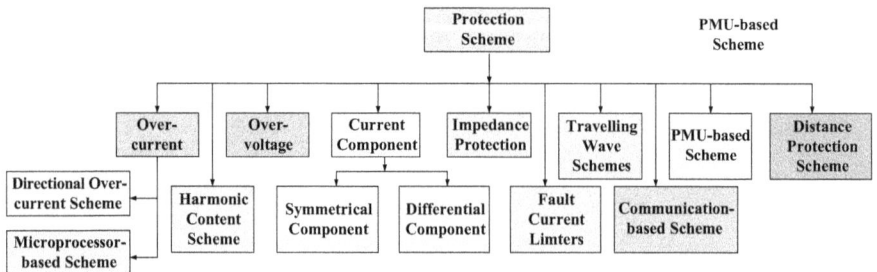

FIGURE 14.6 Protection schemes in microgrid.

14.4.1 Overcurrent-based Protection Scheme

Overcurrent protection is widely employed as a prevalent form of protection in distribution systems. The overcurrent relay (OCR) operation relies on the current magnitude that passes through it. The distribution and sub-transmission electrical networks frequently use directional overcurrent relays (DOCRs) for overcurrent protection. The operating time of an overcurrent relay is influenced by several key parameters, including the time multiplier setting (TMS), plug setting (PS), and relay characteristic coefficients. The primary objective of relay coordination is to minimize the total operating time of relays while ensuring proper coordination among them. The relay coordination problem is typically formulated as an optimization problem, incorporating specified non-linear constraints. The relay coordination problem is broadly classified into three categories: mixed-integer non-linear problem (MINLP), non-linear programming (NLP) and linear programming (LP). The objective function in the LP approach is exclusively expressed in terms of TMS only while all other parameters are kept as fixed [14–17]. Whereas, in the NLP technique, the relay coordination problem is formulated considering both the TMS and PS parameters [18–24]. In Ref. [25], an interior point method (IPM) based optimization technique was introduced, involving a two-phase solution. The first phase treated the PS of EM relays as a continuous variable, converting the MINLP problem into an NLP problem. In the second phase, the lower and upper values of PS were constrained to their nearest discrete values.

14.4.2 Voltage-based Protection

These schemes are employed in microgrids to safeguard against various types of faults using measurements of the output voltage from DG sources. In this scheme, the DG's output voltage is converted into a DC quantity using the d-q reference frame. Any disturbances in the utility input voltage result in corresponding disturbances in the d-q values. From these disturbed d-q values, V_{dist} is derived to represent the deviation from a given reference. The V_{dist} signal undergoes processing through a low-pass filter and a hysteresis comparator, as illustrated in Figure 14.7 [4]. This technique is capable of differentiating between internal and external faults by using a communication channel. The protective zone is determined by comparing the measured voltage between any two relays in a microgrid to their average value [26, 27]. After identifying the fault zone, it is tripped when the fault voltage surpasses the

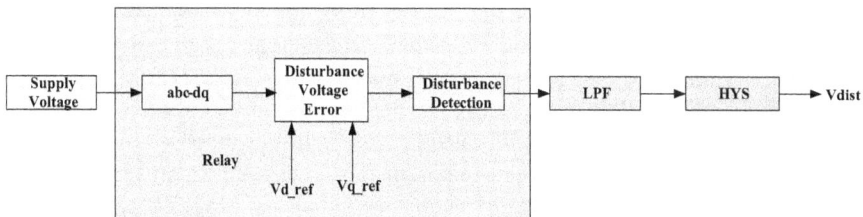

FIGURE 14.7   Block diagram of voltage-based protection scheme.

threshold associated with the specific fault type [28]. The location and type of the fault were determined in Ref. [29] by using the positive sequence component of the fundamental voltage. The performance of over/under voltage relays and rate of change of frequency relays in detecting islanding conditions in microgrids was investigated in Ref. [30].

14.4.3 SYMMETRICAL AND DIFFERENTIAL CURRENT COMPONENT-BASED PROTECTION SCHEME

A microgrid protection strategy that uses differential and symmetrical current components to detect line-to-ground (L–G) and line-to-line (L–L) faults by segmenting the microgrid into separate protection zones has been studied in Refs. [31, 32]. The upstream fault detection employs the differential current component, while the downstream fault detection utilizes the symmetrical current component. By taking into account both synchronous-based DG (SBDG) and inverter-based DG (IBDG), [33] expanded this method to accommodate both grid-connected and island modes of operation. Furthermore, in Ref. [34], this scheme was combined with communication-assisted digital relays employing microprocessor-based relays and phasor measurement unit (PMU) to detect various fault types, including high impedance fault (HIF).

14.4.4 FAULT CURRENT LIMITER-BASED PROTECTION SCHEME

Due to varying fault current levels in microgrid systems, it is difficult, especially with IBDG, to establish a single relay configuration that can be utilized in both grid-connected and island modes of operation. Figure 14.3 illustrates the importance of approximately equal fault current levels in both modes. To mitigate this issue and reduce current levels in grid-connected mode, a series-connected device known as a fault current limiter (FCL) is employed [35]. The FCL provides low impedance during normal operation and high impedance during fault conditions. However, the large size and high cost of FCLs hinder the widespread implementation of this scheme. Various types of FCLs are found in the literature, including resistive FCLs, superconducting FCLs (SFCL), inductive FCLs, solid-state FCLs, hybrid FCLs, and many more, as depicted in Figure 14.8 [36].

14.4.5 CURRENT TRAVELING WAVE-BASED PROTECTION SCHEME

The methodology for locating faults is predicated by the measurement of the propagation time of a wave from the point of fault to the terminal designated for measurement. Accurately determining the location of a fault requires crucial information regarding the speed of wave propagation. The method can use data from either one end of the line [37, 38] or both ends of the line [39, 40] to identify the exact location of the fault. A local data-based protection mechanism is proposed by the author in Ref. [41], which eliminates the requirement of communication channel. The scheme identifies the faulty feeder by analyzing the current traveling wave and bus bar voltage. A current transducer is used to calculate the current travelling wave, and multi-resolution wavelet analysis is used to decompose the signal. This scheme

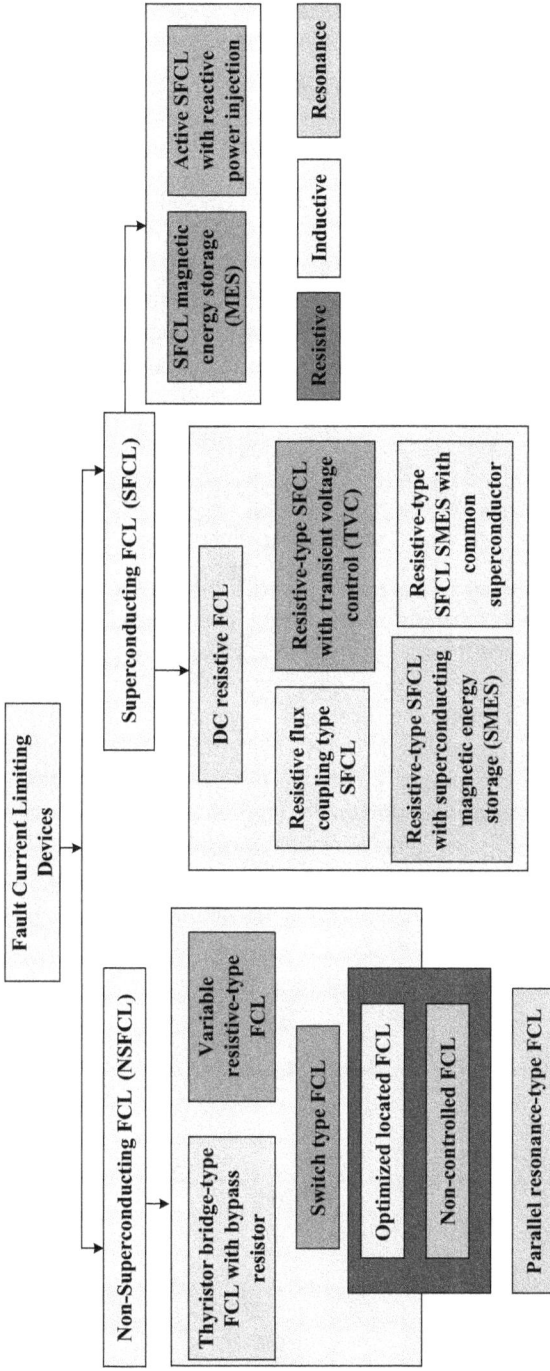

FIGURE 14.8 Classification of FCL.

is applicable across different operating modes of the microgrid. By comparing the original propagating waves, the faulty feeder can be identified.

14.4.6 HARMONIC CONTENT-BASED PROTECTION SCHEME

During a fault, IBDGs can introduce harmonics into the network. The present scheme involves the continuous monitoring of the inverter's output voltage for the purpose of ascertaining the total harmonic distortion (THD) magnitude. In the event of a fault, the tripping of the inverter is triggered by a relay when the THD exceeds a specific threshold. This scheme consists of two steps: 1) identifying the type of fault; and 2) differentiating between faulty zones using THD measurements [42]. In Ref. [43], the author takes advantage of the flexibility of microprocessor-based relays in the islanded mode of a microgrid, without relying on a communication channel. Fault detection in this study is proposed by injecting a small amount of fifth harmonic current into the fault current, and inverter output voltage is increased by the I_f–Z_f droop control when faults are located closer to the inverter.

14.4.7 IMPEDANCE AND ADMITTANCE-BASED PROTECTION SCHEME

The detection of faults by OCRs in the islanded mode of a microgrid with IBDGs is challenging due to the low fault current level. Consequently, impedance or admittance relays have emerged as a dependable alternative to protect power transmission lines. The distance protection scheme operates by calculating the impedance or admittance at the fault point. The measured impedance is compared by the relay to a predetermined value known as the relay reach. Figure 14.9 illustrates the distance protection scheme for microgrid protection [4]. Distance protection is commonly classified into two types: 1) current and voltage measurements are taken only at the main substation [44–46]; and 2) measurements are taken at both DG location and the substation [47–49]. However, these schemes have limitations as they do not consider fault resistance and are suitable only for low DG penetration.

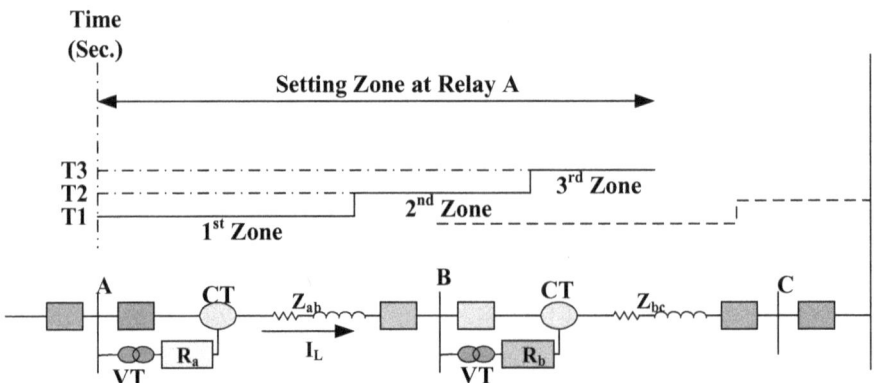

FIGURE 14.9   Reach of distance protection.

In cases of low fault current, an inverse-type relay based online admittance has been proposed for protecting the electric distribution network considering only IBDGs [50]. In Ref. [51], distance and OCR coordination is achieved by optimizing relay characteristics coefficients, TMS, and PS using genetic algorithm (GA). In Ref. [52], a mho characteristic-based distance protection scheme for microgrid protection is introduced, demonstrating good selectivity and sensitivity across different operating modes. The under- or over-reach problem associated with distance relays when a DG is connected between the measurement point and fault point is discussed in Refs. [53, 54]. To evaluate fault distance in the presence of infeed using a real-time information-based algorithm, a communication-based adaptive protection system is recommended in Ref. [55]. In Ref. [56], the author combines distance relays with overcurrent relays and fuses, demonstrating that impedance-based distance relays simplify relay settings and are less affected by changes in network topology compared to OCR. The author of Ref. [57] proposes a line protection approach that defines the fault location based on the phase current shift at both ends of the distribution line.

14.4.8 DIFFERENTIAL PROTECTION SCHEME

In this scheme, the currents are measured at input and output terminals of a differential relay, and the data is transmitted through various communication channels such as power line carrier, pilot wire, wireless networks, and fiber optic as depicted in Figure 14.10 [4]. A fault is detected within the protected zone if the difference between the input and output currents exceeds a pre-set threshold. This scheme offers several advantages, including immunity to power swings, cost-effectiveness, selectivity, high sensitivity, and improved efficiency in detecting HIF due to advancements in communication technology. An advantageous aspect of employing differential protection methods in microgrids is their ability to operate independently from the fault current level and direction of power flow.

For the purpose of detecting symmetrical faults during power swings, a differential power-based fault detection method is proposed in Ref. [58]. The differential power is determined by comparing the anticipated values (derived through auto regression method) with the real measurements of voltage and current samples.

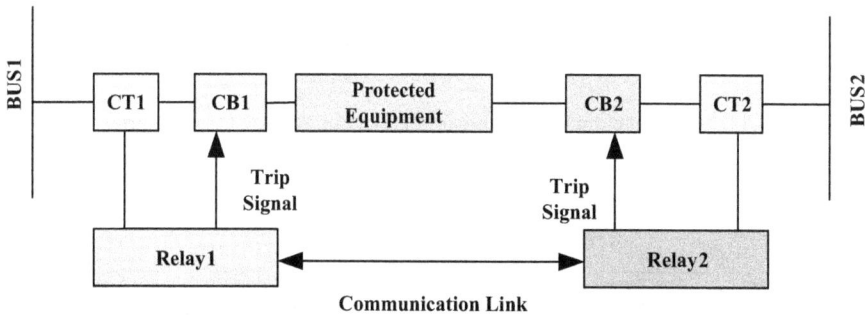

FIGURE 14.10   Block diagram of differential protection.

In Ref. [59], the author utilizes Global Positioning System (GPS) for data synchronization in a feeder protection system employing a differential relay. In situations where GPS signal loss occurs, a reliable technique using symmetrical elements is introduced to enhance the stability of current differential protection.

14.4.9 PMU AND COMMUNICATION-BASED PROTECTION SCHEME

In this scheme, sensors S_1 and S_2 gather voltage and current information and share it with the communication and control center, as depicted in Figure 14.11 [4]. The control center then commands relays R_1 and R_2 to protect the line. In Refs. [60, 61], a combination of intelligent electronic devices (IEDs), communication channels, and directional microprocessor-based relays is utilized to minimize relay operating time. The process of optimizing relay settings is accomplished by taking into account the $(N-1)$ contingency and both operational modes of the microgrid, employing an interior point optimization solver. The IEDs are employed to provide updated information for relay setting updates.

One of the key benefits of PMUs is their ability to achieve time synchronization through GPS, enabling precise measurement of system phase angles and magnitudes. In a PMU, the receiver station receives GPS signals, which are then transferred to an analog to digital converter. The process involves converting the discrete-time signal into a set of complex numbers that correspond to the phasors of the sampled waveform. The positive sequence component can be derived by using the phasors of the three phases. The Discrete Fourier Transform block is utilized to measure the voltage and current signals, as depicted in Figure 14.12 [4].

The utilization of micro-PMUs (μPMUs) for OCR coordination is explored in Refs. [62, 63], where PMUs detect any uncertainties such as line outages and send

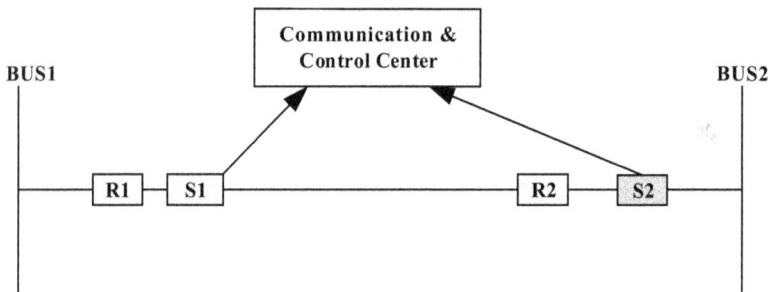

FIGURE 14.11   General communication-based protection scheme for microgrids.

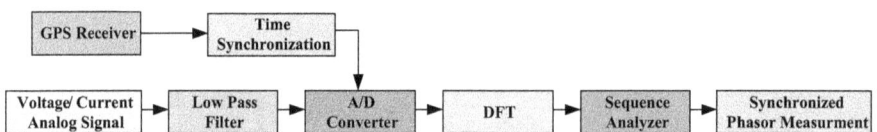

FIGURE 14.12   Block diagram of PMU.

TABLE 14.2

Comprehensive Analysis of Distinct Protection Schemes in a Microgrid

Protection Schemes	Advantage	Disadvantage
Overcurrent protection scheme [14–21, 65]	(i) Economical and easy to implement. (ii) Only current magnitude is required for operation	(i) Not accurate for small short-circuit fault current especially in islanded mode (ii) Cannot work in case of heavy over loading
Voltage-based protection scheme [26–30]	(i) Protection against any type of fault either internal or external to the zone of protection. (ii) Separate protection is used for DG's	(i) Suitable only for islanded mode (ii) For low impedance, fault does not work (iii) Does not work for single phase tripping and symmetrical faults.
Current component scheme [31–34]	(i) Symmetrical and differential protection gives full protection	(i) Tripping of the single phase is not considered (ii) Three phase fault impact has not been considered
Fault current limiter-based protection scheme [35, 36]	(i) Coordination can be maintained even during the faulty condition (ii) Coordination margin can be enhanced	(i) The optimal size of FCL is another optimization problem (ii) Not applicable for IBDGs (iii) Cost of FCL is high
Traveling wave-based protection scheme [37–41]	(i) Independent from the direction of power flow, unbalanced loading (ii) Irrespective of the operational mode of the grid	(i) In case of error during measurement of local information, the fault location can be wrongly identified
Harmonic based protection scheme [42, 43]	(i) Continuous assessment of THD helps in improving the power quality	(i) Cannot work for several dynamic loads (ii) Used only for IBDG

the data to a phasor data concentrator (PDC). The authors of Ref. [64] have utilized a compensation theory-based approach to identify the origin of events within a distribution network by utilizing μPMUs. Table 14.2 presents a comparative analysis of diverse protection schemes employed in microgrids.

14.5 PROTECTION COORDINATION STRATEGIES IN MICROGRIDS

In order to optimize the overall operating time of relays, it is imperative to perform TMS and PS calculations for effective OCR coordination. Numerous strategies for protection coordination have been suggested in scholarly works to improve coordination, as depicted in Figure 14.13. The intricacy of relay coordination is impacted by the nature of the OCR employed, specifically whether it is an EM relay or a digital relay. In the case of EM relays, TMS is continuous, whereas PS can only have

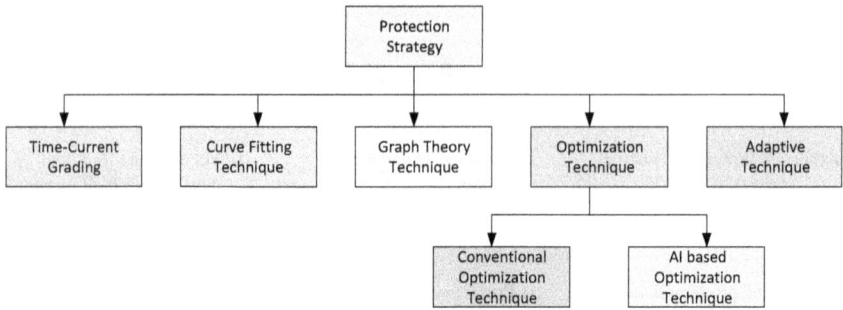

FIGURE 14.13 Protection coordination strategies in microgrid.

discrete values. On the other hand, digital relays allow both TMS and PS to be treated as continuous variables.

14.5.1 PROTECTION COORDINATION USING CURVE FITTING TECHNIQUE

This method involves modeling and storing the relay characteristics in a computer system to facilitate relay coordination. Researchers have explored various approaches to model relay characteristics effectively. The curve fitting method is commonly employed to determine the optimal representation of relay characteristics. Mathematical construction of relay characteristics using polynomial curve fitting has been proposed in Refs. [66, 67]. However, a notable limitation of this method is the requirement of significant memory storage when dealing with multiple relay characteristics.

14.5.2 PROTECTION COORDINATION USING GRAPH THEORY TECHNIQUE

The discernment of clockwise and anticlockwise loops is a pivotal factor in ascertaining the MBPS relays within the realm of graph theory. However, the task of identifying MBPS relays in a large-scale power system with numerous lines and buses can be complex and challenging. The study presented in Ref. [68] utilizes GA to obtain the optimal values of TMS, and a methodology based on graph theory is employed to establish primary-backup relay pairs. Prior research has examined the utilization of graph theory methodologies for the purpose of devising protection coordination strategies [69–72].

14.5.3 PROTECTION COORDINATION USING TIME-CURRENT GRADING

This scheme involves the coordination of primary and backup relay pairs using time delay discrimination, where the two relays have different operating times with a minimal CTI value. In the event of a failure of the primary relay, the corresponding backup relay will be triggered in a sequential manner at predetermined time intervals to effectively isolate the fault [73]. Figure 14.14 illustrates the time grading scheme for a multi-source protection system in a radial distribution network [9]. The conventional

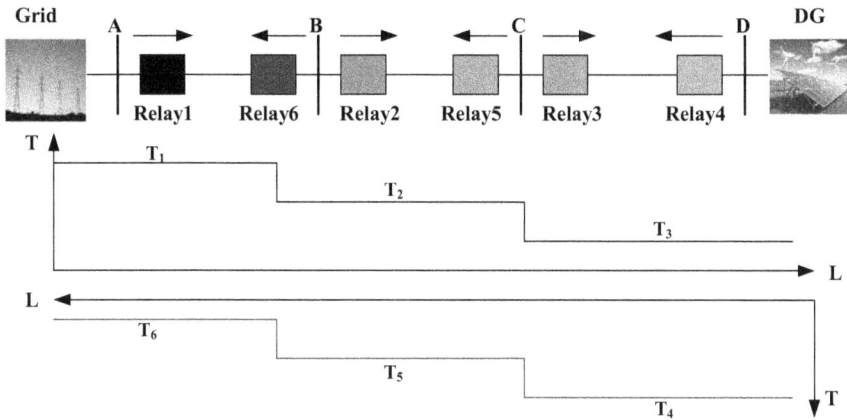

FIGURE 14.14 Time grading scheme for radial distribution multi-source protection system.

method of overcurrent protection is employed in the grid-connected mode to detect faults within a predetermined time frame and restrict their impact. In the event that the fault is not identified within the designated timeframe, the microgrid undergoes a transition from its grid-connected state to an islanded state. During this transition, voltage measurement techniques are utilized to detect the fault. However, in Ref. [74], it is highlighted that voltage measurement protection coordination may not be a coherent technique. To overcome this limitation, microprocessor-based relays are utilized in Ref. [75], where the microgrid's protective unit is coordinated using a time grading system for both reverse and forward fault current directions, considering the different modes of microgrid operation. Conventional protection methodologies have the drawback of requiring predetermined fault assessments, unusual working conditions, and system contingencies [76]. One limitation of this scheme is its inability to account for relay response in unidentified conditions that were not included in the initial assessment.

14.5.4 PROTECTION COORDINATION USING OPTIMIZATION TECHNIQUE

The conventional method of relay coordination in large distribution power system networks becomes impractical due to the lengthy computation time and suboptimal results obtained. As a result, optimization techniques are more suitable for large power systems. In the optimization approach, the use of MBPS relays and graph theory applications are not necessary [77].

14.5.4.1 Protection Coordination Using Conventional and Artificial Intelligence (AI)-based Optimization Technique

A scholarly article, [78], introduced a digital coordination approach for the over and under frequency relays protection in an isolated microgrid that has a significant penetration of DGs. In this study, to mitigate the large frequency variation caused by the nearly zero inertia of solar and wind plants, the author introduced a digital PID controller based on a mapping technique. In Ref. [79], the author proposed a two-step protection scheme for microgrids considering variable penetration levels of PV modules, where the first step involved modifying relay characteristics and the second step focused on identifying mis-coordination without the need for a communication channel. In Ref. [80], the author introduced a new relay characteristic based on a logarithmic function dependent on voltage and current magnitude, utilizing microprocessor-based relays along with IBDG and SBDG. In this study, the values of TMS, PS, and a variable K, which regulates the impact of voltage on relay operating time, were determined through the utilization of sequential quadratic programming (SQP) optimization technique. The introduction of a new objective function incorporating operating time differences (Δt_{mb}) for continuous and discrete TMS in a fixed network topology was proposed in Ref. [81] and optimized using GA. For a fixed network topology, distinct variants of particle swarm optimization (PSO) were described in Ref. [82] to address protection coordination issues. In Ref. [83], the author formulated the protection coordination problem as a MINLP considering various network topologies caused by ($N-1$) contingency, and a hybrid GA-LP algorithm was used to find the solution. In reference [65], an investigation was conducted to analyze the impact of DG control mode on microgrid protection coordination utilizing dual-setting overcurrent relays. This study utilized GA to determine the optimized TMS values for both current control mode (CCM) and voltage control mode (VCM). The utilization of a GA approach was also employed in references [84, 85] to determine the optimal values of TMS and PS, while incorporating DG into a distribution system. In Ref. [86], the author determined a common relay setting (TMS and PS) along with a third variable curve setting using GA, which was applicable in both operating modes without FCL. To relax the protection coordination problem, Ref. [87] introduced the concept of broken constraints and weighting factors determined by expert rules, solved using GA. The Grey Wolf optimization (GWO) algorithm is known for its slow convergence speed. To address this issue, an improved GWO (IGWO) was proposed in Ref. [88] for finding optimal relay settings with a fixed topology. The utilization of omega as a search agent in lieu of the conventional GWO approach of following the top three candidates was introduced by IGWO. In situations where bidirectional current flow occurs in a line, it is crucial to disconnect the line from both sides to prevent the spread of fault current and potential damage to equipment. In order to address this issue, a protection scheme utilizing a differential search algorithm was proposed in Ref. [89]. The purpose of this scheme is to minimize the discrepancy between the operating times of the primary relay for the near and far ends fault.

An alternative optimization algorithm, referred to as advanced teaching-learning-based optimization (ATLBO), was introduced in Ref. [90] as a variant of the original TLBO method. In comparison to the traditional TLBO algorithm, it intended to identify optimal relay settings by gradually lowering the upper bound of TMS in each iteration and thereby shortening the search space and computing time. The protection coordination problem was addressed in a study by utilizing an opposition-based chaotic differential evolution algorithm [91]. The utilization of chaotic scale factor and opposition-based learning was incorporated into the algorithm. In reference [92], the effect of DGs on the coordination of overcurrent relays in microgrids was elucidated. The study employed differential evolution to derive the optimal settings for the relays.

14.5.4.2 Protection Coordination Using Fuzzy and Neural Network-based Techniques

In Ref. [93], a novel approach is presented where the author combines a fuzzy decision-making tool with multi-objective PSO to address the coordination problem of OCRs. In order to improve the efficacy of differential protection, Ref. [94] has suggested a methodology that employs a fuzzy-based approach to leverage the differences between the power at the sending and receiving ends of a transmission line. The utilization of the Hilbert space theory-based approach facilitates the prompt response of the protective unit, within a span of two complete cycles, subsequent to the incidence of a short circuit. Furthermore, the detection of errors becomes independent of threshold values. In Ref. [95], the fault location detection considering the impact of DG penetration is addressed using a feed-forward neural network. The aforementioned technique facilitates precise localization of faults in the DG integrated electrical networks.

In Ref. [96], the author introduces the utilization of a backpropagation neural network to model non-conventional characteristic curves of OCR. This approach eliminates the dependence on standard characteristic curves, enabling more flexible and adaptive protection settings. A novel protective mechanism for microgrids has been suggested in Ref. [97]. The proposed methodology integrates intelligent electronic devices, wavelet transform, and neural networks to effectively identify the fault phase linked with HIFs. This advanced approach enhances the reliability and effectiveness of microgrid protection.

14.5.5 ADAPTIVE PROTECTION COORDINATION SCHEME

The adaptive protection scheme involves the dynamic adjustment of relay settings in response to variations in the system configuration. This involves optimizing the relay settings to accommodate specific factors such as fault location, microgrid operating mode, the number of DG units, DG size, and DG placement. The proposed methodology involves the substitution of EM relays with microprocessor-based relays, and the relay parameters are modified through the utilization of either a communication channel or local data on current and voltage. The primary limitation of this particular scheme pertains to the requirement of costly communication channels and microprocessor-based relays. However, it presents a significant benefit by removing all the constraints associated with EM relays.

In Ref. [98], a meta-heuristic approach using the differential evolution (DE) algorithm is proposed for online coordination of DOCRs considering different network topology variations due to element outages. In order to tackle the issue of significant voltage transients that occur during close-in faults, Ref. [99] proposes an adaptive technique for differential overcurrent relay that utilizes positive and negative sequence superimposed current for the protection of microgrids, without the need for voltage data. In Ref. [100], relay characteristics are updated using local voltage and current information, and a state detection algorithm facilitates the transition of settings between islanded and grid-connected modes.

In Ref. [101], a hybrid adaptive differential overcurrent protection scheme is presented, where differential relays protect the feeder while OC relays safeguard the

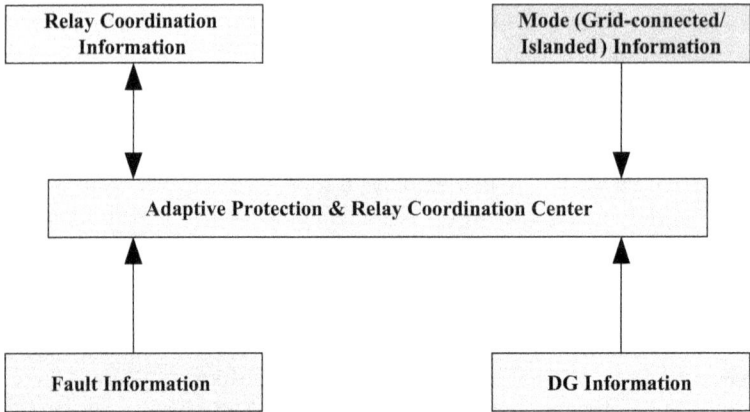

FIGURE 14.15 Adaptive microgrid protection technique.

load points. The adjustment of relay settings is contingent upon the operating mode of the microgrid. An agent-based protection scheme is described in Ref. [102] for a DG integrated distribution network. In this study, the system has been partitioned into distinct sections, and relay agents located at the intersection points of different sections use wavelet coefficient-based techniques to determine the fault direction. Figure 14.15 illustrates a general block diagram of the adaptive microgrid protection technique [4]. A comprehensive study and review of different methods for coordination implementation using conventional, optimization, and adaptive schemes can be found in Table 14.3.

TABLE 14.3
Advantaged and Disadvantages of Protection Coordination Strategies

Protection Coordination Schemes	Advantages	Disadvantages
Curve fitting technique [66, 67]	Used to determine the best suitable function for the representation of data	Large memory is required for relay characteristics storage
Graph theory technique [68–72]	Provides the basis of relay coordination	(i) Determination of MBPS relay is required (ii) Applicable only for fixed topology network
Time-current grading [73–769]	Cheap and economical as communication channel is not required	(i) Complicated and time-consuming method (ii) Applicable only for fixed topology network (iii) Contingency is not considered (iv) Determination of MBPS relay is required

(Continued)

TABLE 14.3 (Continued)

Protection Coordination Schemes	Advantages	Disadvantages
Optimization technique [65, 77–97]	(i) Determination of MBPS relay is not required. (ii) Optimal relay settings are obtained (iii) Less time-consuming	(i) Relay setting obtained by different optimization techniques is different. (ii) Applicable only for fixed topology network
Adaptive protection technique [98–102]	(i) According to the system configuration, the relay settings are updated (ii) An unknown contingency is considered (iii) Fast and more accurate (iv) High speed communication enhances the selectivity, speed, and reliability of the scheme	(i) Expensive due to costly communication channel and equipment (ii) Only numerical relays can be used

14.6 CONCLUSION AND FUTURE SCOPE

This article provides a comprehensive analysis of various protection schemes and strategies that are suitable for AC microgrid systems. All these schemes and strategies seem to be competent and reliable to satisfy the requirements of any advanced protection system. Based on an extensive review of the literature, it is evident that the effectiveness of protection schemes and coordination strategies relies heavily on the system's topology. This is an effort to present all the protection techniques employed in microgrid with their respective pros and cons. It is strict to clear here that limitations concerned with the protection technique are the research gaps for the researchers to make it more effective against all kinds of faults including HIF. The major drawback of adaptive protection techniques is the requirement of the communication link, therefore, there is a great scope of work to enhance the quality of communication network and provide protection against cyberattacks for the secure and safe functioning of future micro smart grid.

REFERENCES

1. Askarian, A. H., Al-Dabbagh, M., Kazemi, K. H., Hesameddin, H. S., and Abul Jabbar Khan, R. (2002). A new optimal approach for coordination of overcurrent relays in interconnected power systems. *IEEE Power Engineering Review*, 22, 60.
2. Beheshtaein, S., Cuzner, R., Savaghebi, M., and Guerrero, J. M. (2019). Review on microgrids protection. *IET Generation, Transmission and Distribution*, 13(6), 743–759.
3. Haron, Ahmad Razani, Mohamed, Azah, and Shareef, Hussain (2012). A review on protection schemes and coordination techniques in microgrid system. *Journal of Applied Sciences*, 12(2), 101–112.
4. Shahzad, U., and Asgarpoor, S. (2017). A comprehensive review of protection schemes for distributed generation. *Energy and Power Engineering*, 9(8), 430–463.

5. Gajbhiye, R. K., De, A., Helwade, R., and Soman, S. A. (2005). A simple and efficient approach to determination of minimum set of break point relays for transmission protection system coordination. *Proc. Int. Conf. Future Power Systems*, 1–5.

6. Salam, M. A., Abdallah, A., Kamel, R., and Hashem, M. (2017). Improvement of protection coordination for a distribution system connected to a microgrid using unidirectional fault current limiter. *Ain Shams Engineering Journal*, 8(3), 405–414.

7. Hooshyar, A. and Iravani, R. (2017). Microgrid protection. *Proceedings of the IEEE* 105(7), 1332–1353.

8. Vasavi, G., Kaviya, V., and Geethamani, R. (2017). Safety assured reliable microgrid. *International Journal of Innovative Research in Science, Engineering and Technology*, 6(14), 81–90.

9. Ram, B., and Vishwakarma, D. N. (2001) *Power System Protection & Switch Gear*. Tata McGraw-Hill Education.

10. Brahma, S. M. and Girgis, A. A. (2002). Microprocessor based reclosing to coordinate fuse and recloser in a system with high penetration of distributed generation. *IEEE Power Eng. Soc. Win. Meeting*, vol. 1, pp. 453–458.

11. Yue, Q., Lu, F., Yu, W., and Wang, J. (2006). A novel algorithm to determine minimum break point set for optimum cooperation of directional protection relays in multiloop networks. *IEEE Transactions on Power Delivery*, 21(3), 1114–1119.

12. Elmitwally, A., Gouda, E., and Eladawy, S. (2016). Restoring recloser–fuse coordination by optimal fault current limiters planning in DG-integrated distribution systems. *International Journal of Electrical Power & Energy Systems* 77(5), 9–18.

13. Choudhary, N. K., Mohanty, S. R., and Singh, R. K. (2014). A review on microgrid protection. *Int. Electr. Eng. Congr. iEECON*, 1–4.

14. Gupta, A., Swathika, O. V. G., and Hemamalini, S. (2015). Optimum coordination of overcurrent relays in distribution systems using big-M and dual simplex methods. *2015 International Conference on Computational Intelligence and Communication Networks (CICN)*, 1540–1543.

15. Wadood, A., Farkoush, S. G., Khurshaid, T., Kim, C. H., Yu, J., Geem, Z. W., and Rhee, S. B. (2018). An optimized protection coordination scheme for the optimal coordination of overcurrent relays using a nature-inspired root tree algorithm. *Applied Sciences*, 8(9), 1664.

16. Stp, S., Verma, P. P., and Swarup, K. S. (2019). A Novel convexified linear program for coordination of directional overcurrent relays. *IEEE Transactions on Power Delivery*, 34(2), 769–772.

17. Tiwari, R., Singh, R. K., and Choudhary, N. K. (2019). Performance analysis of optimization technique for protection coordination in single and multi-loop distribution systems. 2019 International Conference on Electrical, Electronics and Computer Engineering (UPCON), Aligarh, India, pp. 1–6.

18. Sharaf, H. M., Zeineldin, H. H., and El-Saadany, E. (2018). Protection coordination for microgrids with grid-connected and islanded capabilities using communication assisted dual setting directional overcurrent relays. *IEEE Transactions on Smart Grid*, 9(1), 143–151.

19. Yazdaninejadi, A., Nazarpour, D., and Talavat, V. (2018). Optimal coordination of dual-setting directional over-current relays in multi-source meshed active distribution networks considering transient stability. *IET Generation Transmission and Distribution*, 13(2), 157–170.

20. Noghabi, A. S., Sadeh, J., and Mashhadi, H. R. (2009). Considering different network topologies in optimal overcurrent relay coordination using a hybrid GA. *IEEE Transactions on Power Delivery*, 24(4), 1857–1863.

21. Sookrod, P. and Wirasanti, P. (2018). Overcurrent relay coordination tool for radial distribution systems with distributed generation. *2018 5th Int. Conf. Electr. Electron. Eng. ICEEE* 2018, 13–17.

22. Tiwari, R., Singh, R. K., and Choudhary, N. K. (2020). Optimal coordination of dual setting directional over current relays in microgrid with different standard relay characteristics. *2020 IEEE 9th Power India International Conference (PIICON)*, Sonepat, India, pp. 1–6.

23. Tiwari, R., Singh, R. K., and Choudhary, N. K. (2020). A comparative analysis of optimal coordination of distance and overcurrent relays with standard relay characteristics using GA, GWO and WCA. 2020 IEEE Students Conference on Engineering & Systems (SCES), Prayagraj, India, pp. 1–6.

24. Tiwari, R., Singh, R. K., and Choudhary, N. K. (2022). Coordination of dual setting overcurrent relays in microgrid with optimally determined relay characteristics for dual operating modes. *Protection and Control of Modern Power Systems* 7, 6.

25. Alam, M. N., Das, B., and Pant, V. (2016). An interior point method-based protection coordination scheme for directional overcurrent relays in meshed networks. *Electrical Power and Energy Systems*, 81, 153–164.

26. Gopalan, S. A., Sreeram, V., and Iu, H. H. C. (2014). A review of coordination strategies and protection schemes for microgrids. *Renewable and Sustainable Energy Reviews*, 32, 222–228.

27. Al-Nasseri, H., Redfern, M., and Li, F. (2006). A voltage-based protection for microgrids containing power electronic converters. *Proc. IEEE Power Eng. Soc. Gen. Meeting*, 1–7.

28. Al-Nasseri, H., Redfern, M. A., and O'Gorman, R. (2008). Protecting micro-grid systems containing solid-state converter generation. *Proc. Int. Conf. Future Power Systems*, 5.

29. Hou, C. and Hou, X. H. (2009). A study of voltage detection-based fault judgement method in microgrid with inverter interfaced power source. *Int. Conf. Electr. Eng.*, 1–5.

30. Ndou, R., Fadiran, J. I., and Chowdhury, S. (2013). Performance comparison of voltage and frequency-based loss of grid protection schemes for microgrids. *Power and Energy Society General Meeting*, 1–5.

31. Nikkhajoei, H., and Lasseter, R. H. (2006). Microgrid fault protection based on symmetrical and differential current components. *Wisconsin Power Electron. Res. Cent. Dep. Electr. Comput. Eng. Univ.*, 1–72.

32. Casagrande, E., Woon, W. L., Zeineldin, H. H., and Svetinovic, D. (2014). A differential sequence component protection scheme for microgrids with inverter-based distributed generators. *IEEE Transactions on Smart Grid*, 5(1), 29–37.

33. Zeineldin, Hatem H. et al (2006). Distributed generation micro-grid operation: control and protection. *Power Systems Conference: Advanced Metering, Protection, Control, Communication, and Distributed Resources*, 105–111.

34. Sortomme, E., Venkata, S. S., and Mitra, J. (2010). Microgrid protection using communication-assisted digital relays. *IEEE Transactions on Power Delivery*, 25(4), 2789–2796.

35. Miveh, Mohammad, Gandomkar, Majid, Mirsaeidi, Sohrab, and Nasiban, Hosein. (2012). Microgrid protection by designing a communication-assisted digital relay. *American Journal of Scientific Research* (AJSR). 62–68.

36. Naderi, S. B., Davari, P., Zhou, D., Negnevitsky, M., Blaabjerg, F. (2018). A review on fault current limiting devices to enhance the fault ride-through capability of the doubly-fed induction generator-based wind turbine. *Applied Sciences*, 8, 2059–2083.

37. Ando, M., Schweitzer, E., and Baker, R. (1985). Development & field data evaluation of single end fault locator for two terminal HVDC transmission lines Part I: Data collection system & field data. *IEEE Transactions on Power Apparatus and Systems*, PAS-104, 3524–3530.

38. Ando, M., Schweitzer, E., and Baker, R. (1985). Development & field data evaluation of single end fault locator for two terminal HVDC transmission lines Part II: Algorithm & evaluation. *IEEE Transactions on Power Apparatus and Systems*, PAS-104, 3531–3537.

39. Sawai, S., Pradhan, A. K., and Naidu, O. D. (2017). Travelling wave-based fault location of multi-terminal transmission lines. *Proc. IEEE PES Asia-Pac. Power Energy Eng. Conf.*, 1–6.

40. Dewe, M. B., Sankar, S., Arrillaga, J. (1993). The application of satellite time references to HVDC fault location. *IEEE Transactions on Power Delivery*, 8(3), 1295–1302.

41. Shenxing, Shi, Bo, Jiang, Xinzhou, Dong, and Zhiqian, Bo (2010). Protection of microgrid. *10th IET Int. Conference on Developments in Power System Protection (DPSP2010), Managing the Change*, 1–4.

42. Al-Nasseri, H. and Redfern, M. A. (2008). Harmonics content-based protection scheme for micro-grids dominated by solid state converters. *12th Int. Middle East Power Syst. Conf. MEPCON –2008*, 50–56.

43. Chen, Z., Pei, X., Yang, M., Peng, L., and Shi, P. (2018). A novel protection scheme for inverter-interfaced microgrid (IIM) operated in islanded mode. *IEEE Transactions on Power Electronics*, 33(9), 7684–7697.

44. Nunes, J. U. N. and Bretas, A. S. (2010). Impedance-based fault location formulation for unbalanced primary distribution systems with distributed generation. *Proc. Int. Conf. Power Syst. Technol.*, 1–7.

45. Penkov, D. et al. (2005). DG impact on three phase fault location. DG use for fault location purposes. *Proc. Int. Conf. Future Power Systems*, 4, 1–6.

46. Nunes, J. U. N. and Bretas, A. S. (2011). An impedance-based fault location technique for unbalanced distributed generation systems. *Proc. IEEE PES Trondheim PowerTech Power Technol. a Sustain. Soc. POWERTECH 2011*, 1–7.

47. Brahma, S. M. (2011). Fault location in power distribution system with penetration of distributed generation. *IEEE Transactions on Power Delivery*, 26(3), 1545–1553.

48. Henao, C. O., Florez, J. M., and Londono, S. P. (2012). A robust method for single phase fault location considering distributed generation and current compensation. *2012 Proc. 6th IEEE/PES Transm. Distrib. Lat. Am. Conf. Expo.*, 1–7.

49. Cadena, A. B., Henao, C. O., and Florez, J. M. (2012). Single phase to ground fault locator for distribution systems with distributed generation. *Proc. 2012 6th IEEE/PES Transm. Distrib. Lat. Am. Conf. Expo.*, 1–7.

50. Dewadasa, M., Ghosh, A., and Ledwich, G. (2009). An inverse time admittance relay for fault detection in distribution networks containing DGs. *IEEE Reg. 10 Annu. Int. Conf. Proc./TENCON*, 1–6.

51. Chabanloo, R. M., Abyaneh, H. A., Kamangar, S. S. H., and Razavi, F. (2011). Optimal combined overcurrent and distance relays coordination incorporating intelligent overcurrent relays characteristic selection. *IEEE Transactions on Power Delivery*, 26(3), 1381–1391.

52. Lin, H., Liu, C., Guerrero, J. M., and Vasquez, J. C. (2015). Distance protection for microgrids in distribution system. *IECON 2015 – 41st Annu. Conf. IEEE Ind. Electron. Soc.*, 731–736.

53. Baran, M. and El-Markabi, I. (2004). Adaptive over current protection for distribution feeders with distributed generators. *Proc. IEEE PES Power Syst. Conf. Expo. 2004*, 715–719.

54. Voima, S. and Kauhaniemi, K. (2015). Using distance protection in smart grid environment. *Proc. IEEE PES Innov. Smart Grid Technol.*, 1–6.

55. Biswas, S. and Centeno, V. (2017). A communication-based infeed correction method for distance protection in distribution systems. *Proc. North Amer. Power Symp. (NAPS)*, 2–6.

56. Uthitsunthorn, D. and Kulworawanichpong, T. (2010). Distance protection of a renewable energy plant in electric power distribution systems. *Proc. Int. Conf. Power Syst. Technol. Technol. Innov. Mak. Power Grid Smarter, POWERCON–2010*, 1–6.

57. Habib, H. F., Youssef, T., Cintuglu, M. H., and Mohammed, O. A. (2017). Multi-agent-based technique for fault location, isolation, and service restoration. *IEEE Transactions on Industry Applications*, 53(3), 1841–1851.

58. Rao, J. G. and Pradhan, A. K. (2012). Differential power-based symmetrical fault detection during power swing. *IEEE Transactions on Power Delivery*, 27(3), 1557–1564.

59. Li, H. Y. et al. (1997). A new type of differential feeder protection relay using the global positioning system for data synchronization. *IEEE Transactions on Power Delivery*, 12(3), 1090–1097.

60. Alam, M. N. (2019). Adaptive protection coordination scheme using numerical directional overcurrent relays. *IEEE Transactions on Industrial Informatics*, 15(1), 64–73.

61. Voima, S., Kimmo, K., Laaksonen, H. (2011). Novel protection approach for MV microgrid. *21st Int. Conf. Electr. Distrib.*, 0430, 1–4.

62. Ghalei Monfared Zanjani, M., Mazlumi, K., and Kamwa, I. (2018). Application of µPMUs for adaptive protection of overcurrent relays in microgrids. *IET Generation Transmission and Distribution*, 12(18), 4061–4068.

63. Lin, H., Guerrero, J. M., Vasquez, J. C., and Liu, C. (2015). Adaptive distance protection for microgrids. *IECON 2015 - 41st Annu. Conf. IEEE Ind. Electron. Soc.*, 725–730.

64. Farajollahi, M., Shahsavari, A., Stewart, E. M., and Rad, H. M. (2018), Locating the source of events in power distribution systems using micro-PMU data. *IEEE Transactions on Power Apparatus and Systems*, 33(6), 6343–6354.

65. Choudhary, N. K., Mohanty, S. R., and Singh, R. K. (2017). Impact of distributed generator controllers on the coordination of overcurrent relays in microgrid. *Turkish Journal of Electrical Engineering and Computer Sciences*, 25(4), 2674–2685.

66. Albrecht, R. E., Nisja, M. J., Feero, W. E., Rockefeller, G. D., and Wagner, C. L. (1964). Digital computer protective device co-ordination program I-general program description. *IEEE Transactions on Power Apparatus and Systems*, 83(4), 402–410.

67. Zocholl, S. E., Akamine, J. K., Hughes, A. E., Sachdev, M. S., Scharf, L., and Smith, H. S. (1989). Computer representation of overcurrent relay characteristics. *IEEE Transactions on Power Delivery*, 4(3), 1659–1667.

68. Singh, M., Panigrahi, B. K., and Abhyankar, A. R. (2011). Optimal overcurrent relay coordination in distribution system. *Proc. 2011 Int. Conf. Energy, Autom. Signal, ICEAS – 2011*, 2, 822–827.

69. Rao, B. and Rao, S. (1988). Computer aided coordination of directional relay: Determination of break points. *IEEE Transactions on Power Delivery*, 3(2), 545–548.

70. Brown, K. A. and Parker, J. M. (1988). A personal computer approach to overcurrent protective device coordination. *IEEE Transactions on Power Delivery*, 3(2), 509–513.

71. Ramaswami, R. and Mc-Guire, P. F. (1992). Integrated coordination and short circuit analysis for system protection. *IEEE Transactions on Power Delivery*, 7(3), 1112–1119.

72. Jenkines, L., Khincha, H., Shivakumar, S., and Dash, P. (1992). An application of functional dependencies to the topological analysis of protection schemes. *IEEE Transactions on Power Delivery*, 7(1), 77–83.

73. Nikkhajoei, H. and Lasseter, R. H. (2007). Microgrid protection. *IEEE Power Engineering Society General Meeting'07*, 1–6.

74. Zamani, M. A., Sidhu, T. S., and Yazdani, A. (2011). A protection strategy and microprocessor-based relay for low-voltage microgrids. *IEEE Transactions on Power Delivery*, 26(3), 1873–1883.

75. Peng, F. Z., Li, Y. W., and Tolbert, L. M. (2009). Control and Protection of Power Electronics Interfaced Distributed Generation Systems in a Customer-Driven Microgrid. In *Proc. IEEE Power Energy Soc. Gen. Meet.*, 1–8.

76. Abdelaziz, A. Y., Talaat, H. E. A., Nosseir, A. I., and Hajjar, A. A. (2002). An adaptive protection scheme for optimal coordination of overcurrent relays. *Electric Power Systems Research*, 61 (1), 1–9.

77. Martin, K. T., Marchesan, A. C., de Araújo, O. C. B., Cardoso, G., and da Silva, M. F. (2023). Mixed integer linear programming applied to adaptive directional overcurrent protection considering N–1 contingency. *IEEE Transactions on Industry Applications*, 59(3), 2807–2821.

78. Mohamed, E. A., Magdy, G., Shabib, G., Elbaset, A. A., and Mitani, Y. (2018), Digital coordination strategy of protection and frequency stability for an islanded microgrid. *IET Generation Transmission and Distribution*, 12(15), 3637–3646.

79. Fani, B., Bisheh, H., and Sadeghkhani, I. (2017). Protection coordination scheme for distribution networks with high penetration of photovoltaic generators. *IET Generation Transmission and Distribution*, 12(8), 1802–1814.

80. Saleh, K. A., Zeineldin, H. H., Al-Hinai, A., and El-Saadany, E. F. (2015). Optimal coordination of directional overcurrent relays using a new time-current-voltage characteristic. *IEEE Transactions on Power Delivery*, 30(2), 537–544.

81. Razavi, F., Abyaneh, H. A., Al-Dabbagh, M., Mohammadi, R., and Torkaman, H. (2008). A new comprehensive genetic algorithm method for optimal overcurrent relays coordination. *Electric Power Systems Research*, 78(4), 713–720.

82. Deep, K. and Bansal, J. C. (2009). Optimization of directional overcurrent relay times using Laplace crossover particle swarm optimization (LXPSO). *2009 World Congr. Nat. Biol. Inspired Comput. NABIC 2009 – Proc.*, 288–293.

83. Saleh, K. A., Zeineldin, H. H., and El-Saadany, E. F. (2017). Optimal protection coordination for microgrids considering N-1 contingency. *IEEE Transactions on Industrial Informatics*, 13(5), 2270–2278.

84. Bedekar, P. P., Bhide, S. R., and Kale, V. S. (2009). Optimum coordination of overcurrent relays in distribution system using genetic algorithm. *Proc. Int. Conf. Power Syst. ICPS '09*, 1–6.

85. Chakor, S. V. and Date, T. N. (2016). Optimum coordination of directional overcurrent relay in presence of distributed generation using genetic algorithm. 2016 10th International Conference on Intelligent Systems and Control (ISCO), 1–5, Coimbatore, India.

86. Alam, M. N. (2019). Overcurrent protection of AC microgrids using mixed characteristic curves of relays. *Computers and Electrical Engineering*, 74, 74–88.

87. Mohammadi, R., Abyaneh, H. A., Rudsari, H. M., Fathi, S. H., and Rastegar, H. (2011). Overcurrent relays coordination considering the priority of constraints. *IEEE Transactions on Power Delivery*, 26(3), 1927–1938.

88. Jamal, N. Z., Sulaiman, M. H., Aliman, O., Mustaffa, Z., and Mustafa, M. W. (2018). Improved grey wolf optimization algorithm for overcurrent relays coordination. *2018 9th IEEE Control Syst. Grad. Res. Colloquium, ICSGRC 2018 – Proceeding*, no. August, 7–12.

89. Singh, M. and Panigrahi, B. K. (2014). Minimization of operating time gap between primary relays at near and far ends in overcurrent relay coordination. *2014 North Am. Power Symp. NAPS 2014*, 1–6.

90. Kalage, A. A. and Bhuskade, A. (2018). Optimum coordination of directional overcurrent relays using advanced teaching learning-based optimization algorithm. *Proc. IEEE Glob. Conf. Wirel. Comput. Networking, GCWCN 2018*, 2, 187–191.

91. Chelliah, T. R., Thangaraj, R., Allamsetty, S., and Pant, M. (2014). Coordination of directional overcurrent relays using opposition based chaotic differential evolution algorithm. *International Journal of Electrical Power & Energy Systems*, 55, 341–350.

92. Sharma, A. and Panigrahi, B. K. (2018). Phase fault protection scheme for reliable operation of microgrids. *IEEE Transactions on Industry Applications*, 54(3), 2646–2655.

93. Baghaee, H. R., Mirsalim, M., Gharehpetian, G. B., and Talebi, H. A. (2018). MOPSO/ FDMT-based Pareto-optimal solution for coordination of overcurrent relays in interconnected networks and multi-DER microgrids. *IET Generation Transmission and Distribution*, 12(12), 2871–2886.

94. Abdulwahid, A. H. and Wang, S. (2016). A new differential protection scheme for microgrid using Hilbert space-based power setting and fuzzy decision processes. *Proc. IEEE 11th Conf. Ind. Electron. Appl. ICIEA 2016*, 6–11.

95. Rezaei, N. and Haghifam, M. R. (2008). Protection scheme for a distribution system with distributed generation using neural networks. *International Journal of Electrical Power & Energy Systems*, 30(4), 235–241.

96. Muy, T., Anang, T., Ardyono, P., Hery, M. P. (2017). Overcurrent relay modeling using artificial neural network. *International Electrical Engineering Congress (iEECON)*, 2(3), 1–4.

97. Yang, M. T., and Chang, L. F., Blakely, J. N. (Ed.) (2013). Optimal protection coordination for microgrid under different operating modes. In *Mathematical Problems in Engineering*, Taiwan: Hindawi Publishing Corporation, 1–15.

98. Shih, M. Y., Enriquez, A. C., and Trevino, L. M. T. (2014). Online coordination of directional overcurrent relays: Performance evaluation among optimization algorithms. *Elsevier Electric Power System Research*, 110, 122–132.

99. Muda, H. and Jena, P. (2017). Superimposed adaptive sequence current based microgrid protection: A new technique. *IEEE Transactions on Power Delivery*, 32,(2), 757–767.

100. Mahat, P., Chen, Z., Jensen, B. B., and Bak, C. L. (2011). A simple adaptive overcurrent protection of distribution systems with distributed generation. *IEEE Transactions on Smart Grid*, 2(3), 428–437.

101. Wheeler, Keaton A., Faried, Sherif O., Elsamahy, Mohamed (2017). A microgrid protection scheme using differential and adaptive overcurrent relays. *2017 IEEE Electrical Power and Energy Conference (EPEC)*, 1–6.

102. Perera, N. and Rajapakse, A. D. (2006). Agent-based protection scheme for distribution networks with distributed generators. *Proc. IEEE Power Eng. Soc. Gen. Meet. 2006*, Montreal, Canada.

15 A Grid Planning and Optimization Tool for Substation Overload Reduction

Aydin Zaboli, Kuchan Park, Junho Hong, and Wencong Su
University of Michigan-Dearborn, Dearborn, USA

15.1 INTRODUCTION

Recently, as energy issues have been emphasized and the problem of climate change has arisen, many countries have implemented policies such as making renewable energy mandatory or regulating the pollution level of power generation sources. Reducing CO_2 emissions through the installation of PV-BESS and renewable energy resources (RER) is an effective strategy to mitigate climate change and transition to a more sustainable energy system. It is important to note that the effectiveness of CO_2 emission reduction depends on several factors, including the scale of the installation, geographical location, energy demand, and overall energy management practices. Nonetheless, investing in PV-BESS systems and utilizing renewable energy resources are crucial steps toward a greener and more sustainable future (Secchi and Barchi, 2019). In this global trend, the penetration of distributed energy resources (DERs) (e.g., PV, BESS, wind turbines, and EV charging stations) has been increasing recently (Abdmouleh et al., 2015). Since the installation of additional DERs causes uncertainties, low power quality, and cost problems, various considerations, such as upgrading the substation, are required. Over time, as demand and policy for RER increase, the manufacturing capacity of RER will expand, and market prices will decrease. Moreover, the efficiency of a PV system will be increased according to technological advancements (Kenneth et al., 2014). However, since the performance of PV is highly dependent on time and irradiation, an increase in PV penetration causes problems such as the duck curve. That is, power is oversupplied while the sun is up, and power is scarce at night (Hou et al., 2019). Therefore, it is necessary to match supply and demand by limiting the energy produced by PV, which is a disadvantage in terms of energy efficiency. Furthermore, several strategies can help facilitate a better alignment between supply and demand as follows:

DOI: 10.1201/9781003481836-15

- Interconnected grids and energy markets: Interconnecting regional or national grids and establishing energy markets can help balance supply and demand over larger geographic areas.
- Forecasting and advanced analytics: Accurate forecasting of renewable energy generation and demand patterns is crucial for optimizing the match between supply and demand. Leveraging advanced analytics and machine learning algorithms can help improve predictions, enabling more effective planning and management of renewable energy resources (Zaboli et al., 2023).
- Flexible pricing and incentives: Implementing time-of-use pricing or dynamic pricing structures can encourage consumers to adjust their energy consumption based on supply availability and price signals. Incentives (e.g., tax credits) can also promote the adoption of renewable energy technologies (Nouicer et al., 2023).
- Energy storage installation: Implementing energy storage systems (e.g., batteries, pumped hydro storage, or compressed air energy storage) allows excess energy generated during high supply periods to be stored and used during periods of high demand. Energy storage helps balance the intermittent nature of renewable sources by bridging the gap between supply and demand (Pfeifer et al. 2018; Aranizadeh et al., 2019).

Combining PV and BESS gives the reliability and stability of a PV system (Mejia-Giraldo et al., 2019). For example, Hoon and Tan (2020) verified that the duck curve could be relaxed by charging early and discharging at peak times through load shifting. Porzio and Scown (2021) proposed that installing PV-BESS can help achieve carbon reduction goals and can solve energy problems. In addition, new and large-capacity loads such as EV charging stations are steadily increasing in a distribution system (Liu et al., 2012). The installation of these EV chargers can create an overload and voltage drop issue in the distribution system (Mahadeo et al., 2017). However, because of the cost of BESS, it is important to determine BESS size during the planning stage. Knap et al. (2016) considered BESS for the frequency response of a power system. In Khezri et al. (2023), machine learning-based optimal sizing of BESS was proposed. In addition, several research studies have focused on maximizing the benefits of the system. The economic effect of combining PV generation and BESS was demonstrated in Nottrott et al. (2013). However, BESS sizing was not considered. Zhou et al. (2018) proposed a battery control strategy considering the market conditions of the power system. In addition to BESS, research on loads is also steadily increasing. New and large-capacity loads are steadily increasing in a distribution system (Liu et al. 2012). The remaining capacity of the transmission or distribution system must be considered for the installation of facilities such as new PV or EV charging stations. In this case, overloading the facility should be avoided. If capacity is insufficient, the transmission grid or substation should be upgraded. Since these upgrades are expensive and time-consuming, BESS can be installed to replace them. In Kaldellis et al. (2010), the BESS relieves the overload of the substation caused by the PV power plant connection. In Haozhong et al. (2016), scholars verified an economic effect of PV-BESS structures. Nagarajan and Ayyanar (2015) proposed a PV-BESS design methodology considering PV variability, battery life cycle, and substation transformer loss. Therefore, utilities should consider solutions (e.g., substation upgrades or PV-BESS installations) to be prepared for overload issues along with these changes (Yoo et al. 2020). Therefore, a cost-benefit analysis is required to make the best selection.

This chapter proposes a tool for users to facilitate the transition to a grid that includes renewable energy resources. In order to increase overall grid stability with the penetration of new loads (e.g., EV chargers), a substation upgrade is considered. The main contributions of this chapter are listed below:

- To provide a solution to substation overloads based on cost-benefit analysis, a comparison between two scenarios, including a typical substation upgrade and PV-BESS installation, is proposed.
- To develop the GPOT, the environmental impacts of power sources in the form of CO_2 emissions are considered. It finds the installation of PV-BESS as defined by NPV and the optimal substation upgrade. Moreover, GPOT compares them in terms of annual and lifetime NPV, CO_2 emissions, and grid demand.

The rest of this chapter is organized as follows: Section 15.2 represents the substation upgrade and PV-BESS installation in general. Section 15.3 describes the proposed methodology, which explains the proposed GPOT, optimization process, and mathematical modeling. The case studies, including results and discussions, are mentioned in Section 15.4, and finally, the chapter is deduced in Section 15.5.

15.2 A GENERAL OVERVIEW OF SUBSTATION UPGRADE AND PV-BESS INSTALLATION

A high-level diagram based on different parts of the proposed tool is shown in Figure 15.1. As can be observed, there are three types of functions. The first function, called Solar in this figure, takes solar generation and loads data as inputs, and then determines the net demand and the financial benefit. If the amount of solar power generation exceeds the amount of the load, it is curtailed to the maximum load for that hour. The net demand calculated in the first function becomes the input to the second function, called BESS. Other inputs of the second function are battery parameters such as battery capacity, energy capacity, discharging/charging threshold of the battery, and the power rating of the substation. This function is related to the battery. It determines the output power of the battery, the amount of energy stored, the net demand with solar and battery output taken into account, the total cycle, and profits from load shifting. In other words, it acts as a charge/discharge algorithm based on threshold and substation ratings. The third is overload calculation. When a substation upgrade is considered, it uses information about the substation rating, environmental charges, available capacity to extend, grid load information, and net demand with combined solar and BESS outputs to calculate the overload over time, duration, and intensity that the substation can withstand. The results of the GPOT can be shown in three cases of the output as follows: 1) no change in the substation; 2) conventional substation upgrade; and 3) PV-BESS installation. The first case means that the

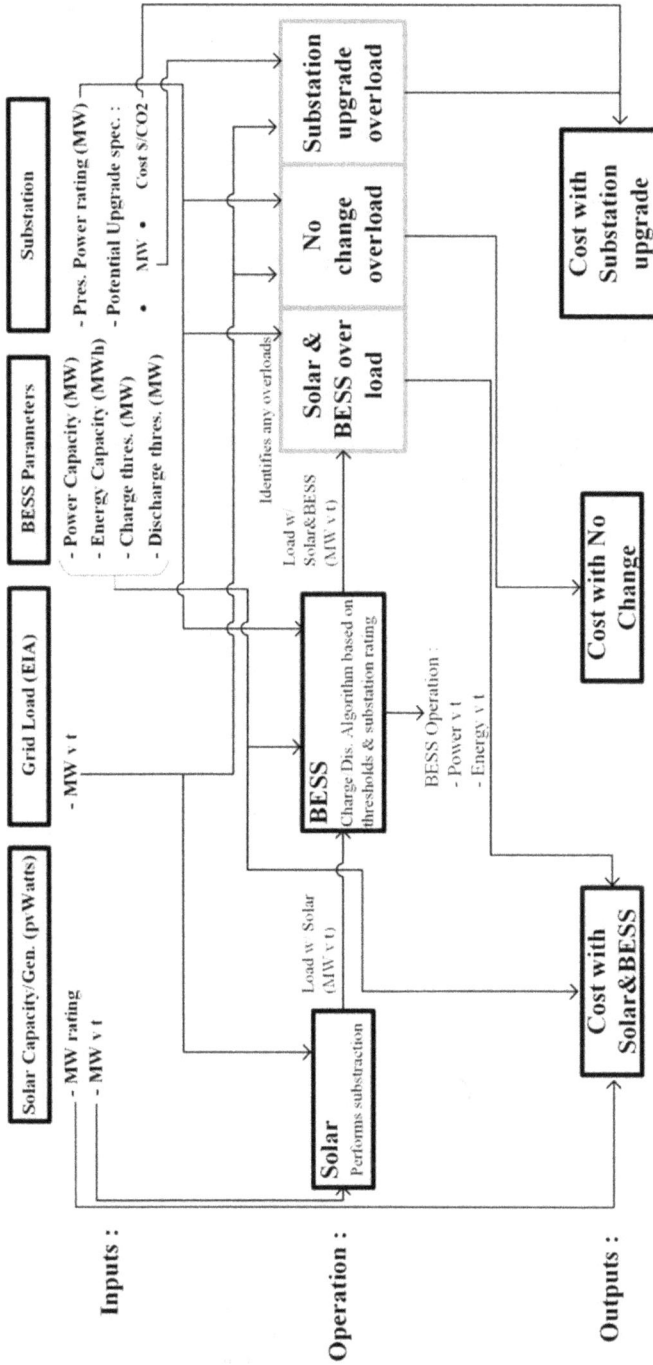

FIGURE 15.1 A high-level framework of the performance of the proposed GPOT.

substation does not require an additional upgrade. The meaning of the second case is the necessity of the substation upgrade because of insufficient capacity. The third case means that although the substation capacity is sufficient, additional installation of PV-BESS is required because the capacity of power generation facilities is insufficient. The cost for each case is calculated using the results obtained through the above three functions (cost-benefit analysis).

15.3 PROPOSED METHODOLOGY

15.3.1 Optimization Procedure and Mathematical Modeling

The performance of this tool based on the optimization process is depicted in Figure 15.2. Data for the initial power grid, such as costs of overloads or energy usage, is created once the scenarios have been provided. The cost of substation overload, in other words, the cost imposed on a substation by overload, can vary depending on several factors, including the severity and duration of the overload, the size and capacity of the substation, and the specific circumstances of the electrical grid involved. Then, the tool will simulate each scenario, and then, once the system's lifetime has elapsed (considered as 30 years), it records the NPVs. Finally, the NPV of the possible outcomes is evaluated, and the maximum values for the PV-BESS and substation upgrade are determined. Moreover, this tool performs a re-simulation to compare further outcomes after analyzing the two optimum outcomes using iterations across other scenarios (Hou et al., 2019; Hoon and Tan, 2020; Ghayoor et al., 2022). Processing solar and BESS datasets, overload computations, and conducting cost-benefit assessments are the key functions employed in the modeling process.

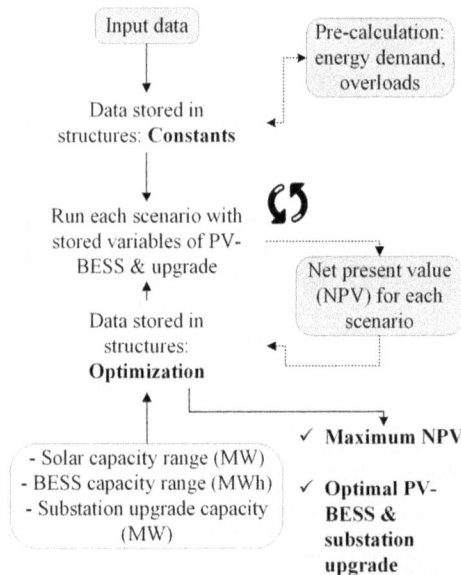

FIGURE 15.2 A flowchart for operational capabilities of an alternative option for substation upgrade by an optimization process.

15.3.2 SOLAR DATA ASSESSMENT

The modeling of the solar part can provide mathematical formulations of the output power of the solar array (MW), P_{pv}, net load, L_{pv} by only the solar array, total energy, E_{pv}, revenue R_{pv} by a solar system in the 1st year, DC capacity, C_{pv}, and AC output power per unit of 1 MW DC of solar, $P_{MW,pv}$ as shown in Eqs. (15.1)–(15.4). A definition of different parameters is mentioned in the Nomenclature section. Solar power is expected to be restricted to a maximum of the load for that hour if it ever surpasses the baseline substation load (Mejia-Giraldo et al., 2019).

$$P_{pv} = P_{MW,pv} \times C_{pv}, \tag{15.1}$$

$$L_{pv} = L_o - P_{pv}, \tag{15.2}$$

$$E_{pv} = \sum P_{pv} \times \Delta t, \tag{15.3}$$

$$R_{pv} = \sum P_{pv} \times F \times \Delta t. \tag{15.4}$$

15.3.3 BESS DATA ASSESSMENT

BESS data processing measures output power (MW), P_{bess}, total energy in 1st year, E_{bess}, net substation load (MW), L_{bess}, number of cycles in 1st year, and revenue from shifting load from peak to off-peak period R_{bess}. Also, N_{bess} shows the logical array for local minimum in E_{bess} with at least 50% discharge. This operation can be managed by a load- or time-based control algorithm. The load-based method charges and discharges the BESS depending on a proportion of the seven-day average load. If the load reaches 110% of the average value, the BESS discharges, and if it falls below 90%, it charges. The discharge factor in the load-based algorithm reduces BESS output when it discharges (a discharge factor of 100% causes the BESS to discharge to reduce the load). The time-based algorithm uses daily periods, a minimum state of charge for overload mitigation, and a "charge-via-solar" ratio. This algorithm will operate within the provided time frames to discharge to the acceptable minimum state of charge (5%) and charge to maximum capacity. The BESS will charge if solar power exceeds the charge-via-solar factor.

Both strategies discharge the BESS to avoid overloads and load ramping (Liang et al. 2017). BESS data processing can be formulated using Eqs. (15.5)–(15.8):

$$L_{bess} = L_{pv} - P_{bess} \tag{15.5}$$

$$E_{bess} = \sum P_{bess} \times \Delta t \tag{15.6}$$

$$R_{bess} = \sum P_{bess} \times F \times \Delta t \tag{15.7}$$

$$N = \sum N_{bess} \tag{15.8}$$

DC–AC and AC–DC efficiencies are expected to be equal for a bidirectional inverter. "PVWatts" (measures the output energy of a PV array over time) sets to 96%

efficiency (PVWatt Calculator, 2022). Lithium-ion batteries charge with 86% round-trip efficiency and 98% DC–DC converter efficiency. Solar energy charging is more economical because of its better efficiency.

15.3.4 Overload Assessment

Eqs. (15.9)–(15.12) determine overloads, duration, and intensity as a percent of sub-station rating, peak time index values, and total energy of the overload. Substation overloads are expected to be treated as regional blackouts (Danish et al., 2020). A detrimental overload happens when the intensity reaches a threshold and lasts for a given number of hours as illustrated in Figure 15.3).

As an example, the substation overload limitations can be defined as greater than 235 and 100% of substation capacity (intensity of overload) for any time duration and 16 hours, respectively.

$$OL = L - C_s, \tag{15.9}$$

$$I_{ol} = \frac{OL_{max}}{C_s} \times 100, \tag{15.10}$$

$$E_{ol,n} = \sum OL \times \Delta t, \tag{15.11}$$

$$E_{ol,d} = \sum OL_d \times \Delta t. \tag{15.12}$$

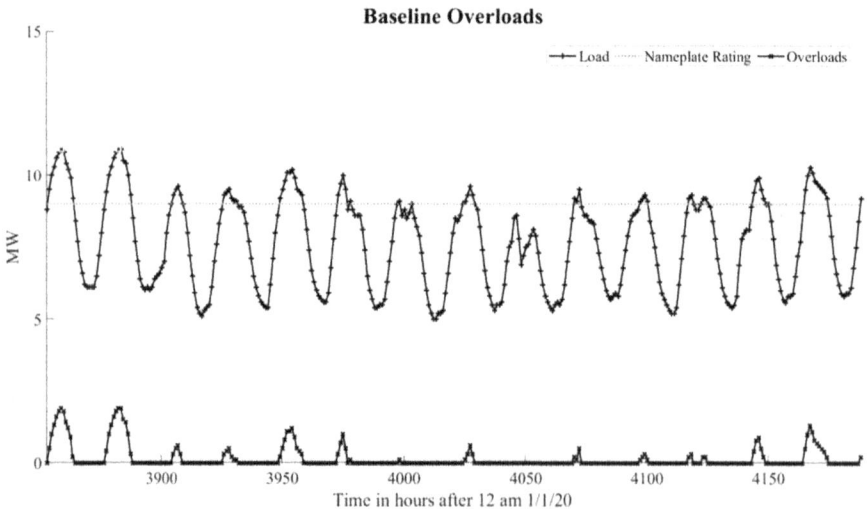

FIGURE 15.3 Overload changes based on the substation nameplate rating.

15.3.5 Cost-benefit Analysis

Yearly CO_2-equivalent emissions, net emissions, annual returns, and net present value (NPV) are the primary metrics computed in the cost-benefit analysis stage. All costs and benefits are in relation to the existing grid. Therefore, a zero-sized installation should have an NPV close to zero. The cost-benefit analysis takes into account the following benefits: decreased overloads, revenue from solar power sales, the solar investment tax credit (applicable exclusively to solar) (Investment Tax Credit, 2022), load shifting through BESS, and lower CO_2-eq emissions. Replacement costs (for batteries and hardware) consist of upfront implementation costs, yearly O&M for all installations, positive CO_2-eq emissions, and excessive ramping in the load. They are all taken into account in the cost-benefit analysis. All of the parameters are considered in the modeling, and details of NPV are described in Eq. (15.13) in which NPV_n shows the NPV at nth year, r is an interest rate, C_{up} is the upfront cost of the project, and CR_{fr} is the annual financial return.

$$NPV_n = -C_{up} + \sum_{k=1}^{n} \frac{CR_{fr}^{(k)}}{(1+r)^k} \qquad (15.13)$$

15.4 CASE STUDIES AND DISCUSSIONS

Case studies have been carried out to find an optimal strategy in particular geographical areas and show a comparison between substation upgrades & PV-BESS installations. It includes the capability of the GPOT in two cities with macro- and micro grid-scales applications for overload relief. Micro and macro grid-scales have been considered as Homestead, FL and Upper Great Plains West (UGPM), and MT regions, respectively. The solar and grid demand datasets based on "PVWatts" and the US Energy Information Administration (EIA) (US Energy Information Administration, 2022) have been used. The constant values, including charge-discharge parameters, PV-BESS and upgrade limitations in MW, discharge factors, charging time, and state of charge (SOC), are not mentioned in this chapter.

15.4.1 A General Description of GPOT

Alternative options for substation upgrades were created using MATLAB in addition to a graphical user interface. This tool assists users in the decision-making process between PV-BESS installation and substation upgrade through an assessment of their comparative costs and benefits. Based on these scenarios, it determines the optimal scenario that can maximize NPV. Several categories containing Input Data, System Specifications, Cost Variables, and Optimization contain the tool's key features are illustrated in Figure 15.4. The "Input Data" tab has options that enable users to enter solar generation and grid load data from the csv format. The "System Specifications" tab has inputs that are utilized by the BESS functionalities, such as permitting the user to select the BESS algorithm that will be used and determining the lifetime of

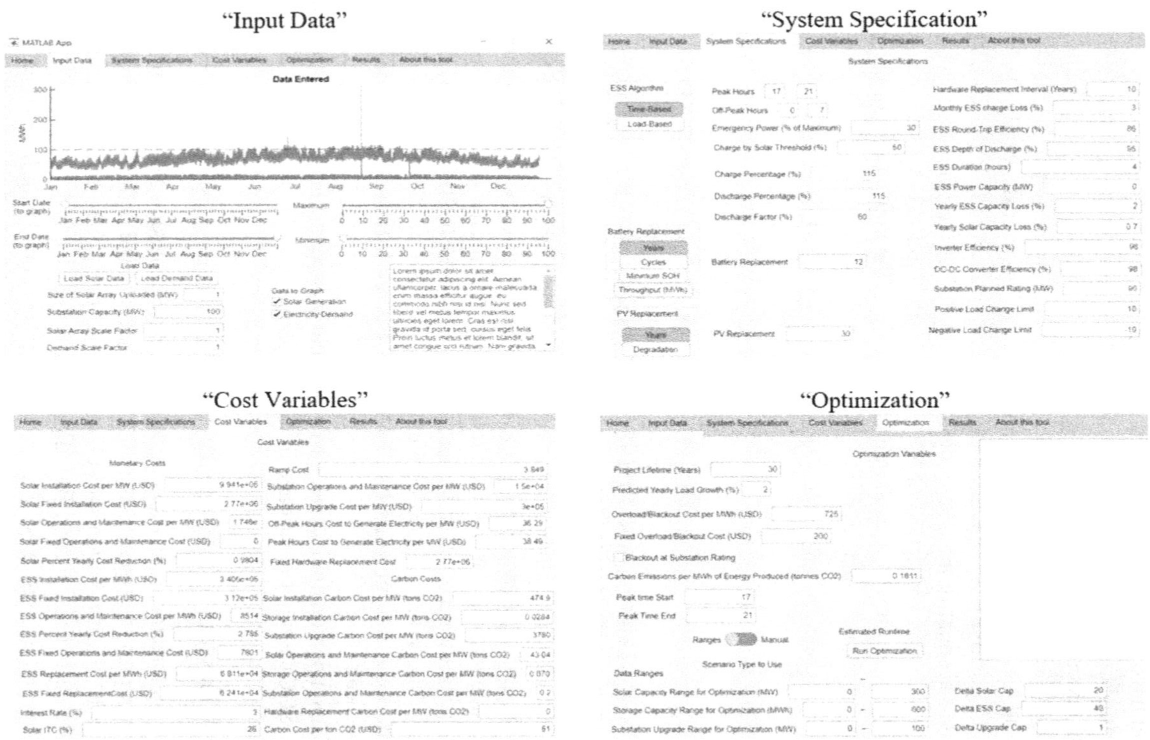

FIGURE 15.4 A demonstration of the proposed graphical user interface along with different taps for data inputs.

the PV and BESS components. The inputs needed to determine the NPV and CO_2-eq costs of PV-BESS and substation upgrade can be found on the "Cost Variables" tab.

Under the "Optimization" tab, there are two sub-tabs in which "Manual Entry" contains data entry tables that allow the user to specify precise individual capacities and corresponding installation, operations, and maintenance costs for PV, BESS, and substation upgrades. The "Optimization Variables" tab includes variables relevant to the optimization, comprising the lifetime and the limitations for PV, BESS, and substation capacity. Another sub-tab, named "Run Optimization," allows the user to start an optimization.

The "Results" section has different parts related to NPV graphs based on the substation upgrade and PV-BESS installation scales. Additionally, it includes multiple sets of values for the best-case scenario for both PV-BESS and substation upgrading, including lifetime NPV, annual O&M expenses, and average annual overload costs.

The "Returns and Carbon Emissions" sub-tab offers graphs for net CO_2-eq and NPV cost. The "Demand Graphs" sub-tab enables the user to plot a number of metrics over the duration of the project's first year of operation. Users can visualize and evaluate overloads from potential scenarios including substation upgrades, PV-BESS installations, and no infrastructure change using the "Overloads" sub-tab. Most GPOT data input capabilities include troubleshooting and pre-processing procedures. These options open the user's default file entry window, which can import the .csv files. If solar and demand data have various sizes, the user will be asked to modify the dataset or upload new information.

The decision to choose Homestead, Florida, and Upper Great Plains West (UGPW), Montana as the locations was based on the availability of public data on the 2020 time-series load from the EIA. Homestead, FL was utilized for the first case study, which focused on a micro-scale implementation with reduced costs and smaller installations aimed at alleviating overloads. On the other hand, UGPW, MT was considered for macro-scale implementation. Solar generation data for the two locations was obtained using PVWatts, a tool that provides information on the time-series energy output of a PV array. Table 15.1 presented showcases the variables that were altered in each case study in order to evaluate the performance of the two charge-discharge algorithms. Apart from utilizing all the default constants, one of the key modifications made was specifying the power capacity of the BESS as a 4-hour BESS in both case studies. Furthermore, the assumption was made that blackouts would occur when the load surpassed the capacity of the substation. These modifications and assumptions were incorporated to facilitate a comprehensive analysis of the effectiveness and efficiency of the charge-discharge algorithms under varying conditions.

15.4.2 Case Study 1: A Microgrid Scale Analysis

Homestead, FL considered a micro-case study. The substation's nominal rating was specified at 9 MW, as indicated in Figure 15.3. The maximum capacities of solar arrays, BESS, and substation upgrades are 5 MW, 20 MWh, and 5 MW, respectively. The time-based algorithm, 12 a.m.–7 a.m. charging, and 15% reserved SOC were found to have the optimal charge-discharge characteristics. As shown in Figure 15.5,

TABLE 15.1

Charge–discharge Parameters Used in Both Case Studies

Load-Based Parameters Tested	Statement	Time-Based Parameters Tested	Statement
Discharge factor (0–100%)	Discharge and charge percent were kept at 100%	Charge time 9 p.m.–5 p.m. (all off-peak), 12 a.m.–7 a.m., 6 a.m.–12 p.m.	Reversed SOC kept at 30%. Charge-via-solar factor kept at 50% and discharge time kept at 5 p.m.–9 p.m.
Discharge percent (100–130%)	Charge percent kept at 100% and the best discharge factor used	Reversed SOC (0–30%)	Best charge time case used. Charge-via-solar factor kept at 50%. Furthermore, discharge time is kept at 5 p.m.–9 p.m.
Charge percent (best discharge percent – 30% × best discharge percent)	Best discharge factor and discharge percent used	—	—

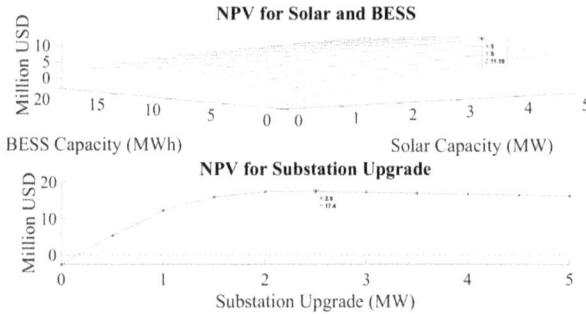

FIGURE 15.5 NPV values for both scenarios for a comparison of financial return after 30 years.

with an NPV of $11.16 million, the PV-BESS installation is less than the substation upgrade's NPV of $17.39 million. Although PV-BESS has a positive NPV within the lifetime of this project (based on Figure 15.6), the most cost-effective investment is the substation upgrade, with $6.23 million greater financial return after 30 years.

The high starting cost of installing a PV array and BESS shows why it has a lower NPV than substation upgrading. The PV-BESS NPV graph always dips to a lower NPV at the outset. Additionally, if only one transformer needs to be replaced rather than an entire substation being upgraded or built from the ground up, the cost per MW of enhanced capacity is only $300,000.

The solar investment tax credit (ITC) is another likely reason, so if enough solar energy is used to charge the BESS, the ITC can be applied to the BESS (Investment Tax Credit, 2022). Finally, the tool may be underestimating the real cost of CO_2

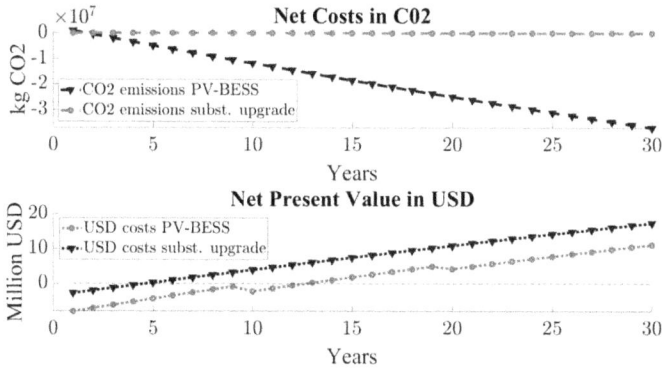

FIGURE 15.6 A comparative cost-benefit analysis of CO_2-equivalent emissions and NPV (case study #1).

emissions and the ensuing climate change. The social cost of carbon is $51/metric ton, although this may not account for all of the potential effects of climate change over the next 30 years.

15.4.3 CASE STUDY 2: A MACROGRID SCALE ANALYSIS

UGPW, MT has been considered a macro-case study and an optimization process. The substation nominal rating was specified at 120 MW because of the bigger scale of this grid. The maximum limitations of solar arrays, BESS, and substation upgrades are 100 MW, 200 MWh, and 50 MW, respectively.

The results show that this case study has a similar pattern with different values for NPV and CO_2 emissions. In both case studies, the maximum NPV for PV-BESS was lower than the NPV for the substation upgrading, i.e., $6.23 and $10.537 million for Case Studies 1 and 2, respectively. Since Homestead, FL, experiences more solar irradiation than Montana, it is conceivable that the first case study was closer to having a greater NPV for PV-BESS. The maximum NPV for Case Study 1 occurs at maximum solar capacity, where the corresponding capacity for Montana is only half of the maximum capacity. The value of the substation upgrade can be seen to be higher in terms of $kgCO_2$ reduction. The load-based BESS method shows the highest NPV when the discharge factor is assumed to be zero. This could be because the BESS conserves charge and discharges only when doing so. It will limit overloads or decrease ramping in the load, both of which are necessary to make a profit.

From the two referenced case studies, it has been observed that employing a load-based BESS algorithm with a zero percent discharge factor optimally enhances the NPV. This phenomenon can potentially be attributed to the BESS's strategy of preserving charge, releasing energy only to counteract overloads and decrease load ramping, both crucial for ensuring positive returns. Consequently, a null discharge factor means that BESS's charging and discharging are unaffected by the discharge percent, leaving the NPV unchanged. Both case studies exhibited this trend.

PV-BESS's NPV is notably less than that of a substation upgrade. A plausible explanation for this difference is the significant initial financial outlay needed to establish a comprehensive solar panel array combined with a battery BESS, evidenced by the notably lower starting point of the NPV curve for PV-BESS.

Further factors that might contribute include the ITC linked to solar energy. In real-world applications, the ITC might extend to the BESS if the energy sourced for BESS charging primarily comes from solar energy, although it was postulated that the ITC would not apply to the BESS. Lastly, there may be a potential underestimation in accounting for emission costs and the ensuing impacts of climate change in the evaluation tool. The methodology of calculating NPV, which incorporates an interest rate, inherently diminishes the perceived influence of future expenses, such as those related to climate change.

15.5 CONCLUSION

PV-BESS installation was compared with substation upgrades in terms of CO_2-eq emissions and NPV using a MATLAB-based tool. Substation upgrade deferral and decreased grid overloads were the main advantages discussed. Time-series data for the load and solar energy and a set of assumptions used mostly for cost estimation all make up the tool's output. It helps to optimize by determining the best capacity to install a PV-BESS and upgrade the substation, maximizing NPVs of both operations. By calculating the NPV over 30 years, it can be found that substation upgrading is the preferable choice for the increased loads in the grids.

It is our purpose to pursue future research on the topic of deferring substations through renewable energy sources and making improvements to the existing tool. First and foremost, the accuracy of the tool, known as GPOT, should be validated by comparing its results with established values of CO_2-eq emissions and NPV for PV-BESS installations and substations that already exist. Additionally, it is important to consider additional costs and benefits in order to reduce assumptions made during the cost-benefit analysis. These considerations may include factors such as overload prices at different time scales (daily, weekly, yearly), the application of ITC to BESS, the impact of high solar penetration on distribution grid integrity (with and without BESS), and taking temperature into account for BESS and solar efficiency. Improving the accuracy of GPOT can be achieved by incorporating more reliable default values for constants. Many of the current values, such as O&M costs for substations and prices for excessive ramping in the load, are arbitrary estimates and can be replaced with more robust data. Furthermore, the optimization algorithm used in the tool can be enhanced by employing a less brute-force approach, leading to improved efficiency. Moreover, it would be beneficial to expand the capabilities of GPOT by optimizing for additional parameters (e.g., the charge-discharge inputs of the BESS). This enhancement would significantly enhance user-friendliness. Furthermore, GPOT could be extended to compare various forms of renewable energy, explore different investments like large-scale power generation infrastructure versus DERs, and consider alternative forms of energy storage (e.g., pumped hydro, thermal storage, and various battery types). In summary, future endeavors should focus on validating the accuracy of GPOT, considering additional costs and benefits, improving default values and optimization algorithms, optimizing for charge-discharge inputs, and expanding the tool's capabilities to incorporate various renewable energy sources, investments, and energy storage options.

ACKNOWLEDGMENT

This work was supported in part by the US National Science Foundation (NSF) under Award No. 1757522. Any opinions, findings, conclusions, or recommendations expressed in this material are those of the author(s) and do not necessarily reflect the views of the US National Science Foundation.

NOMENCLATURE

Δt	Time increment (1 hr)
C_{pv}	DC capacity of the solar array (MW)
C_s	Substation nameplate rating (MW)
E_{bess}	Total energy of the BESS in 1st year (MWh)
$E_{ol,d}$	Total energy of damaging overloads (MWh) to prevent substation damage
$E_{ol,n}$	Total energy of overload (MWh) for blackouts ($L > C_s$)
E_{pv}	Total energy generated by solar in 1st year (MWh)
F	Cost to generate 1 MWh of electricity ($/MWh)
I_{ol}	Intensity of each overload
L	Net BESS/Original load (MW)
L_{bess}	Net PV-BESS load (MW)
l_o	Original load on the substation (MW)
L_{pv}	Net load with solar (No BESS) (MW)
N	Total number of BESS cycles in 1st year
N_{bess}	Logical array for local minimum in E_{bess} with at least 50% discharge
OL/OL_d	Time-series of overloads without/with damaging (MW)
OL_d	Time-series of damaging overloads (MW)
P_{bess}	Output power of the BESS (MW)
$P_{MW,pv}$	AC output power per unit of 1 MW DC of solar
P_{pv}	Output power of the solar array (MW)
R_{bess}	Revenue made by BESS load shifting in 1st year ($)
R_{pv}	Revenue from generated solar energy in 1st year ($)

REFERENCES

Abdmouleh Z., Alammari R., & Gastli A. (2015). Review of policies encouraging renewable energy integration & best practices. *Renewable and Sustainable Energy Reviews*, vol. 45, pp. 249–262.

Aranizadeh A., Zaboli A., Gashteroodkhani O. & Vahidi B. (2019). Wind turbine and ultracapacitor harvested energy increasing in microgrid using wind speed forecasting. *Engineering Science and Technology, An International Journal*, vol. 22(5), pp. 1161–1167.

Danish S. M. S., Ahmadi M., Danish M. S. S., Mandal P., Yona A. & Senjyu T. (2020). A coherent strategy for peak load shaving using energy storage systems. *Journal of Energy Storage*, vol. 32, p. 101823.

Ghayoor F., Ghannadpour S. F. & Zaboli A. (2022). Power network-planning optimization considering average power not supplied reliability index: Modified by chance-constrained programming. *Computers & Industrial Engineering*, vol. 164, p. 107900.

Guide to the federal investment tax credit for commercial solar photovoltaics. Available: https://www.energy.gov/sites/prod/files/2020/01/f70/Guide%20to%20the%20Federal%20Investment%20Tax%20Credit%20for%20Commercial%20Solar%20PV.pdf (Accessed on Jul. 10, 2022).

Haozhong C., Yi Z. & Pinglian Z. (2016). Active distribution network expansion planning integrating dispersed energy storage systems. *IEEE Access*, vol. 10, pp. 638–644.

Hoon J. & Tan R. H. G. (2020). Grid-connected solar PV plant surplus energy utilization using battery energy storage system. *2020 IEEE Student Conference on Research and Development (SCOReD)*, pp. 1–5.

Hou Q., Zhang N., Du E., Miao M., Peng F., & Kang C. (2019). Probabilistic duck curve in high PV penetration power system: Concept, modeling, and empirical analysis in china. *Applied Energy*, vol. 242, pp. 205–215.

Kaldellis J. K., Zafirakis D. & Kondili E. (2010). Optimum sizing of photovoltaic-energy storage systems for autonomous small islands. *International Journal of Electrical Power & Energy Systems*, vol. 32(1), pp. 24–36.

Khezri R. Razmi P., Mahmoudi A., Bidram A. & Khooban M. H. (2023). Machine learning-based sizing of a renewable-battery system for grid-connected homes with fast-charging electric vehicle. *IEEE Transactions on Sustainable Energy*, vol. 14(2), pp. 837–848.

Knap V., Chaudhary S. K., Stroe D., Swierczynski M., Craciun B. & Teodorescu R. (2016). Sizing of an energy storage system for grid inertial response and primary frequency reserve. *IEEE Transactions on Power Systems*, vol. 31(5), pp. 3447–3456.

Liang Y., Su J., Xi B., Yu Y., Ji D., Sun Y., Cui C & Zhu J. (2017). Life cycle assessment of lithium-ion batteries for greenhouse gas emissions. *Resources, conservation and recycling*, vol. 117, pp. 285–293.

Liu Z., Wen F. & Ledwich G. (2012). Optimal planning of electric-vehicle charging stations in distribution systems. *IEEE transactions on power delivery*, vol. 28(1), pp. 102–110.

Mahadeo G. E., Bahadoorsingh S. & Sharma C. (2017). Analysis of the impact of battery electric vehicles on the low voltage network of a Caribbean Island. *2017 IEEE Transportation Electrification Conference and Expo (ITEC)*, pp. 364–369.

Mejia-Giraldo D., Velasquez-Gomez G., Munoz-Galeano N., Cano-Quintero J. B. & Lemos-Cano S. (2019). A BESS sizing strategy for primary frequency regulation support of solar photovoltaic plants. *Energies*, vol. 12(2), pp. 1–16.

Nagarajan A. & Ayyanar R. (2015). Design and strategy for the deployment of energy storage systems in a distribution feeder with penetration of renewable resources. *IEEE Transactions on Sustainable Energy*, vol. 6(3), pp. 1085–1092.

Nottrott A., Kleissl J. & Washom B. (2013). Energy dispatch schedule optimization and cost benefit analysis for grid-connected, photovoltaic-battery storage systems. *Renewable Energy*, vol. 55, pp. 230–240.

Nouicer A., Meeus L. & Delaruw E. (2023). The economics of demand-side flexibility in distribution grids. *The Energy Journal*, vol. 44(1), DOI: 10.5547/01956574.44.1.anou

Pfeifer A., Dobravec V., Pavlinek L., Krajacic G. & Duic N. (2018). Integration of renewable energy and demand response technologies in interconnected energy systems. *Energy*, vol. 161, pp. 447–455.

Porzio J. & Scown C. D. (2021). Life-cycle assessment considerations for batteries and battery materials. *Advanced Energy Materials*, vol. 11(33), p. 2100771.

Priye Kenneth A. & Folly K. (2014). Voltage rise issue with high penetration of grid connected PV. *IFAC Proceedings*, vol. 47(3), pp. 4959–4966.

PVWatt Calculator (2022). National Renewable Energy Laboratory (NREL), Golden, CO, United States. Available at: https://pvwatts.nrel.gov/index.php (Accessed on Aug. 10, 2022).

Secchi M. & Barchi G. (2019). Peer-to-peer electricity sharing: maximising PV self-consumption through BESS control strategies. *2019 IEEE International Conference on Environment and Electrical Engineering and 2019 IEEE Industrial and Commercial Power Systems Europe (EEEIC/ICPS Europe)*, pp. 1–6.

U.S. Energy Information Administration. Available: https://www.eia.gov/, (Accessed on Aug. 15, 2022).

Yoo Y., Jang G. & Jung S. (2020). A study on sizing of substation for PV with optimized operation of BESS. *IEEE Access*, vol. 8, pp. 214577–214585.

Zaboli A., Tuyet-Doan V., Kim Y., Hong J. & Su W. (2023). An LSTM-SAE-based behind-themeter load forecasting method. *IEEE Access*, vol. 11, pp. 49378–49392.

Zhou L., Zhang Y., Lin X., Li C., Cai Z. & Yang P. (2018). Optimal sizing of PV and BESS for a smart household considering different price mechanisms. *IEEE Access*, vol. 6, pp. 41050–41059.

16 Heuristic Techniques and Evolutionary Algorithms in Microgrid Optimization Problems

Aykut Fatih Güven
Yalova University, Yalova, Turkey

16.1 INTRODUCTION

In recent years, energy production has fallen below the required level due to significant changes in living standards, increased use of technology in the production sector, and population growth [1]. As traditional energy sources are rapidly depleting, they are not a suitable option to fill this gap. At present, fossil fuels like oil, natural gas, and coal fulfill more than 70% of the world's electricity needs [2]. Over the years, society has continued to consume these limited resources inefficiently. In the near future, many countries will grapple with challenges like the depletion of fossil fuel reserves, risks associated with nuclear energy use, and environmental pollution. To mitigate these issues, countries are emphasizing and investing in clean renewable energy sources (RESs) [3]. As a result, Hybrid Renewable Energy Systems (HRES), which comprise cost-effective, sustainable, and environmentally friendly energy sources, are garnering significant attention. HRES are versatile, suitable for on-grid and off-grid systems, as well as microgrid and smart grid systems [4].

In Turkey, investigations have underscored the substantial capacity for generating electricity using wind and solar resources [5]. Notably, while renewable energy sources present many benefits, they come with the challenge of inherent inconsistencies and periodic variations [6]. Particularly, solar and wind energy contribute to this unpredictability, as they produce electrical energy with significant fluctuations over short periods, from hours to days [7]. This introduces the challenge of managing multiple energy sources. Hybrid systems employ storage devices, like batteries or fuel cells, to balance energy between production sources and consumers [8]. In designing a hybrid system, the aim is to identify the optimal size of the system, considering both economic and technical indicators [9]. While the financial metric provides information about the costs associated with energy generation, the engineering metric sheds light on the reliability of meeting the load [10]. Consequently, the design objective is to identify the optimal size of a HRES that minimizes overall system costs and ensures the desired level of load reliability. To achieve the ideal system size for specific location demands, one must utilize intelligent optimization

DOI: 10.1201/9781003481836-16

methods anchored in robust, rule-based energy management schemes. These schemes aim to minimize costs and maximize reliability by effectively coordinating power flow [11]. Maintaining a consistent energy supply to satisfy load requirements is vital. Identifying the perfect size to cater to the energy needs of a particular area is complex. This intricacy arises from the unpredictable characteristics of energy sources, the difficulties in establishing a reliable cost evaluation framework, and the intensive computational efforts required by optimization algorithms for sizing [12].

Recent literature has extensively focused on both the technological and economic aspects of HRES [13–20]. Researchers have adopted diverse HRES modeling techniques, power management strategies, and optimization methodologies for scaling. Components like Photovoltaic Panels (PV), Wind Turbines (WT), Diesel Generators (DG), and Batteries (BT) offer multiple combination possibilities in renewable energy systems. In this vein, a technical-economic and environmental analysis was conducted for a significant commercial load in Saudi Arabia [21]. Among all the different system configurations considered, the Solar PV/grid setup emerged as the most cost-effective, achieving the lowest total net present cost (TNPC) of $14.2 million and an affordable levelized cost of energy (LCOE) at $0.115/kWh. The system's renewable energy proportion stood at 44%, correlating with a total emission of 3,370 tons. To achieve the optimal design of a stand-alone HRES in Egypt's Farafra region, Kharrich and his team utilized an advanced variant of the Improved Archimedes Optimization Algorithm (IAOA). They conducted a comparative analysis against five distinct algorithms, aiming to evaluate the proficiency of these algorithms in optimizing HRES design. This assessment considered metrics like convergence speed, solution quality, computational efficiency, and robustness. Numerical results from the simulations highlighted the IAOA's superiority over the other tested algorithms, emphasizing its standout performance [22]. Samy and colleagues [23] devised a power supply strategy for a tourist location in Hurghada, Egypt, using a Hybrid HRES with PV, WT, and fuel cell (FC) units. This strategy considered the financial implications of electricity procurement and its resale to the grid. The system's optimization was achieved using a hybrid firefly and harmony search approach, benchmarked against the particle swarm optimization (PSO) method. The results showcased the system's capability to manage grid buying and selling capacities of 4 GW and 3 GW, respectively. In a related study, Güven et al. [24] assessed the optimal sizing of a stand-alone PV/WT/DG/BT energy system. They employed tools like HOMER, the Ant Colony Optimizer, and the Jaya algorithms. Interestingly, while the system produced an excess of 1.8431×10^6 kWh, the potential to sell this surplus back to the grid wasn't explored. In another investigation, Güven and Samy [25] evaluated stand-alone PV, WT, biomass gasifier (BG), and FC systems for the same location. They harnessed additional wind and solar energy for hydrogen storage and used various algorithms, including the genetic algorithm and the sine–cosine algorithm. Their findings indicated an overproduction of 8.0280×10^5 kWh of energy at a rate of 1.3416 $/kWh. However, the resale of this surplus wasn't considered. It's evident that modern scholarly works are delving deep into the nuances of HRES.

While many studies employ software tools, these often entail longer computation times than traditional optimization techniques. Conversely, metaheuristic techniques tend to yield faster and more effective results. However, the limited support for user-defined functions in many software packages curtails their applicability. Given the complexity of HRES design and the adaptability of metaheuristic techniques, researchers often favor the latter for HRES design solutions.

This study primarily aims to ascertain the optimal size and components of a grid-connected HRES, balancing economic and environmental factors, and offers a comparative performance analysis using metaheuristic algorithms. The study focuses on an energy system comprising PV, WT, DG, and BT to fulfill the grid's demand. We employed the Snake Optimization Algorithm (SOA) for HRES sizing and benchmarked its performance against GWO, RSA, FFA, BRO, and ACO algorithms.

The investigation is structured into five distinct sections to provide a comprehensive analysis. Section 16.2 provides an in-depth examination of the HRES and its individual components, highlighting the intricacies of each part. Section 16.3 offers a detailed overview of the employed methodology, encompassing key aspects such as system modeling, in-depth load demand analysis, and the implementation of an advanced optimization algorithm. Each aspect is meticulously examined to ensure clarity in the methodology. In Section 16.4, the Snake Optimization Algorithm is discussed. Section 16.5 presents clear simulation results for the PV/WT/DG/BT system, followed by an in-depth discussion that explores the nuances and implications of the simulations. Finally, Section 16.6 conducts a thorough evaluation of the results, critically assessing their validity and significance. It also suggests potential avenues for future research, aiming to further the field and provide a holistic analysis of the topic.

16.2 MATERIALS AND METHOD

This section is shedding light on the numerical data measurements, the mathematical representations of power from various energy sources, and the equations being utilized for economic analysis. To navigate the intricacies of optimization, appropriate constraints and an objective function are being chosen. A more comprehensive description of the optimization process follows. The research methodology is illustrated in a process flowchart, as shown in Figure 16.1. All elements of the HRES to be installed are being collated in a database, which includes techno-economic parameters, meteorological data, and load demand data. The project is presumed to have a lifespan of 20 years. The interest rate (IR) is incorporated into calculations to ensure financial accuracy. Metaheuristic optimization algorithms in the MATLAB environment are being used, and the optimal size of the HRES is being determined, taking into account the objective function and constraints. After this sizing process, comprehensive information about the system's energy profile for every second of a year is being collected, as represented by extensive energy data and graphs.

16.2.1 Mathematical Modeling of Hybrid System

To fine-tune an HRES, the initial step involves creating a mathematical representation of the system. This entails the use of equations that characterize the operations of various system components, including PV, WT, DG, and BT. It also includes the

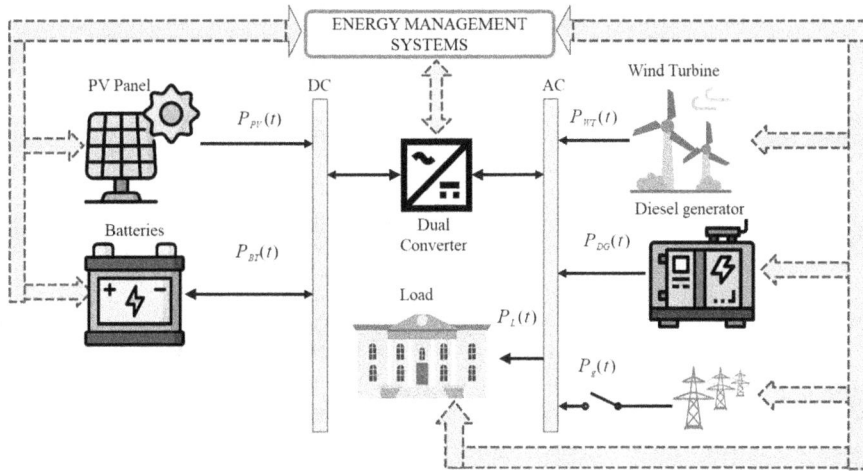

FIGURE 16.1 The recommended on-grid PV/WT/DG/BT HRES model for the study area.

delineation of objective functions and the articulation of the system's energy management strategy. The model must also consider the electrical load that the system is designed to serve. Figure 16.1 presents the block diagram of the suggested grid-connected PV, WT, DG, BT system.

16.2.2 Photovoltaic System Modeling

The power output from photovoltaic (PV) sources is contingent on factors such as solar radiation, temperature, and the properties of the PV array. The researchers use Eq. (16.1) to estimate the power produced by the PV panels (P_{PVout}) [26].

$$P_{\text{pv}_{\text{out}}}\left(t\right) = P_{\text{PV}_{\text{rated}}} \times \frac{G\left(t\right)}{1000} \times \left[1 + \alpha_t \left(\left(T_{\text{amb}} + \left(0.0256 \times G_t\right)\right)\right) - T_{\text{C\_STS}}\right] \quad (16.1)$$

Here, $P_{\text{pv}_{\text{out}}}(t)$ represents the PV module output power (W), $G_{(t)}$ represents the value of solar radiation (W/m^2), $P_{\text{(PV}_{\text{rated}})}$ represents nominal PV power under standard conditions, α_t represents -3.7×10^{-3} (1/°C) as the temperature coefficient, while $T_{\text{C\_STS}}$ and T_{amb} are the temperature values for cell at STC (°C) and the ambient (°C), respectively.

The photovoltaic panel features a rated capacity of 0.345 kW, a temperature coefficient of −0.390, and an operational temperature of 44°C. It also boasts an efficiency rate of 17.8%, a lifespan of 20 years, and a capital cost of $650/kW. The replacement cost is also set at $650/kW, and the annual maintenance and operation costs amount to $50/year.

16.2.3 Wind Turbine Modeling

The power generated by the WT can be estimated using Eq. (16.4), as outlined below [27]:

$$P_{WT} = \begin{cases} 0 & v(t) \le v_{cut-in} \text{ or } v(t) \ge v_{cut-out} \\ P_r \times \dfrac{v(t) - v_{cut-in}}{v_r - v_{cut-in}} & v_{cut-in} \le v(t) \le v_r \\ P_r & v_r \le v(t) \le v_{cut-out} \end{cases} \tag{16.2}$$

P_r represents the WT nominal power (kW), $v(t)$ represents the wind speed (m/s), $v_{cut-out}$ represents the low shear speed of WT (m/s), v_r represents the WT nominal speed (m/s), and v_{cut-in} represents the high shear speed values of WT (m/s).

$$v(t) = v_{ref}(t) \times \left(\frac{H}{H_{ref}} \right)^{a_h} \tag{16.3}$$

Eq. (16.3) articulates a relationship that ties together the wind speed at the reference hub height ($v(t)$) and the wind speed at the anemometer height (v_{ref}). Additionally, it factors in the heights of the anemometer (H_{ref}) and the WT hub (H), along with the values for the exponential power law (a_h).

The coefficient a_h, which is determined by elements like surface texture and the stability of the environment, has a usual span from 0.05 to 0.5. In the context of the locations under consideration, we've adopted an a_h value of 0.14, based on the findings in Ref. [28].

A standard WT boasts a rated power of 1 kW, a tower height of 17 m, a capital cost of $2000/kW, and a replacement cost also at $2000/kW. It incurs an operation and maintenance cost of $200/year and is expected to have a lifespan of 20 years.

16.2.4 Battery Storage Modeling

Given the probabilistic nature of PV cells and WTs outputs, the modeling of battery storage becomes a stochastic process. Thus, accurately determining the appropriate battery size is pivotal for effectively meeting the load demand. The battery's state of charge serves as the primary variable for regulating both overcharging and overdischarging. Overcharging may arise from excessive power generation by the hybrid system or during periods of low load demand. In order to avoid overcharging, the management system takes action and discontinues the charging process once the battery's state of charge reaches its maximum value (E_{BSS_max}). Conversely, when the state of charge hits its minimum value (E_{BSS_min}), the management system curtails the load to safeguard the battery's operational lifespan. The battery's status at any given hour relies entirely on its previous state of charge, playing a pivotal role in harmonizing load demand and supply within the energy management system. The fluctuations in energy production, consumption, and battery charge status are denoted by the hour t and the previous hour ($t - 1$). In scenarios where the PV, WT, and DG sources generate surplus power surpassing the load demand, the excess power is stored in the battery bank, resulting in its charging. The Eq. (16.4) presents an effective method for computing the accessible capacity of the battery bank at hour t without compromising accuracy [29].

$$E_{\text{BSS,ch}}(t) = E_{\text{BSS}}(t-1) \times (1-\sigma) + \left[E_{\text{WT}}(t) + E_{\text{PV}}(t) - \frac{E_L(t)}{\eta_{\text{Inv}}} \right] \times \eta_{\text{BC}} \quad (16.4)$$

When the combined energy output from multiple sources doesn't meet the load requirements, the system will tap into the battery bank, discharging its stored energy. As such, Eq. (16.5) provides an accurate determination of the battery bank's initial available capacity at the start of the discharge process, denoted as hour 1.

$$E_{\text{BSS,disch}}(t) = E_{\text{BSS}}(t-1) \times (1-\sigma) + \left[\frac{E_L(t)}{\eta_{\text{Inv}}} - \left(E_{\text{WT}}(t) + E_{\text{PV}}(t) \right) \right] \times \eta_{\text{BD}} \quad (16.5)$$

Within this framework, the battery's accessible capacity at hour t (measured in kWh) is represented by $E_{\text{BSS}}(t)$, while $E_{\text{BSS}}(t-1)$ signifies the battery's capacity at hour $(t-1)$. The efficiency of battery charging is captured by η_{BC}, the efficiency of battery discharging is symbolized by η_{BD}, and η_{Inv} encapsulates the converter efficiency. The battery's rate of self-discharge is symbolized by σ. E_{WT} highlights the energy output from the WT generator at the specific hour t (in kWh), while E_{PV} points to the energy accumulated by the PV module during the same timeframe. The energy demand at hour t (in kWh) is defined by E_{Load}. When it comes to battery efficiency, η_{BC} stands for its charging phase, η_{BD} for its discharging phase, and the efficiency of the converter is denoted by η_{Inv}.

In the aforementioned equations, η_{BC} and η_{BD} represent the charging efficiency and discharging efficiency of the battery, respectively. These efficiencies are contingent upon the charging current at various stages, playing a crucial role in determining the overall performance of the battery system. It is crucial to acknowledge that the efficiency of both the battery charging and discharging processes can exhibit variability. For the purposes of this study, a constant value of 90% was assumed for the battery's charging efficiency.

In situations where renewable resources generate an abundance of power, the surplus energy finds its sanctuary within the battery banks. However, it is imperative to impose a restraining criterion upon the battery capacity. The reservoir of stored power within the battery bank cannot reach excessive levels. Thus, if the battery attains full charge, any surplus power must be carefully released. To ensure the longevity of the battery, it is crucial to establish a predefined threshold known as the maximum permissible depth of discharge (DOD), which is expressed as a percentage. This threshold serves as a safeguard for the battery's lifespan. In this particular study, the DOD value was held at an auspicious 80%. This signifies that the battery cannot be drained entirely, as the DOD value illustrates the extent to which it may be discharged. The determination of the minimum battery capacity is meticulously executed through the implementation of Eq. (16.6) in the following manner [30].

$$E_{\text{BSS\_min}} = (1 - \text{DOD}) \times E_{\text{BSS\_max}} \quad (16.6)$$

Moreover, the restriction on the battery capacity at each hour is explicitly defined by means of Eq. (16.7).

$$E_{BSS\_min} \leq E_{BSS}(t) \leq E_{BSS\_max} \tag{16.7}$$

Within this framework, DOD designates the upper threshold of permissible battery discharge (expressed as a percentage), while E_{BSS_max} and E_{BSS_min} represent the apex and nadir of allowable battery storage capacity, correspondingly.

This specific battery exhibits a voltage of 600 V, and it boasts a nominal capacity of 100 kWh along with a peak capacity of 167 Ah. It also demonstrates a round trip efficiency of 90%. Notably, its maximum charging current stands at 167 A, it has a minimum state of charge pegged at 20%, and the maximum discharging current it can endure is 500 A. Predicted to have a serviceable life of a decade, this battery carries both a capital cost and a replacement charge of $550.00 per kW.

16.2.5 DIESEL GENERATOR MODELING

In an HRES, a DG plays the role of a compensatory mechanism, activating when the combined power yield from the RESs such as wind and PV, along with the BSS, falls short of the required supply. The fuel intake of this DG is intrinsically tied to its power output, a relationship that is encapsulated by Eq. (16.8) [31].

$$q(t) = a \times P_{DG}(t) + b \times P_r \tag{16.8}$$

In this scenario, $P_{DG}(t)$ stands as the embodiment of power furnished by the DG at any given hour t, measured in kilowatts (kW). $q(t)$ portrays the fuel consumption rate, quantified in liters per hour (L/h). P_r signifies the mean power generation from the DG. The constants a and b, expressed in L/kW, serve as standard benchmarks in fuel consumption parameters, holding respective values of 0.246 and 0.08415.

This DG features a substantial capacity of 1500 kW, accompanied by a replacement expense valued at $175 per kW. It is subject to an operation and maintenance cost of $30 per kWh, and carries an upfront capital expenditure of $175 per kW. The fuel cost is a modest $1 per liter. With a projected lifespan of 10 years, this DG represents a noteworthy investment in the realm of energy production.

16.2.6 INVERTER MODELING

The inverter plays a pivotal role within the system as an electronic entity meticulously designed to convert the direct current (DC) power harnessed from RESs into alternating current (AC) power. Its primary function lies in enabling the seamless and efficient transformation and transmission of electrical power to meet the system's demands.

The efficacy of the inverter, denoted by the symbol (η_{Inv}), represents its ability to convert DC power into AC power while minimizing energy losses. It serves as a crucial metric, quantifying the inverter's efficiency in terms of power conversion. The precise value of (η_{Inv}), can be determined by utilizing Eq. (16.10), as referenced in the literature [32]. The mentioned equation offers a comprehensive framework for assessing and evaluating the efficiency of the inverter, thereby providing valuable insights into its performance within the specific system configuration.

$$P_{\text{inv}}(t) = \frac{P_L(t)}{\eta_{\text{inv}}} \qquad (16.9)$$

Within this framework, $P_L(t)$ and η_{inv} serve as respective representations for load power and inverter efficiency. A system converter inverter, boasting a capacity of 1 kW, is employed in this research. The financial considerations include a capital investment and a replacement cost of $300 each, along with an annual operating and maintenance expense of $50. Notably, the inverter, with an impressive efficiency rating of 95%, is expected to perform reliably over a lifespan of 15 years.

16.2.7 GRID MODELING

A grid operates as a seemingly inexhaustible energy channel, adept at both generating and absorbing power. In circumstances where the amalgamated energy yield from the PV, WT, and BT doesn't fulfill the electrical load demand at a particular juncture, the energy deficit is offset by harnessing power from the grid, while the DG serves as a backup. Eq. (16.10) is employed to assess the revenue accrued from selling excess energy back to the grid.

$$R_{\text{grid}} = \sum_{t=1}^{8760} r_{\text{feed}\_\text{in}} \times E_{\text{grid}\_\text{s}}(t) \qquad (16.10)$$

Here, $r_{\text{feed}\_\text{in}}$ represents the feed-in tariff guarantee rate, which is provided as 0.01 $/kWh and is the actual value. The anticipated power purchase from the grid is computed utilizing Eq. (16.11).

$$C_{\text{grid}} = C_p \sum_{t=1}^{8760} E_{\text{grid}\_p}(t) \qquad (16.11)$$

Here, C_p represents the estimated 1 kW electricity purchase from the grid for the central campus, which is designated as the operating area. This value is provided by the University's Department of Facilities Management at a rate of 0.25 $/kWh, which is the actual value. Generally, the grid availability (GA) can be expressed as Eq. (16.12) [33].

$$GA(\%) = \frac{\text{Grid available in period}}{\text{Hours in the period}} \times 100 \qquad (16.12)$$

16.2.8 ECONOMIC PARAMETERS

During the process of financial analysis, it is vital to give careful thought to the effects of specific variables that influence the economic aspects of HRES. The simulation was conducted using a real interest rate of 2.07% for the year 2022. Moreover, the study set the system's operational duration at 20 years. The constructor/distributor of HRES played a pivotal role by providing up-to-date prices for each individual component, which were seamlessly integrated into the comprehensive simulation process.

16.2.9 LOAD PROFILE OF STUDY AREA

The investigation predominantly emphasized collecting detailed load measurements specific to a location, which is the heart of a university campus positioned at latitude 40°39.2′N and longitude 29°13.2′E in Turkey. Climatic details, covering hourly wind patterns, sunlight intensity, and prevailing temperatures, were obtained from Turkey's Meteorological Authority. By analyzing the annual records of these variables in conjunction with the load data, a comprehensive profile of the research area was carefully constructed. The load profiles of the university campus are illustrated in Figure 16.2, while Figure 16.3 visually presents the profiles of solar radiation, wind speed, and average temperature for the designated research area.

16.3 METHODOLOGY

This chapter presents an overview of the methodologies used to address the research questions of the study. The subsequent subsections provide detailed information on HRES sizing, energy management strategies, objective functions, and optimization algorithms employed in the analysis.

The study's primary focus is on optimizing a grid-connected HRES, comprising PV, WT, BT, and DG. The optimization process is performed using the SOA algorithm, aiming to minimize the annual cost of the system (ACS). The hybrid system is located in Turkey, where the primary RESs are PV and WT, complemented by BT and DG. In optimal sizing, the pivotal step is system modeling. This involves various component models, like PV modules, WT, BT, and DG, as discussed in the previous chapter. Power formulas are employed to gauge energy production considering varying levels of solar exposure and wind velocities. The energy derived from these calculations is subsequently integrated into the system's simulation model. Figure 16.4 provides an outline of the methodology and process in use.

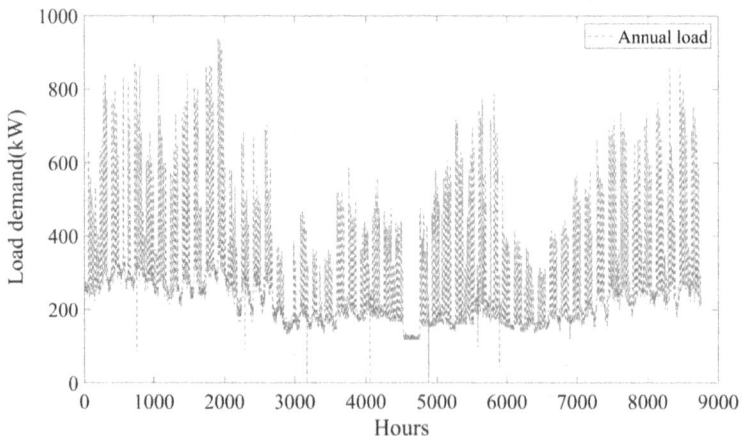

FIGURE 16.2 Annual electricity consumption patterns of the research area.

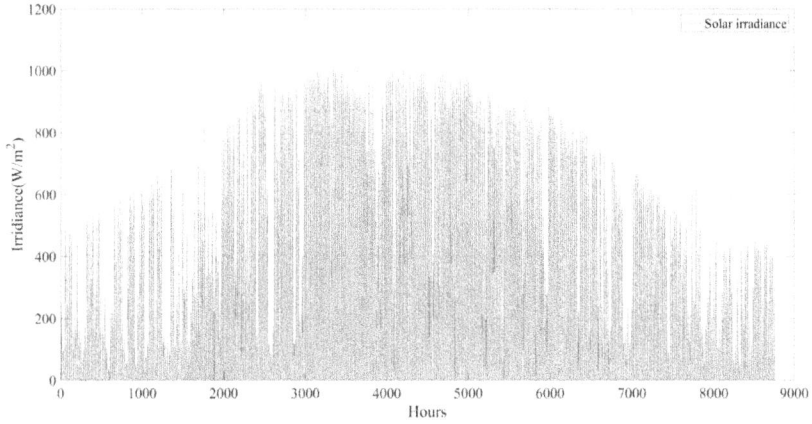

(a)

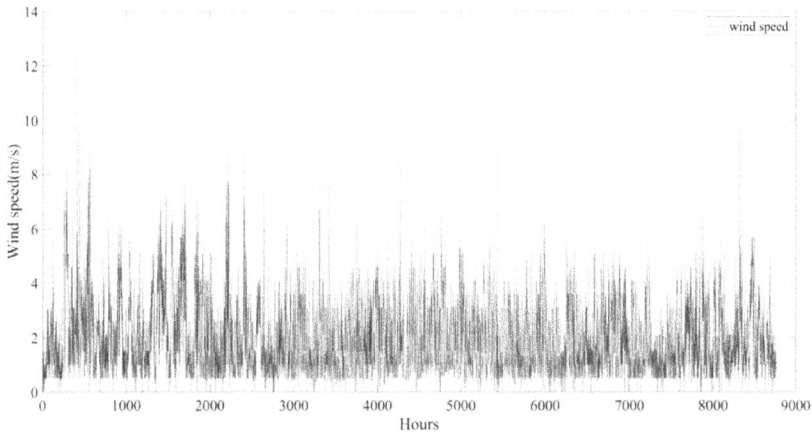

(b)

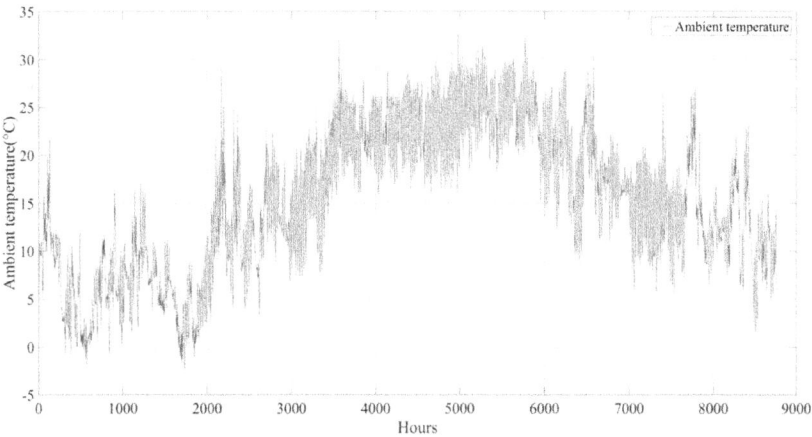

(c)

FIGURE 16.3 The meteorological parameters recorded in the study area (a) solar radiation (W/m²), (b) wind speed (m/s), (c) ambient temperature (˚C).

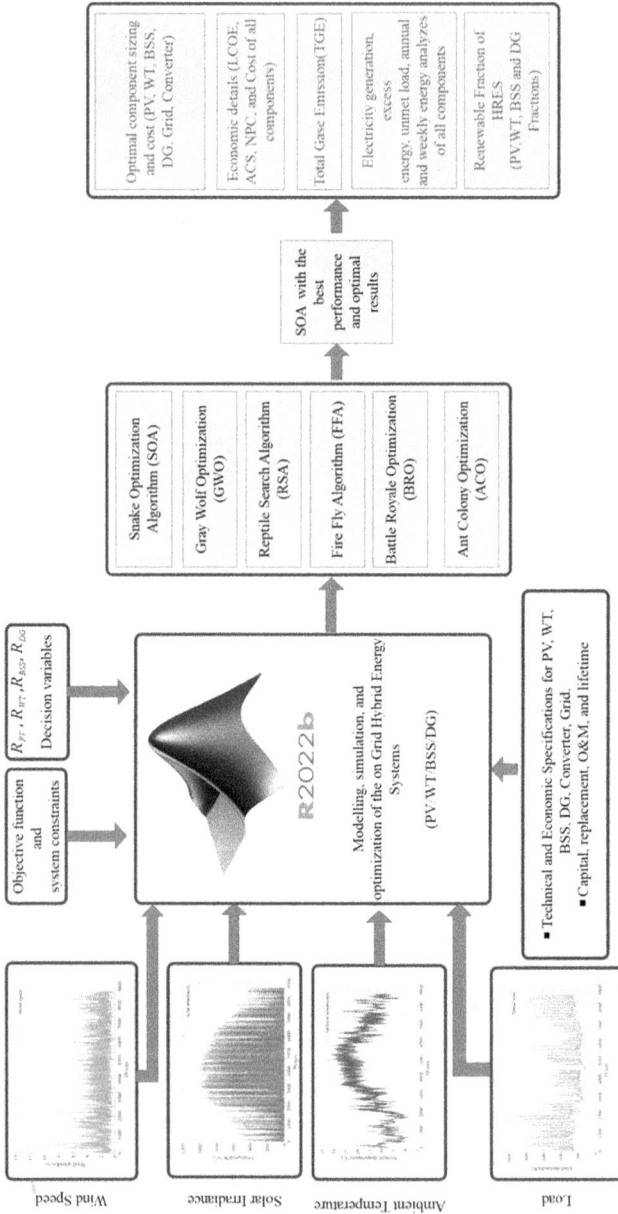

FIGURE 16.4 The methodology and process for sizing hybrid energy systems.

16.3.1 Energy Management Strategy

An energy management strategy is essential for optimizing a grid-connected HRES. The primary aim of energy management is to optimize the utilization of energy harnessed from renewable sources, like PV and WT, to fulfill energy demands while reducing the total system expenditure. In this study, a tiered control approach is adopted to regulate the energy distribution among various components of the hybrid system, encompassing the grid, renewable energy sources, and energy storage mechanisms.

At the top level of the control hierarchy, a supervisory controller is used to set the reference power for each component based on the energy demand and the availability of the renewable sources. The reference power is then distributed to the lower-level controllers, which are responsible for regulating the power flow to and from the different components. For instance, a bidirectional converter is used to manage the power flow between the battery bank and the grid. The converter is controlled using a proportional-integral (PI) controller, which adjusts the charging and discharging rates of the battery based on the reference power and the battery state of charge (SOC).

To enhance the system's stability and reliability, a frequency control strategy is implemented, ensuring the grid frequency remains within an acceptable range. The frequency controller modifies the power output of the DG to counterbalance variations in renewable energy output and load demand. Furthermore, based on current electricity market prices, an economic dispatch algorithm determines the optimal power output for each component of the hybrid system. This algorithm aims to minimize the system's overall cost while satisfying energy demand. Fundamentally, the energy management strategy is essential in optimizing a grid-connected HRES, ensuring its operation is efficient, reliable, and cost-effective. The Energy Management System (EMS) controller operates in four distinct modes:

- **Case 1**: The energy harvested from the PV and WT sources is sufficient to meet the energy requirements. Any excess energy generated will be directed toward charging the battery bank, ensuring that it remains within its maximum capacity.
- **Case 2**: If the renewable energy generated exceeds the energy demand and the battery bank is fully charged, any extra energy will be supplied to the grid.
- **Case 3**: In the event that the energy generated by the PV and WT resources fails to meet the energy demand, the battery bank will come into play to bridge the gap, provided it remains above the minimum battery capacity.
- **Case 4**: When both PV and WT outputs cannot fulfill the energy needs and the battery bank is exhausted, the system has the option to either tap into a DG or pull energy directly from the grid. The choice between these alternatives will be determined by the optimization algorithm, which will consider factors such as cost-effectiveness and feasibility.

Figure 16.5 illustrates the flow algorithm for the EMS, highlighting the system's complex interactions. Additionally, Figures 16.6 and 16.7 elucidate comprehensive flowcharts that illustrate the strategic methodologies employed for charge and discharge operations, respectively.

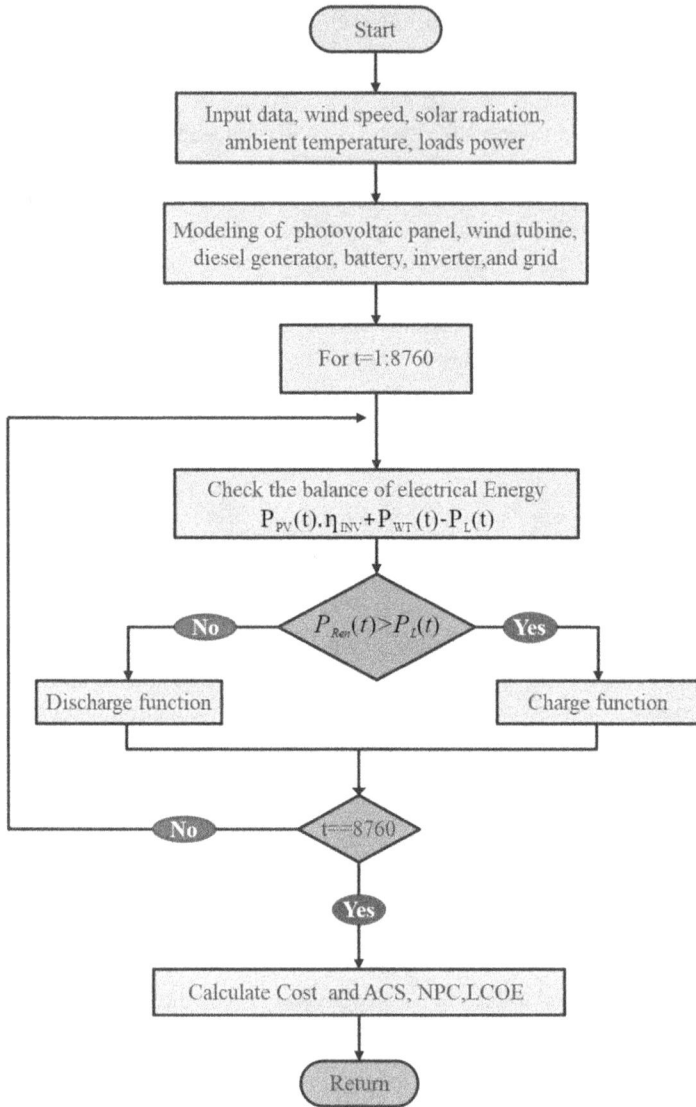

FIGURE 16.5   The flowchart for designing a grid-connected HRES.

16.3.2   OBJECTIVE FUNCTION OF HYBRID ENERGY SYSTEM

In the assessment of the microgrid system, the ACS stands out as the principal objective function. The overarching aim is to refine the system, guaranteeing both a steadfast and dependable power supply and concurrently reducing costs to their bare minimum. The optimal configuration is determined by four pivotal decision factors: WT power, PV power, BT number, and DG power. For economic considerations, the ACS serves as a benchmark, suggesting that the configuration with the minimal ACS

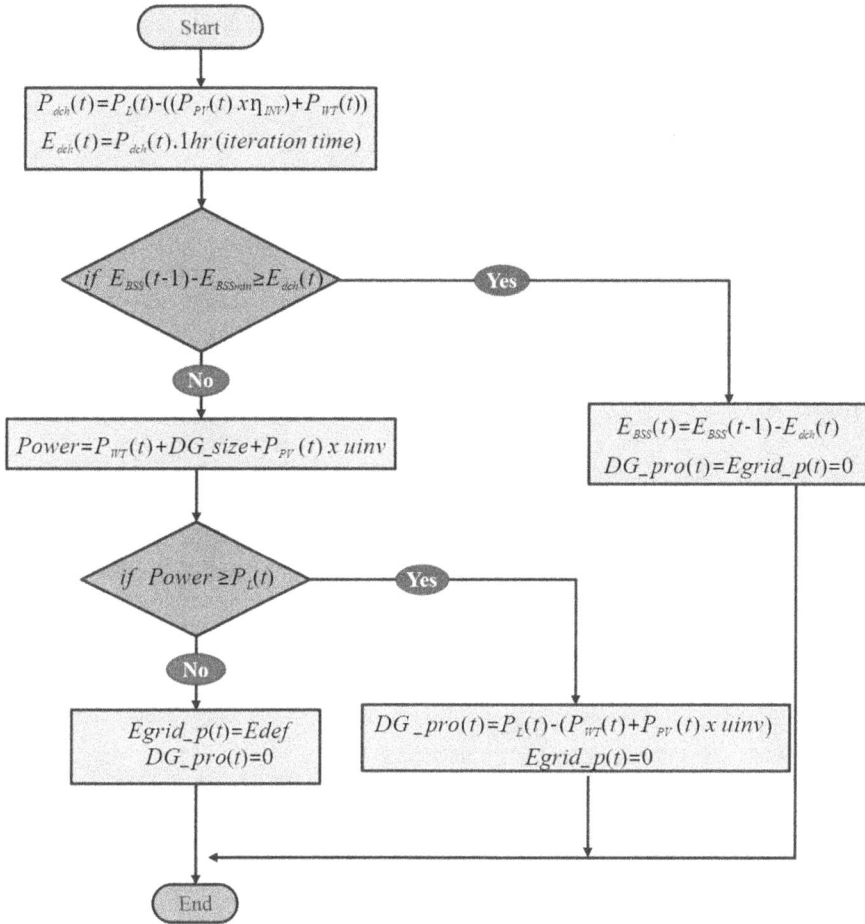

FIGURE 16.6 The flowchart for discharge function.

is deemed optimal, as long as it adheres to all other constraints and parameters. The goal function primarily addresses the comprehensive system expenditure, integrating initial capital outlay, renewable energy-associated costs, and the routine operational and maintenance charges attributed to the system components. The capital cost also accounts for the installation and construction expenses of the components. The salient components of this optimization endeavor encompass ACS, LCOE, TNPC, REF, PV, WT, DG, BT, and the converter's capacity. ACS's objective function is articulated in Eq. (16.13).

$$\text{Objective Function} = \min\left(\text{ACS},\text{TNPC},\text{LCOE}\right)\left\{R_{\text{WT}},R_{\text{PV}},R_{\text{BT}},R_{\text{DG}}\right\} \quad (16.13)$$

The LCOE hinges on the NPC, encompassing investment, operation, and maintenance, as well as replacement costs [34]. In this research, the considered expenses are those associated with PV, WT, BT, and DG. The LCOE, viewed as an essential

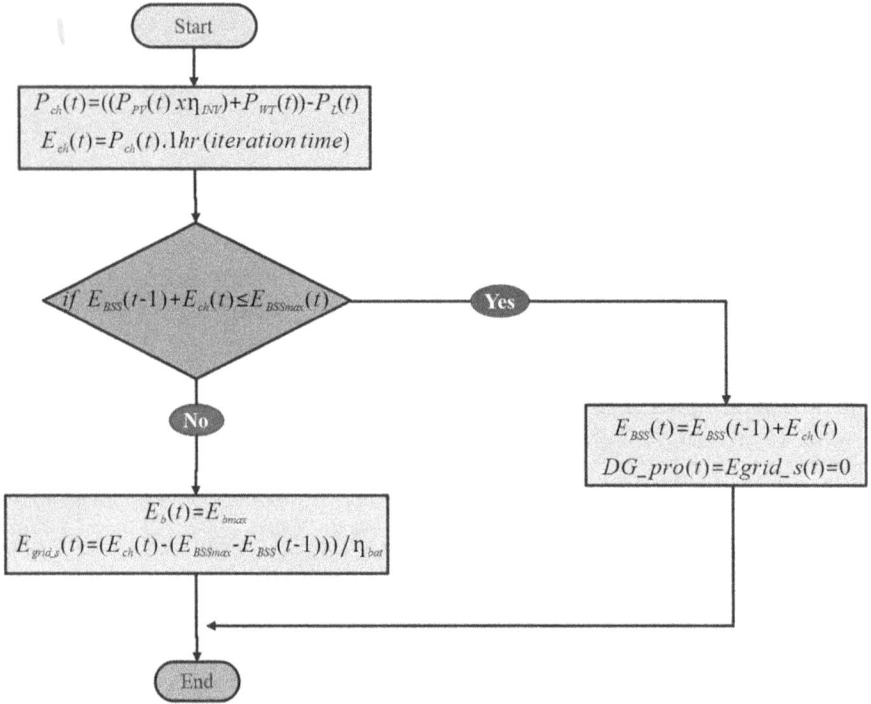

FIGURE 16.7 The flowchart for charge function.

metric for economic assessment targeted for minimization, is determined through the subsequent equations:

$$C_t^{PV} = N_{PV} \left(C_C^{PV} + C_{O\&M}^{PV} \times \left(\frac{(1+i_r)^n - 1}{i_r (1+i_r)^n} \right) \right) \tag{16.14}$$

$$C_t^{WT} = N_{WT} \left(C_C^{WT} + C_{O\&M}^{WT} \times \left(\frac{(1+i_r)^n - 1}{i_r (1+i_r)^n} \right) \right) \tag{16.15}$$

$$C_t^{BT} = C_C^{BT} + C_{O\&M}^{BT} \times \left(\frac{(1+i_r)^n - 1}{i_r (1+i_r)^n} \right) + C_R^{BT} \times \sum_{j=1}^{\left(\frac{n}{n_{BT}}\right)-1} \left(1 + \frac{1}{(1+i_r)^{jn_{BT}}} \right) \tag{16.16}$$

$$C_t^{DG} = C_C^{DG} + C_{O\&M}^{DG} \times \left(\frac{(1+i_r)^n - 1}{i_r (1+i_r)^n} \right) + C_R^{DG} \times \sum_{j=1}^{\left(\frac{n}{n_{DG}}\right)-1} \left(1 + \frac{1}{(1+i_r)^{jn_{DG}}} \right) \tag{16.17}$$

The calculation of the net present cost can be derived by utilizing Eq. (16.18).

$$\text{TNPC} = C_t^{\text{PV}} + C_t^{\text{WT}} + C_t^{\text{BT}} + C_t^{\text{DG}} + C_{\text{C}}^{\text{Inv}} + C_{\text{grid}} - R_{\text{grid}} \tag{16.18}$$

N_{PV} and N_{WT} represent the quantities of PV modules and WT units, respectively. The terms C_{C}^{PV}, C_{C}^{WT}, C_{C}^{BT}, C_{C}^{DG}, and $C_{\text{C}}^{\text{Inv}}$ denote the capital expenditures for PV, WT, BT, DG, and inverter, respectively. The terms $C_{\text{O\&M}}^{\text{PV}}$, $C_{\text{O\&M}}^{\text{WT}}$, $C_{\text{O\&M}}^{\text{BT}}$, and $C_{\text{O\&M}}^{\text{DG}}$ represent the operation and maintenance expenditures for PV, WT, BT, and DG, respectively. C_{R}^{BT} and C_{R}^{DG} are the replacement costs of BT and DG, respectively. n_{BT} and n_{DG} are the lifetimes of BT and DG, respectively.

The LCOE is widely recognized as a pivotal metric in assessing the economic viability of a microgrid system. Representing the cost per unit of electricity, expressed in \$/kWh, it is derived using the formula delineated in Eq. (16.19). In this formula, P_{load} represents the hourly power consumption, indicating the quantity of electrical energy used within each respective hour.

$$\text{LCOE}\,(\$\,/\,\text{kWh}) = \frac{\text{TNPC}}{\sum\limits_{t=1}^{8760} P_{\text{load}}} \times \text{CRF} \tag{16.19}$$

Eq. (16.20) is utilized to convert the initial cost into an annualized capital cost. Within this equation, the capital recovery factor (CRF), as outlined by ref. [35], is applied. In this context, i_r signifies the real interest rate (%), while n denotes the projected lifespan of the HRES system in years.

$$\text{CRF}\,(i_r, n) = \frac{\left[\, i_r \times (1 + i_r)^n \,\right]}{\left[\, (1 + i_r)^n - 1 \,\right]} \tag{16.20}$$

16.3.3 THE LPSP-BASED POWER RELIABILITY CONSTRAINT

In microgrid system evaluations, the reliability metric is quantified by examining the potential occurrence of power supply disruptions, technically defined as the loss of power supply probability (LPSP). Given the inherent uncertainties associated with factors like intermittent solar radiation, fluctuating wind speeds, and potential grid outages, conducting a reliability study becomes an indispensable step in the design process of the proposed system. The LPSP serves as a reliability metric, indicating the potential inability of the hybrid model to satisfy the load demand due to insufficient power supply. This parameter is computed using Eq. (16.21) as described in ref. [36].

$$\text{LPSP} = \frac{\sum\limits_{t=1}^{8760} P_{\text{deficit}}(t)}{\sum\limits_{t=1}^{8760} P_{\text{demand}}(t)} \tag{16.21}$$

In this context, $P_{\text{deficit}}(t)$ represents the power supply shortfall at time t, while $P_{\text{demand}}(t)$ indicates the actual power consumption during that specific interval. The LPSP, as a reliability metric, signifies the likelihood of the hybrid model not meeting the load requirements due to an inadequate power supply. As outlined in reference [37], its value ranges between 0 and 1.

16.3.4 RENEWABLE ENERGY FRACTION (REF)

The "Renewable Energy Fraction" (REF) denotes the proportion of total electricity produced by an electricity system that comes from RESs. This metric provides insights into the extent to which RESs, such as solar, wind, and hydroelectric, contribute to the overall electricity production.

The REF serves as an indicator of an electricity system's sustainability and its shift away from fossil fuel dependency. Furthermore, it offers a perspective on the economic competitiveness of RESs in the energy market.

Many countries prioritize the REF in their energy policies. For instance, the European Union has set a goal to achieve a 32% renewable energy contribution by 2030. If a HRES incorporates diesel as one of its energy sources, the REF can be determined using Eq. (16.22).

$$\text{REF}(\%) = \left(1 - \frac{\sum_{t=1}^{8760} P_{\text{DG}}(t)}{\sum_{t=1}^{8760} P_{\text{PV}}(t) + P_{\text{WT}}(t)}\right) \times 100 \qquad (16.22)$$

To maximize the REF, a pivotal objective function, it's imperative to minimize the equation's second component.

To achieve the maximization of REF, an essential objective function, the focus lies on minimizing the second component of the equation. One of the objectives in the optimization process is to minimize the function by taking into account the second part. It is crucial to note that this function is subject to constraints, ensuring it remains below 100%. Consequently, during the optimization procedure, it is essential to maintain the REF value below the desired threshold (ε_{REF}) as stated in Eq. (16.23) [38].

$$\text{REF}(\%) \leq \varepsilon_{\text{REF}} \qquad (16.23)$$

16.3.5 DESIGN OPTIMIZATION VARIABLES

Design variables in optimization are the variables used to determine the characteristics and performance of the system to be designed. These variables are used to make decisions about the design, components, sizes, capacities, operating parameters, and so on. For instance, in designing a solar energy system, design variables might include the number and type of panels, battery capacity and count, and inverter capacity and efficiency. Optimizing these variables helps improve system performance and reduce costs.

The boundary range of design variables in optimization is determined based on the physical properties of the variable, design constraints, and problem objectives. For instance, when solving an engineering problem using optimization, design variables are typically used to specify a lower and upper limit range. This range represents the minimum and maximum values that the design variables can take. Engineers usually determine this range based on their experience, experimental results, or the boundary ranges used in similar problems. Additionally, some optimization software can automatically determine a boundary range or restrict the boundary range of the optimized function. When setting the boundary range, a penalty function can be used for values outside the boundary range of the variables. This minimizes the impact of the design variables outside the boundary range on the optimization result, and ensures that the variables to be optimized take values close to the boundary range.

In this case, the design variables or decision variables considered are the PV power (R_{PV}), WT power (R_{WT}), BT power (R_{BT}), and DG power (R_{DG}). The Eq. (16.24) outlines the prescribed upper and lower limits for the variables:

$$\text{Design Variables} = \begin{cases} 1\,\text{kW} \leq R_{WT} \leq 5000\,\text{kW} \\ 1\,\text{kW} \leq R_{PV} \leq 5000\,\text{kW} \\ 1\,\text{kW} \leq R_{BT} \leq 800\,\text{kW} \\ 1\,\text{kW} \leq R_{DG} \leq 1000\,\text{kW} \end{cases} \tag{16.24}$$

16.4 SNAKE OPTIMIZATION ALGORITHM (SOA)

Snakes are among the most captivating creatures in the reptile kingdom. As depicted in Figure 16.8, they are ectothermic vertebrates characterized by their limbless, elongated bodies. Their jointed skull structures enable them to consume prey much larger than their heads. Literature research indicates that there are 520 genera and 3,600 species of snakes [39]. Their distinctive reproductive behaviors further add to their intrigue. Females have the capability to influence genotypes through mate selection and the dynamics of sperm competition. Conversely, male–male competition exhibits a variety of behaviors, including mate guarding and strategies that mimic female behaviors. The victor in intense confrontations often plays a pivotal role in mate selection [40].

This section contains information about the inspiration and mathematical model of SOA.

16.4.1 SNAKE MATING BEHAVIOR

Snakes display unique mating behaviors influenced by environmental factors, such as temperature fluctuations and food supply. In colder regions, these reptiles tend to mate as spring transitions to summer. While male snakes compete for a female's favor, the female ultimately decides her mate. After mating, she chooses an apt location to lay her eggs, often in nests or burrows, and typically departs once her offspring hatch. Figure 16.9 provides a visual representation of this behavior.

FIGURE 16.8 Snakes in natural habitat (a) The characteristics and behavior of individual snakes, (b) The occurrence of compact bouts between snakes leading to conduction current density, (c) The specific behaviors of snakes during the egg-laying process [39].

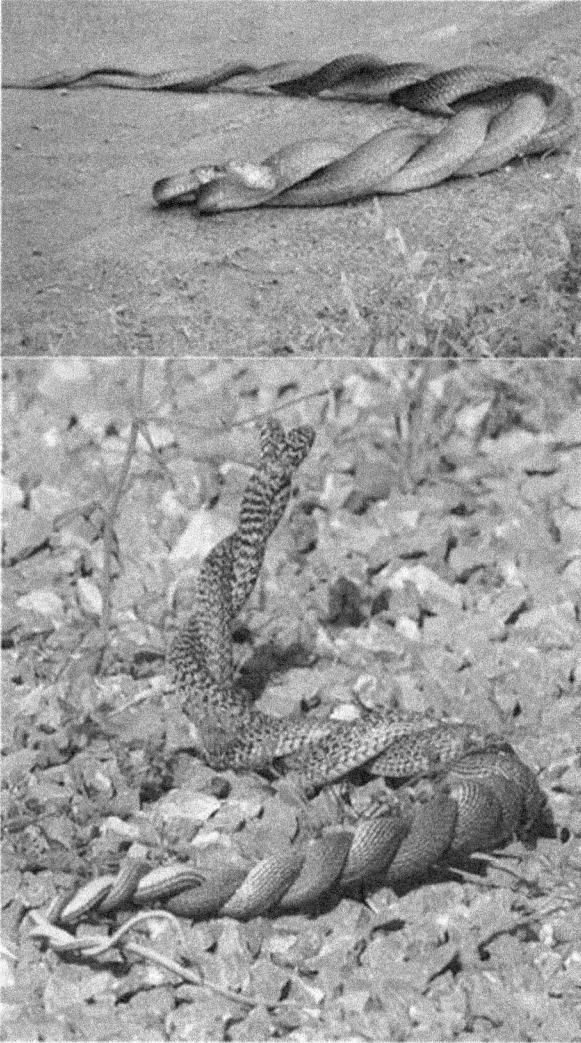

FIGURE 16.9 Captivating sight of snakes engaging in the intricate act of mating [39].

16.4.2 INSPIRATION SOURCE

The SOA, inspired by the intricate mating behavior of snakes, functions based on temperature and food availability. In environments with low temperatures but abundant food sources, mating takes place. Conversely, in cold conditions with limited food, snakes prioritize foraging. The algorithm's search process consists of two main stages: exploration and exploitation. During exploration, snakes focus on nearby food resources, considering factors like cold temperatures and scarce food. In the exploitation stage, several transitional phases enhance the algorithm's global search efficiency. In situations with abundant food and warmer temperatures, snakes prioritize feeding.

When there's ample food in cold conditions, the mating ritual unfolds, divided into two phases: combat and copulation. During combat, every male competes for the dominant female, while each female aims to select the most suitable male. In the copulation phase, mating occurs between pairs, depending on food availability. If mating happens during foraging, the chances of the female laying eggs that produce new snakes increase significantly [41].

16.4.3 THE MATHEMATICAL MODEL OF SOA

The subsequent sections provide a detailed explanation of the mathematical model and algorithm. Like other metaheuristic algorithms, the SOA initiates the optimization process by generating a random population with a uniform distribution. This initial population is derived using the equation provided. The pseudocode for SOA is depicted in Figure 16.10, while its corresponding flowchart can be seen in Figure 16.11. The relevant sections of these figures are elaborated upon in the sections that follow.

```
1: Initialize Problem Setting (Dim, UB, LB, and Pop_Size(N), Max_Iter(T ), Curr_Iter t
2: Initialize the population randomly
3: Divide population N to 2 equal groups N_m and N_f using Eqs. (26) and (27).
4: while (t ≤ T ) do
5:        Evaluate each group N_m and N_f
6:        Find best male f_best,m
7:        Find best male f_best,f
8:        Define Temp using Eq. (28).
9:        Define food Quantity Q using Eq. (29).
10:       if (Q < 0.25) then
11:              Perform exploration using Eqs. (30) and (32)
12:       else if (Q > 0.6) then
13:              Perform exploitation Eq. (34)
14:       else
15:       if (rand > 0.6) then
16:              Snakes in Fight Mode Eqs. (35) and (36)
17:       else
18:              Snakes in Mating Mode Eqs. (41) and (42)
19:              Change the worst male and female Eqs. (43) and (44)
20:              end if
21:       end if
22: end while
23: Return best solution.
```

FIGURE 16.10   The pseudocode for SOA.

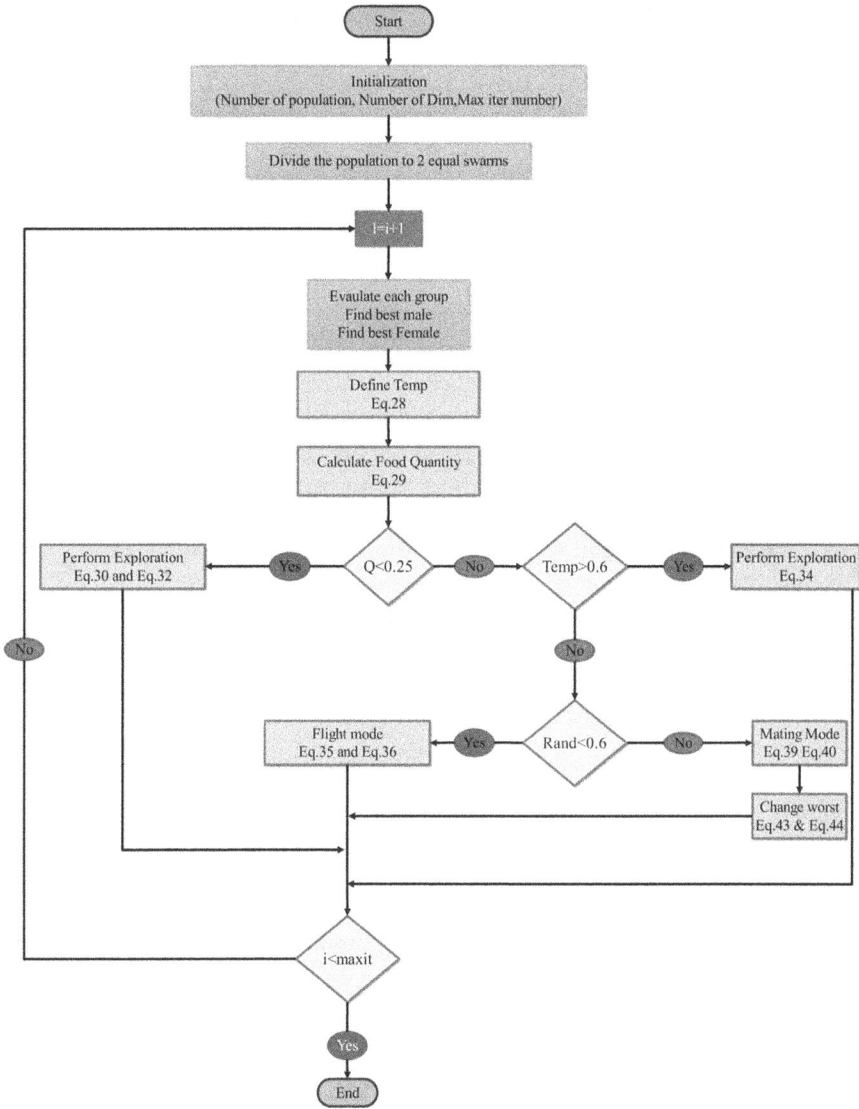

FIGURE 16.11 The flowchart for SOA.

16.4.4 INITIALIZATION

As with all metaheuristic algorithms, SOA begins the optimization algorithm process by creating a random population with a uniform distribution. The initial population is obtained using Eq. (16.25) [42].

$$X_i = X_{\min} + r \times \left(X_{\max} - X_{\min} \right) \tag{16.25}$$

Here, X_i signifies the position of the i-th entity, r is a stochastic number ranging from 0 to 1, and X_{min} and X_{max} stand for the lower and upper bounds of the problem, correspondingly.

- **Partition the group into two equivalent sets: males and females**.
  In this research, we postulate an equal distribution of males and females in the population, with a 50–50 percentage split. The population is bifurcated into two segments: the male cohort and the female cohort. Eqs. (16.26) and (16.27) are employed to segregate the swarm into these dual groups.

$$N_m \approx N/2 \qquad (16.26)$$

$$N_f = N - N_m \qquad (16.27)$$

  Here, N denotes the aggregate number of entities, N_m represents the count of male individuals, and N_f stands for the number of female entities [42].

- **Conduct an evaluation of each cluster, identifying temperature and quantity of sustenance**.
  Identify the optimal individual within each group and secure the finest male (f_{best_m}) and female (f_{best_f}), as well as the location of the food source (f_{food}).

$$\mathrm{Temp} = \exp\left(\frac{-t}{T}\right) \qquad (16.28)$$

  The thermal measure (Temp) is ascertained utilizing Eq. (16.28), wherein t symbolizes the present iteration, and T encapsulates the maximum iteration count. The food quantity (Q) can be deduced from Eq. (16.29), where c_1 denotes a constant value of 0.5 [43].

$$Q = c_1 \times \exp\left(\frac{t-T}{T}\right) \qquad (16.29)$$

- **Exploration Phase – Absence of Sustenance**
  If $Q <$ Threshold (Threshold $= 0.25$), snakes seek nourishment through arbitrary position selection, consequently updating their respective positions. Eqs. (16.30)–(16.33) are defined to model the exploration phase [44].

$$X_{i,m}(t+1) = X_{rand,m}(t) \pm c_2 \times A_m \times (X_{max} - X_{min}) \times \mathrm{rand} + X_{min} \qquad (16.30)$$

  Here, $X_{i,m}$ denotes the position of the i-th male, while $X_{rand,m}$ stands for the position of a randomly chosen male. The term 'rand' is a random number that falls within the range of 0 and 1. A_m nomenclature encapsulates the proficiency of the male individual in the acquisition of sustenance, the computation of which can be carried out as delineated below:

$$A_m = \exp\left(\frac{-f_{\text{rand},m})}{f_{i,m}}\right) \qquad (16.31)$$

Here, $f_{\text{rand},m}$ represents the fitness of $X_{\text{rand},m}$, the position of a random male, and ($f_{i,m}$) represents the fitness of an individual male in the group, and c_2 is a constant equal to 0.05.

$$X_{i,f} = X_{\text{rand},f}\left(t+1\right) \pm c_2 \times A_f \times \left(X_{\max} - X_{\min}\right) \times \text{rand} + X_{\min} \qquad (16.32)$$

In this framework, $X_{i,f}$ signifies the geographical positioning of an identified female entity, while $X_{\text{rand},f}$ captures the location of a randomly selected female counterpart, with 'rand' being a probabilistic value constrained between 0 and 1. Additionally, A_f symbolizes the foraging capacity of a female individual, the computation of which can be illustrated as follows:

$$A_f = \exp\left(\frac{-f_{\text{rand},f})}{f_{i,f}}\right) \qquad (16.33)$$

Here, $f_{\text{rand},f}$ represents the fitness of the randomly selected female, X_{rand}, and $f_{i,f}$ represents the fitness of the i-th individual in the female group [45].

- **Exploration Phase-Food is available**
  If Q > Threshold. If the thermal measure > Threshold (0.6) (hot). The snakes will maneuver solely toward the food source when sustenance is available.

$$X_{i,j}\left(t+1\right) = X_{\text{food}} \pm c_3 \times \text{Temp} \times \text{rand} \times \left(X_{\text{food}} - X_{i,j}\left(t\right)\right) \qquad (16.34)$$

Here, $X_{i,j}$ represents the position of an individual (male or female), X_{food} is the position of the best individuals, and c_3 is a constant equal to 2.
  If the thermal measure < Threshold (0.6) (cold) The snakes will engage in either combat or mating mode.

$$X_{i,m}\left(t+1\right) = X_{i,m}\left(t\right) + c_3 \times \text{FM} \times \text{rand} \times \left(Q \times X_{\text{best},f} - X_{i,m}\left(t\right)\right) \qquad (16.35)$$

Here, $X_{i,m}$ represents the male agent's position, $X_{\text{best},f}$ represents the position of the best female individual in the female group, and FM represents the male agent's fighting ability.

$$X_{i,f}\left(t+1\right) = X_{i,f}\left(t+1\right) + c_3 \times \text{FF} \times \text{rand} \times \left(Q \times X_{\text{best},m} - X_{i,F}\left(t+1\right)\right) \qquad (16.36)$$

Here, $X_{i,f}$ represents the female's position, $X_{\text{best},m}$ represents the position of the best male individual in the male group, and FF represents the female agent's fighting ability.

FM and FF can be calculated using the following Equations (16.37) and (16.38):

$$FM = \exp\left(\frac{-f_{\text{best},f})}{f_i}\right) \tag{16.37}$$

$$FF = \exp\left(\frac{-f_{\text{best},m})}{f_i}\right) \tag{16.38}$$

Here, $f_{\text{best},f}$ is the fitness of the best individual in the female group, $f_{\text{best},m}$ is the fitness of the best individual in the male group, and f_i is the fitness of the agent.

This is the Mating Mode:

$$X_{i,m}(t+1) = X_{i,m}(t) + c_3 \times M_m \times \text{rand} \times \left(Q \times X_{i,f}(t) - X_{i,m}(t)\right) \tag{16.39}$$

$$X_{i,f}(t+1) = X_{i,f}(t) + c_3 \times M_f \times \text{rand} \times \left(Q \times X_{i,m}(t) - X_{i,f}(t)\right) \tag{16.40}$$

Here, the position of the i-th individual within the female group is denoted by $X_{i,f}$, whereas $X_{i,m}$ represents the position of the i-th individual within the male group. Furthermore, the mating ability of the male and female individuals is symbolized by M_m and M_f, respectively. The calculations for these parameters can be derived using the following formulas [46]:

$$M_m = \exp\left(\frac{-f_{i,f})}{f_{i,m}}\right) \tag{16.41}$$

$$M_m = \exp\left(\frac{-f_{i,m})}{f_{i,f}}\right) \tag{16.42}$$

In the event of an egg hatching, the most inferior male and female are identified and replaced [47].

$$X_{\text{worst},m} = X_{\min} + \text{rand} \times \left(X_{\max} - X_{\min}\right) \tag{16.43}$$

$$X_{\text{worst},f} = X_{\min} + \text{rand} \times \left(X_{\max} - X_{\min}\right) \tag{16.44}$$

Herein, $X_{\text{worst},m}$ identifies the least successful member of the male population, while $X_{\text{worst},f}$ pinpoints the least effective participant from the female demographic. The flag direction operator $\pm$, alternatively known as the diversity factor, enables opportunities for amplification or diminution of positional solutions, which ultimately sways the directionality of variables, enhancing a robust exploration of all potential routes within the search field. This factor, produced randomly to present an unexpected dimension, is an integral element in the machinery of any metaheuristic algorithm [48].

16.5   SIMULATION RESULT AND DISCUSSION

In this investigation, we constructed an objective function aimed at minimizing cost metrics—including ACS, TNPC, and LCOE—for a HRES composed of grid-connected PV, WT, DG, and battery storage system (BSS) elements. We deployed state-of-the-art algorithms such as SOA, GWO, RSA, FFA, BRO, and ACO to secure an optimal solution and proceeded to compare their relative efficiencies. Our optimization of HRES took into account numerous constraints, incorporating LPSP_max, REF, SOC, and optimal sizing, in order to ensure the system's utmost reliability in satisfying electricity demand. Furthermore, a sensitivity analysis was performed to assess the impact of varying LPSP_max values on overall system performance. Notably, as we scrutinized configurations with LPSP values of 0.5%, 2%, 5%, and 10%, we observed a marked decrease in ACS, TNPC, and LCOE values. The results of our optimization efforts are articulated in Tables 16.1–16.3.

The energy costs obtained in this study can be compared with previous studies. Güven et al. investigated independent PV/WT/DG/BSS systems in a similar location [24]. In this study, conducted assuming an LPSP value of zero and using the Harmony Search optimization algorithm, the following results were obtained: LCOE of 0.2012 $/kWh, TNPC of 6.5027×106 ($), ACS of 4.0084×105 ($), and Wasted Energy of 1.8431×106 (kWh). Additionally, Güven and Samy [25] conducted a different study based on the techno-economic analysis of WT, PV, BG, and FC systems. Using the HFGA, the optimal system configuration was determined to be 1094.68 kW WT, 2256.17 kW PV, and 775 H2 storage tanks, resulting in an ACS of $2921702.3, TNPC of $2.4639 \times \$107$, and LCOE of $1.3416/kWh. The SOA utilized in this paper notably excelled above other optimization methodologies with respect to production cost, size of renewable resources, and convergence duration. In Table 16.2, when the LPSPmax value was set to 0.5%, the LCOE of HRES was calculated as $0.1511/kWh, with TNPC of 3.3406×10^6 and ACS of 3.8302×10^5, and 100% REF was achieved. Moreover, a revenue of 8.0104×10^3 MWh energy can be obtained by selling it to the grid throughout the year.

The designs of the SOA, GWO, RSA, FFA, BRO, and ACO algorithms in Table 16.1 were applied separately, taking into account all details such as project costs, maintenance and operating costs, location coordinates, project lifespan, and the number and prices of hybrid power system components, to obtain results with high accuracy in minimum time, and the outcomes were evaluated.

According to the optimization results, the proposed system meets the total energy demand through solar and wind energy sources, batteries, and grid support. This way, an optimal cost-effective, technically feasible, environmentally friendly, and socially applicable hybrid system is created. Based on the optimization results, where the lower limit of the WT is set to 1 kW, the WT result is calculated as 1 kW. However, if the lower limit were considered as 0, it could be seen that all energy demand could be met by solar energy sources, batteries, and grid support. Figure 16.12 depicts the monthly energy equilibrium over the span of a year. As indicated in Figure 16.12, solar and wind resources have been evaluated in accordance with the prevailing conditions. During 7 months of the year (January, February, March, April, October, November, and December), the energy demand is met by photovoltaic panels (PV), batteries (BT), and the grid, while in the remaining 5 months, only PV and BT usage are observed.

TABLE 16.1

The Energy Analysis Results of Optimization for Optimal Sizing with a Maximum LPSP of 0.5% Using Diverse Methods Such as SOA, GWO, RSA, FFA, BRO, and ACO

	SOA	GWO	RSA	FFA	BRO	ACO
Time Elapsed (sec)	393.65	314.15	460.25	356.43	377.59	503.58
Power of WT (kW)	1	1.038202	1	1	7.0153	51.4949
Power of PV (MW)	1.9900	1.9901	1.9485	2.3239	2.1236	2.7782
BSS Units	1.5649×10^2	1.5678×10^2	1.1451×10^2	4.1488×10^2	2.1890×10^2	2.5926×10^2
Energy of WT (kWh)	125.5744	130.3717	125.5744	125.5744	880.9451	6.4664×10^3
Energy of PV (MWh)	2.9124×10^3	2.9125×10^3	2.8517×10^3	3.4010×10^3	3.1079×10^3	4.0659×10^3
Energy of DG (kWh)	—	—	—	—	—	99.1936
Wasted Energy (kWh)	—	—	—	—	—	—
Total Load Demand (MWh)	2.5357×10^3	2.5357×10^3	2.5357×10^3	2.5357×10^3	2.5357×10^3	2.5357×10^3
TGE (kg/year)	0	0	0	0	0	69.2
BSS input Energy (kWh)	1.8920×10^5	1.8918×10^5	1.9674×10^5	1.4524×10^5	1.7306×10^5	1.1664×10^5
BSS output Energy (kWh)	7.5061×10^5	7.5067×10^5	7.2927×10^5	8.5873×10^5	7.8880×10^5	9.1409×10^5

TABLE 16.2

The Economic Analysis Results of Optimization for Optimal Sizing with a Maximum LPSP of 0.5% Using Diverse Methods Such as SOA, GWO, RSA, FFA, BRO, and ACO

	SOA	GWO	RSA	FFA	BRO	ACO
LCOE ($/kWh)	0.1511	0.1511	0.1514	0.1528	0.1522	0.1636
Grid sale (MWh)	8.0104×10^3	1.1059×10^3	1.0690×10^3	1.4797×10^3	1.2602×10^3	2.1083×10^3
Grid Purchased (MWh)	1.1058×10^3	8.0095×10^2	8.2973×10^2	6.4013×10^2	7.4264×10^2	5.5090×10^2
TNPC ($)	3.3406×10^6	3.3415×10^6	3.2391×10^6	4.0884×10^6	3.6348×10^6	4.9717×10^6
REF (%)	100	100	100	100	100	1000
NPC of WT ($)	6.5759×10^3	6.8271×10^3	6.5759×10^3	6.5759×10^3	4.6132×10^3	3.3863×10^3
NPC of PV ($)	2.9096×10^6	2.9097×10^6	2.8489×10^6	3.3977×10^6	3.1049×10^6	4.0620×10^6
NPC of DG ($)	433.7446	576.2867	1.0828×10^3	433.7446	1.9350×10^3	9.0363×10^4
NPC of BSS ($)	1.8161×10^5	1.8195×10^5	1.3290×10^5	4.8148×10^5	2.5405×10^5	3.0088×10^5
NPC of Inverter ($)	4.2148×10^4	4.2148×10^4	4.2148×10^4	4.2148×10^4	4.2148	4.2148×10^4
ACS ($)	3.8302×10^5	3.8305×10^5	3.8387×10^5	3.8757×10^5	3.8589×10^5	4.1474×10^5
Cost of WT ($)	404.8825	420.3499	404.8825	404.8825	2.8404×10^3	2.0849×10^4
Cost of PV ($)	1.7914×10^5	1.7915×10^5	1.7541×10^5	2.0920×10^5	1.9117×10^5	2.5010×10^5
Cost of DG ($)	26.7059	35.4823	66.6674	26.7059	119.1373	5.563×10^3
Cost of BSS ($)	1.1182×10^4	1.1203×10^4	8.1826×10^4	2.9645×10^4	1.5642×10^4	1.8525×10^4
Cost of Inverter ($)	3.0598×10^3	3.0598×10^3	3.0598×10^3	3.0598×10^3	3.0598×10^3	3.0598×10^3

TABLE 16.3

Optimization Results of Sensitivity Analysis with LPSP Parameter Changes

LPSP (%)	LCOE ($/kWh)	ACS ($)	NPC($)
0.5	0.1511	3.8302×10^5	3.3406×10^6
1	0.1505	3.8170×10^5	3.3283×10^6
3	0.1484	3.7642×10^5	3.2687×10^6
5	0.1463	3.7108×10^5	3.2297×10^6
10	0.1410	3.5765×10^5	3.1299×10^6
15	0.1357	3.4404×10^5	3.0173×10^6
20	0.1302	3.3023×10^5	2.8814×10^6

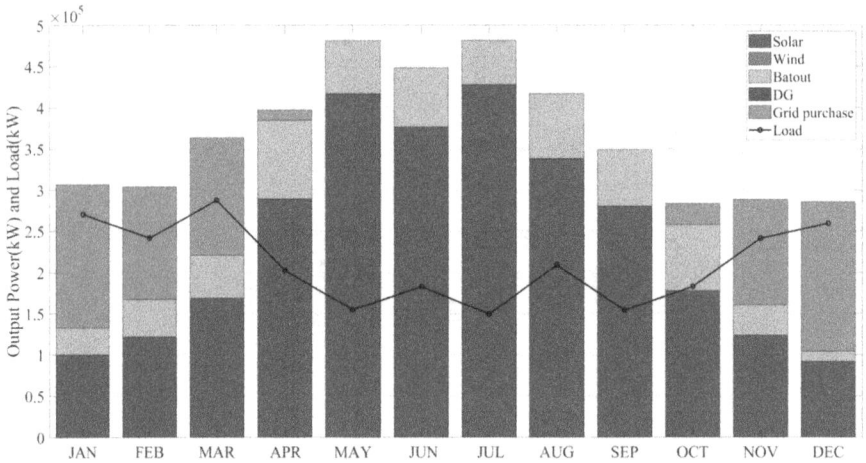

FIGURE 16.12 The optimal configuration balance and energy sharing of HRES components in meeting the load demand.

SOA was applied along with five other metaheuristic algorithms (GWO, RSA, FFA, BRO, and ACO) to estimate the optimal size of the grid-connected HRES, and the results were compared with SOA. During the iteration process, the values of ACS, LCOE, and TNPC decreased, indicating that the algorithm moved toward the optimal size by reducing the objective function. MFOA provided better results than the other algorithms, offering the lowest cost solution. The optimal size of the HRES using SOA was determined to be 1.9900 MW of solar panels, 1 kW of WT, and 1.5649×10^2 batteries at 0.5% LPSP. The LCOE value was calculated to be $0.1511, and the values for ACS and TNPC were 3.8302×10^5 and 3.3406×10^6, respectively. As the analyses were performed on a kW basis, it should be noted that the HRES is dominated by solar panels and batteries. This research corroborates that the SOA and other computation-based optimization techniques can be effectively employed to discern optimal solutions for intricate microgrid design dilemmas.

Figure 16.13 illustrates the daily, seasonal, and annual fluctuations in RESs. These fluctuations impact the efficiency of RESs and require a combination of different

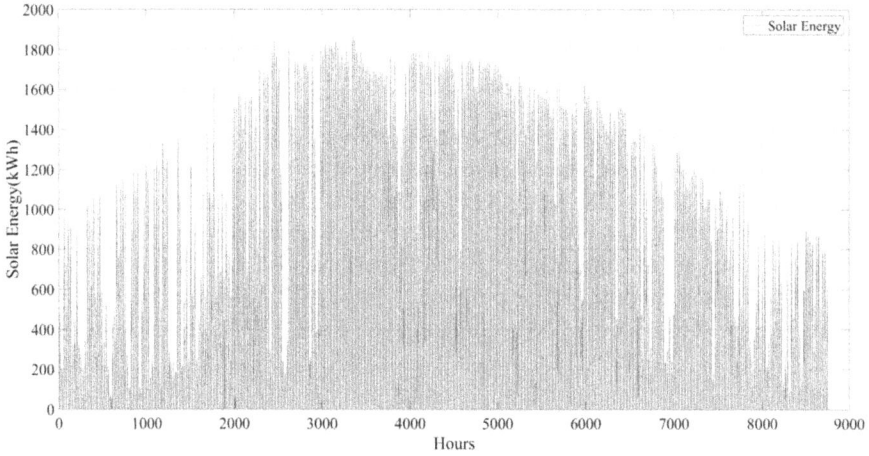

(a)

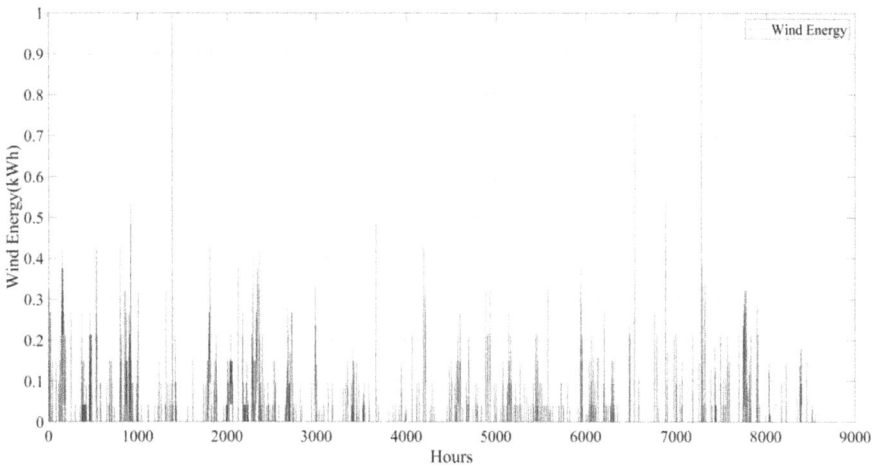

(b)

FIGURE 16.13 The energy profile of RESs (a) Annual solar energy, (b) Annual wind energy.

energy sources to meet the demand of the energy grid. Figure 16.13(a) presents the cumulative energy harvested by a solar panel throughout a year, whereas Figure 16.13(b) exhibits the total energy accrued by a WT during the identical timeframe. While the solar panel produces more energy at certain times of the year, at other times it produces less energy due to variations in exposure to sunlight. The net time intervals for these periods can be observed from the graph. The average wind speed is a useful value for determining the potential energy output of a WT. However, because we set the boundary range for wind between 1 kW and 5000 kW during optimization, the resulting value is evaluated as the lowest value, which is 1 kW. Taking into account the availability of RESs and the costs of components, our energy management strategy evaluates the most cost-effective solution for meeting the demand for load with each component. If cost-effective optimization is not achieved, increasing the minimum interval value for the WT to 1000 kW will result in a higher utilization rate of wind energy, but our annual cost will also increase along with the unit energy cost.

Performance and lifespan of the battery are vital for the efficiency of hybrid energy systems. Therefore, the battery management system must accurately monitor and optimize the SOC of the battery. SOC is a measure of how much of the accessible capacity of the battery has been used. A battery that is completely discharged or fully charged can shorten its lifespan and decrease performance. Figure 16.14(a) and (b) show the annual minimum, maximum, and average SOC values of the battery. The battery's SOC drops below 30% in January, February, March, November, and

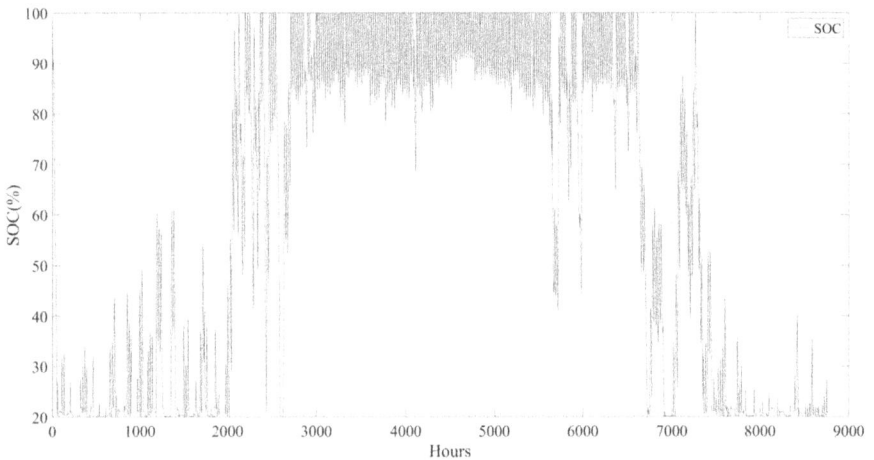

(a)

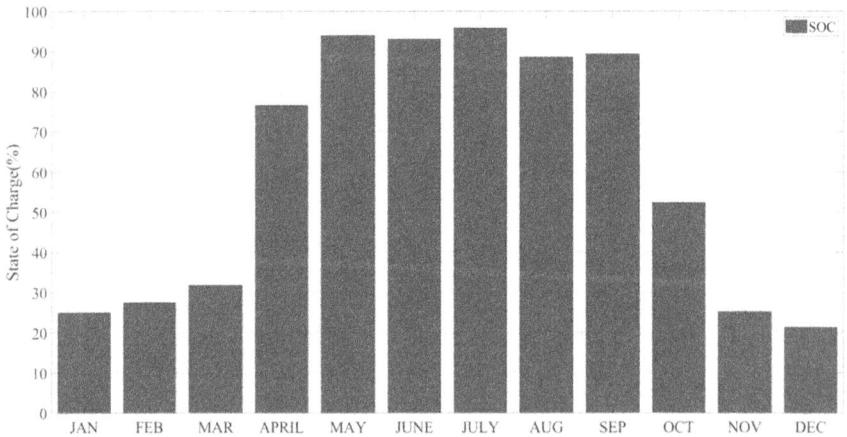

(b)

FIGURE 16.14   Battery profile (a) The hourly variation of the SOC, (b) The average monthly variation of the SOC.

December when natural resources are scarce, but rises above 90% in April, May, June, July, August, and September. However, with some exceptions, the battery's SOC values are generally good. The data obtained from Figure 16.14 illustrates the importance of managing the SOC values of the battery accurately.

Optimizing the charging and discharging processes of a battery is critical for enhancing its lifetime and performance. The data provided in Figure 16.15 shows the total input and output energy of the battery during a specific time period of discharging and charging operations. The Batout data represents the battery's total energy consumption, consumption trends, and how energy demands change over time. The Batin data shows the total energy loaded into the battery, charging trends, and how charging needs change over time. These data points are important for proper battery management. Since the consumption and charging processes have a direct impact on the battery's lifetime and performance, it is essential to consider this information in the design of a battery management system. Particularly, optimizing the charging/discharging processes is critical to extend the battery's lifetime and enhance its performance.

The analyses based on the data reveal that Figure 16.16 is a time series graph showing how energy demand was met throughout a year by different sources, such as hydroelectric, wind, and natural gas. The graph provides detailed and numerical information about the variability of energy supply.

On the other hand, Figure 16.17 illustrates the total energy sold to and purchased from the grid over a year. The graph tracks the total supply and demand of grid energy for a year and provides numerical values, such as 8.0104×10^3 MWh of sold energy and 1.1058×10^3 MWh of purchased energy.

Both graphs provide detailed information about how energy demand and supply changed over a year, serving as an important tool for planning and managing energy sector activities. These data could be useful for managing energy resources in the future.

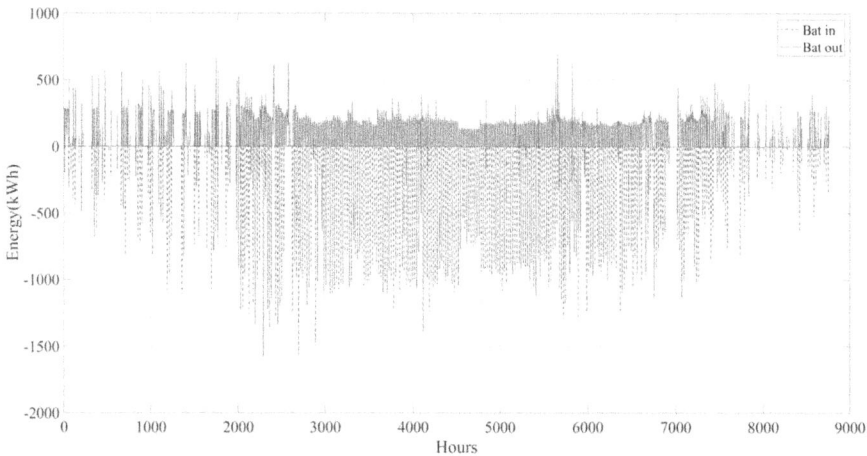

FIGURE 16.15   The hourly fluctuation of the input and output energy of the battery.

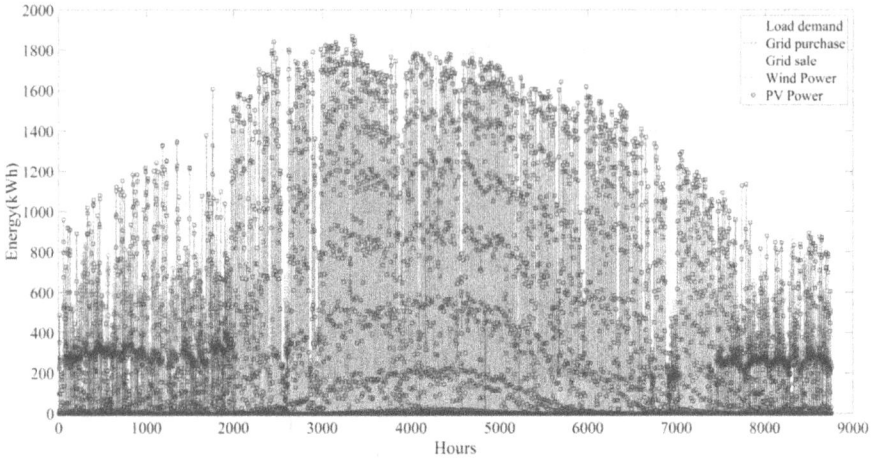

FIGURE 16.16 Comparison of electricity demand for one year with energy sources.

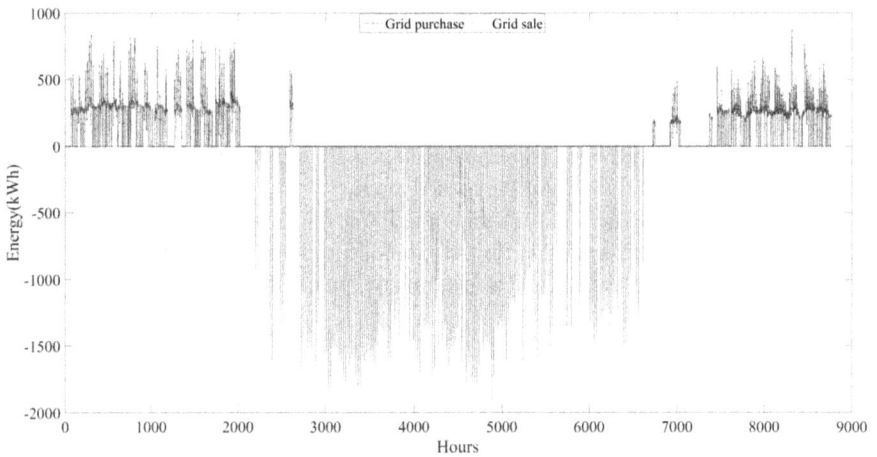

FIGURE 16.17 Comparison of the energy balance of purchased from and sold to the grid in an HRES.

Throughout the year, on March 21st, 2022 (at 19:07), the highest electricity demand of 938.498 kW was recorded. Figure 16.18 is a weekly graph showing the balance analysis met by HRES during the time interval of 1907–2074 hour. In Figure 16.18(a) and (b), it can be seen that the battery was discharged during the period of 1907–1984 hour, and the electricity demand was met by RESs and power purchased from the grid. Figure 16.18(c) represents the weekly profile of the battery's input and output energy. As seen in Figures 16.18(a) and (b), it was confirmed that the required power was provided from the grid and RESs for approximately 77 hours during the first week of March 21st, 2022. As observed, the power fluctuations of renewable sources were balanced through grid energy consumption and battery charge-discharge management to constantly meet the electricity demand.

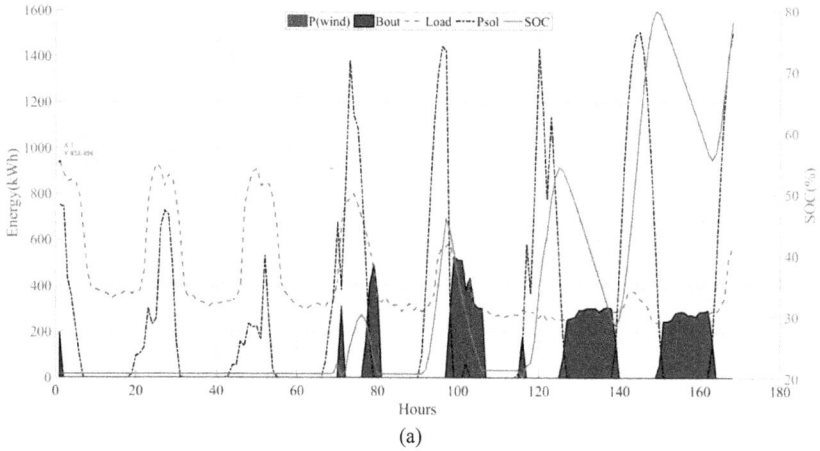

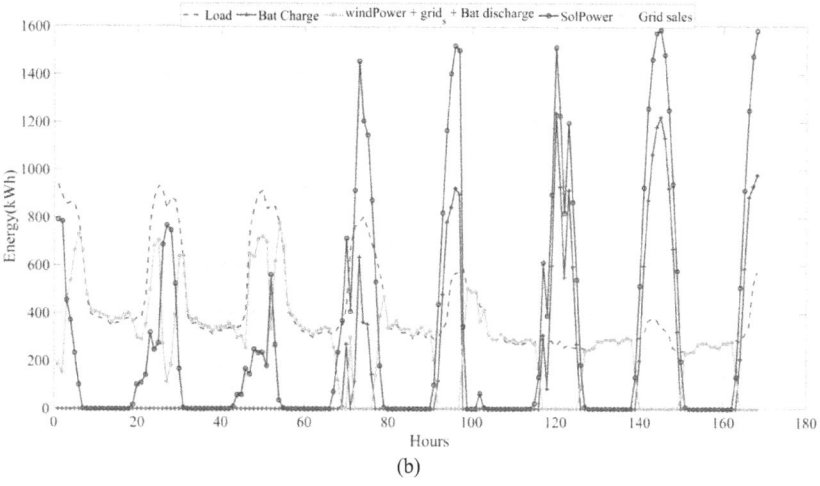

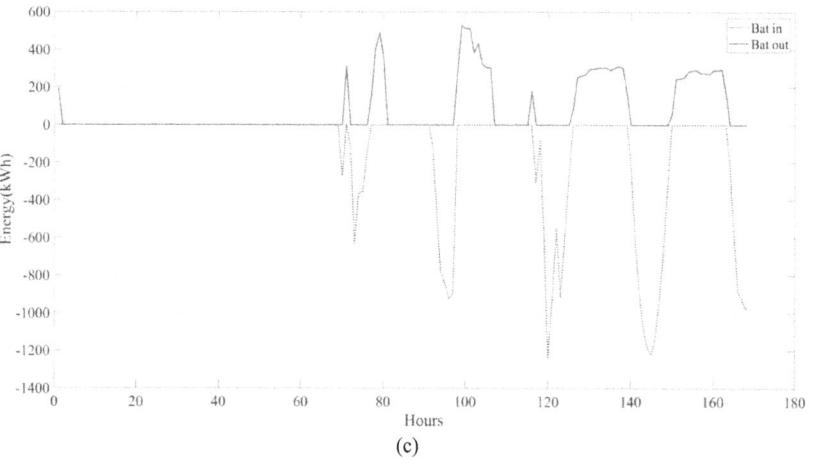

FIGURE 16.18 Energy balance analysis of HRES during the time interval of 1907–2074 hour against the highest electricity demand of the year.

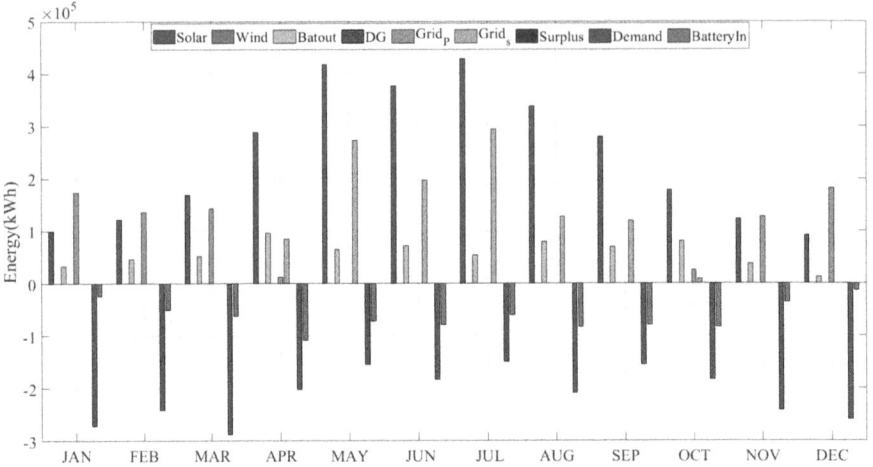

FIGURE 16.19 Monthly average energy balance analysis of HRES.

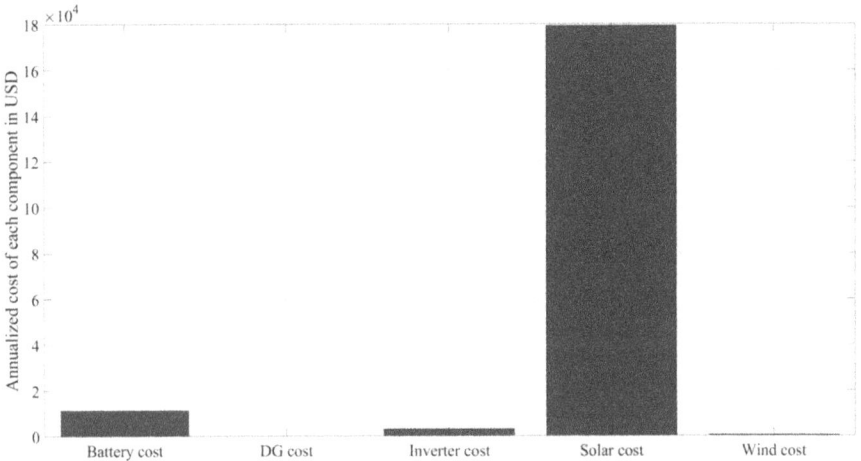

FIGURE 16.20 The cost distribution of HRES components as a result of optimization.

The monthly average energy balance analysis of HRES is shown in a different way for 8760 hours. This graph, seen in Figure 16.19, displays the monthly energy distribution balance of HRES and the total average power values of its components, making it easier to evaluate the overall situation.

Figure 16.20 presents a cost analysis of the components of the optimized HRES, displaying the optimal values obtained using the proposed SOA method. The annual costs for the WT, PV, BT, inverter, and the total are $404.8825, 1.7914×10^5,

$1.1182 × 10⁴, $3.0598 × 10³, and $3.8302 × 10⁵, respectively. These results could serve as a guide for cost-effective HRES installations.

A sensitivity analysis was conducted to guide optimal sizing decisions. Dimensioning algorithms were simulated using various LPSP_max values ranging from 0.5% to 20%, and the results were graphically depicted. In a study by Güven et al., optimization was based on an LPSP value of 0, leading to an LCOE value of $0.2012/kWh. In contrast, this study reported an LCOE value of $0.1511/kWh. With the inclusion of battery storage efficiency, the grid-supported HRES achieved an LCOE value nearly 33% lower than the off-grid system. In summary, when considering the impact of system reliability on cost, it's evident that a significant decrease in LPSP increases both TNPC and LCOE values. The results presented are consistent with previously published literature.

The sensitivity analysis of HRES is used to investigate the effects of variations in renewable energy source parameters on system performance, energy production costs, carbon emissions, and other metrics. This analysis helps determine how the system would perform best under various weather conditions and energy demand scenarios. Additionally, it assists in optimizing the system by identifying the impact of changes in component characteristics on system performance. The effects of changes in the LPSP on the system are presented in Table 16.3 and Figure 16.21. There is an inverse relationship between TNPC, ACS, LCOE values, and LPSP.

Moreover, a sensitivity analysis has been carried out to comprehend the influence of interest rate fluctuations on system costs. Based on the findings of the sensitivity analysis depicted in Figure 16.22(b), as interest rates rise, the ACS value increases, while the TNPC value demonstrates a declining trend. As depicted in Figure 16.22(a), the LCOE value escalates as interest rates rise, while the TNPC value simultaneously experiences a decline.

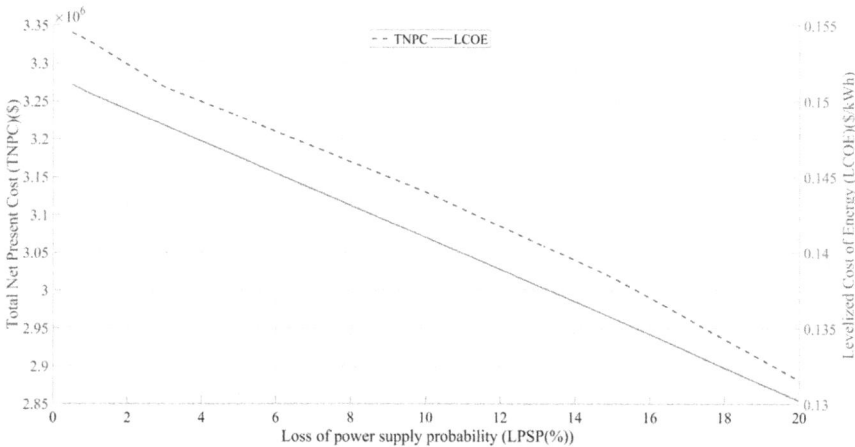

FIGURE 16.21 Changes in LPSP and their effects on TNPC and LCOE.

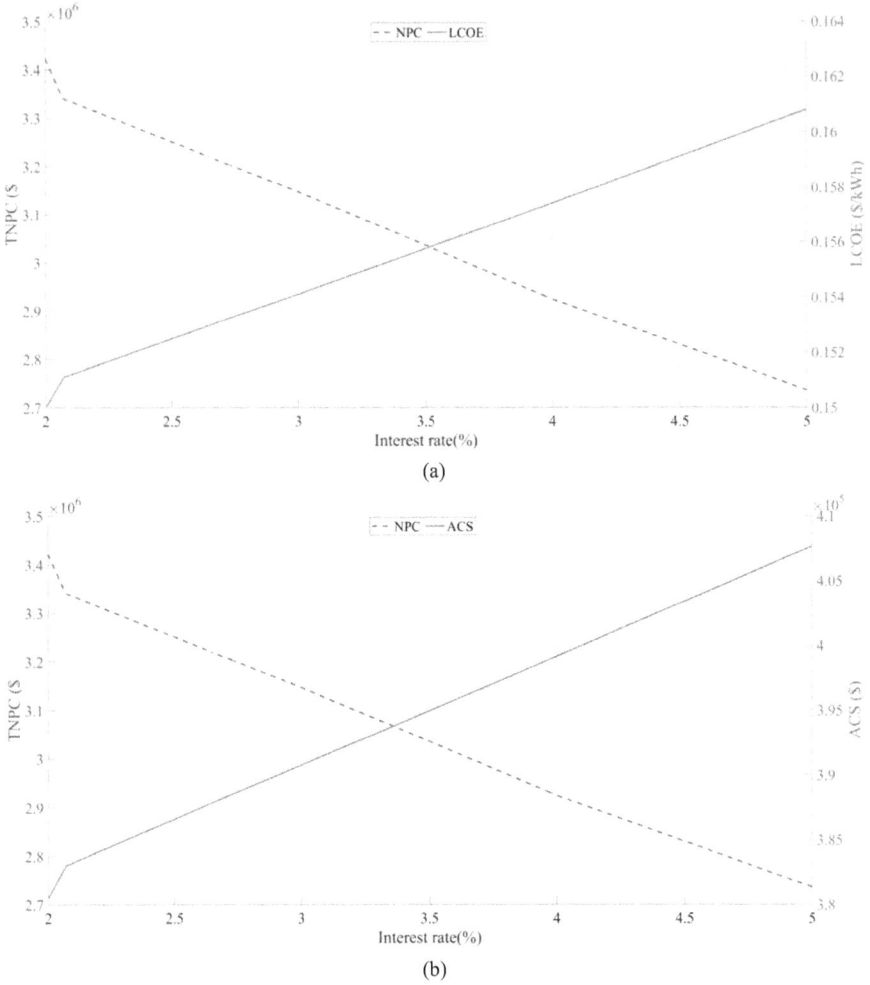

FIGURE 16.22 Changes in interest rate (a) Effects on LCOE and TNPC, (b) Effects on TNPC and ACS.

As illustrated in Figure 16.23(a), the GWO algorithm begins to converge at the 4th iteration, FFA at the 7th, and SOA at the 20th iteration. After a thorough fitness evaluation, GWO remains ahead of SOA until the 32nd iteration. However, from the 33rd iteration onwards, SOA overtakes GWO, quickly arriving at the optimal solution. While the results are closely matched, Figure 16.23(b) shows that SOA outperforms all other algorithms, concluding the optimization process with the lowest ACS, NPC, and LCOE values.

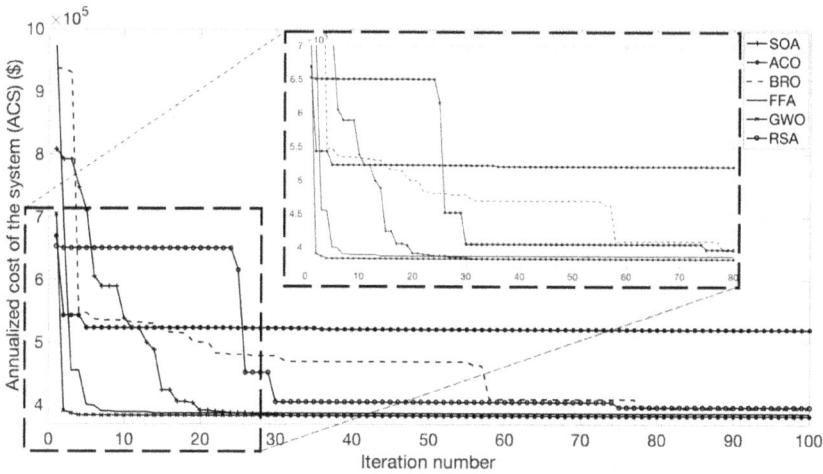

(a)

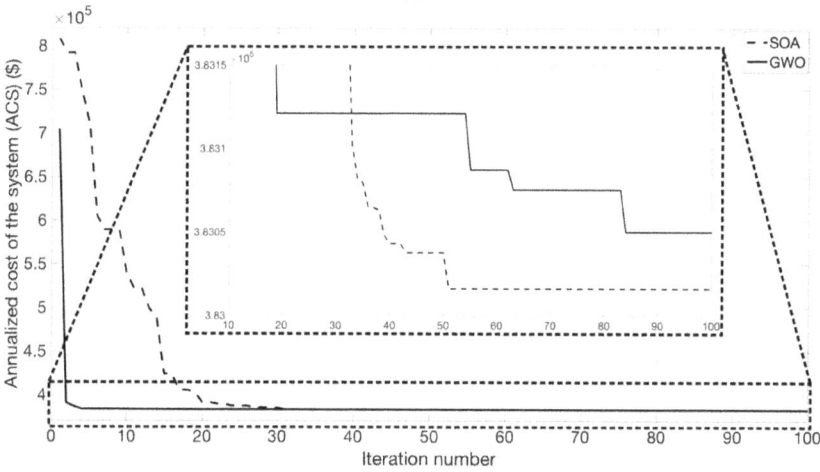

(b)

FIGURE 16.23 Comparison of convergence curves for (a) SOA, GWO, RSA, FFA, BRO, and ACO algorithms, (b) SOA, FFA, and GWO.

16.6 CONCLUSION

In this chapter, an optimization method using SOA for grid-connected HRES with PV/WT/DG/BSS components is presented. The objective is to ascertain the ideal capacity of renewable energy resources by minimizing the system's ACS, LCOE, and TNPC values, whilst considering the LPSP as a constraint to ensure system reliability. The optimal design results obtained using SOA, considering the specified values for the probability of power supply shortage, show superior performance in terms of unit energy cost, annual system cost, dimensions of renewable energy resources, and algorithm convergence time compared to other optimization methods such as GWO, RSA, FFA, BRO, and ACO.

As a result of the optimization process, it has been determined that the existing demand can be fulfilled using a combination of a 1 kW wind turbine, a 1.99 MW solar panel array, a battery storage system with 1.5649×10^2 kW capacity, and by drawing 1.1058×10^3 MWh of energy from the grid. The TNPC value of the HRES is $\$3.3406 \times 10^6$, the ACS value is $\$3.8302 \times 10^5$, and the LCOE value is $\$0.1511/$ kWh. According to the sensitivity analysis performed for different LPSP targets, as the LPSP value increases, there is a decrease in ACS, NPC, and LCOE values. Throughout the year, 8.0104×10^3 MWh of excess energy generated from RESs is sold to the grid, generating revenue. The presented HRES meets load demands at all times of the year with a 100% renewable energy factor, and due to the diesel generator not operating at all during the year, the total gas emission value is zero. Shifting toward renewable energy investments and installations is of great importance in combating energy crises and increasing carbon emissions.

All algorithms used in the optimization process are found to be applicable for HRES sizing optimization, and each algorithm possesses the ability to approach optimal solutions. In this case, the search space density can be adaptively controlled; the search space can be narrowed when the algorithm is close to the optimal solution or expanded to encounter new search points when far from the optimal solution. In this context, the proposed SOA method provides highly competitive results compared to other algorithms in terms of HRES sizing problem. The algorithm is expected to be used in complex and challenging problems in the future.

RESs such as solar and wind energy, although not selected in all optimal systems due to their inherent variability, are indispensable power generators for all optimal systems, thanks to the stable energy supply provided by the grid, BT, and DG. The state of energy sources and loads collaboratively govern the optimal architecture of HRES; the load delineates the demands for the aggregate energy that needs to be supplied, while the configuration of the generative apparatus sharing this energy dispensation is contingent on the conditions of the energy sources.

Moreover, the proportion between energy sources and storage devices is dictated by both the conditions of the energy sources and the loads. Consequently, the methodology elucidated in this study can offer an efficient framework for the design and evaluation of HRES. Further, considering local resource conditions and load data, this comprehensive research can be leveraged to predict the performance of any system in a different city, as well as to pinpoint the most suitable system for that particular location. In future investigations, given that the load utilized is akin to the type of load specified in this study (university campus, commercial building, hospital, etc.), and the community is of a comparable size, the data on load and climatic conditions can be directly applied when available.

REFERENCES

[1] Makhdoomi S., & Askarzadeh A. (2023). Techno-enviro-economic feasibility assessment of an off-grid hybrid energy system with/without solar tracker considering pumped hydro storage and battery. *IET Renewable Power Generation*, 17(5), 1194–1211.

[2] Li C., Zhang L., Qiu F., & Fu R. (2022). Optimization and enviro-economic assessment of hybrid sustainable energy systems: The case study of a photovoltaic/biogas/diesel/battery system in Xuzhou. China. *Energy Strategy Reviews*, 41(March), 100852.

[3] Wimalaratna Y.P., Afrouzi H.N., Mehranzamir K., Siddique M.B.M., Liew S.C., & Ahmed J. (2022). Analysing wind power penetration in hybrid energy systems based on techno-economic assessments. *Sustainable Energy Technologies and Assessments*, 53(PB), 102538.

[4] Kirim Y., Sadikoglu H., & Melikoglu M. (2022). Technical and economic analysis of biogas and solar photovoltaic (PV) hybrid renewable energy system for dairy cattle barns. *Renewable Energy*, 188, 873–889.

[5] Karayel G.K., Javani N., & Dincer I. (2022). Green hydrogen production potential for Turkey with solar energy. *International Journal of Hydrogen Energy*, 47(45), 19354–19364.

[6] Eslami S., Noorollahi Y., Marzband M., & Anvari-Moghaddam A. (2022). District heating planning with focus on solar energy and heat pump using GIS and the supervised learning method: Case study of Gaziantep. Turkey. *Energy Conversion and Management*, 269(May), 116131.

[7] Ali S., Ur Rehman A., Wadud Z., Khan I., Murawwat S., Hafeez G., Albogamy F.R., Khan S., & Samuel O. (2022). Demand response program for efficient demand-side management in smart grid considering renewable energy sources. *IEEE Access*, 10, 53832–53853.

[8] Anusha U., Jagadeesh S., Rao K.V., & Umavani M. (2022) Energy Management of an Off-Grid and Grid Connected Hybrid Renewable Energy Source Micro-Grid System for Commercial Load. *IEEE 2nd Mysore Sub Section International Conference*.

[9] Sun H., Ebadi A.G., Toughani M., Nowdeh S.A., Naderipour A., & Abdullah A. (2022). Designing framework of hybrid photovoltaic-biowaste energy system with hydrogen storage considering economic and technical indices using whale optimization algorithm. *Energy*, 238, 121555.

[10] Naderipour A., Abdul-Malek Z., Nowdeh S.A., Kamyab H., Ramtin A.R., Shahrokhi S., & Klemeš J.J. (2021). Comparative evaluation of hybrid photovoltaic, wind, tidal and fuel cell clean system design for different regions with remote application considering cost. *Journal of Cleaner Production*, 283, 1–20.

[11] Ghadimi N., Sedaghat M., Azar K.K., Arandian B., Fathi G., & Ghadamyari M. (2023). An innovative technique for optimization and sensitivity analysis of a PV/DG/BESS based on converged Henry gas solubility optimizer: A case study. *IET Generation, Transmission Distribution*, 1–15.

[12] Thirunavukkarasu M., Sawle Y., & Lala H. (2023). A comprehensive review on optimization of hybrid renewable energy systems using various optimization techniques. *Renewable and Sustainable Energy Reviews*, 176, 113192.

[13] Merrington S., Khezri R., & Mahmoudi A. (2023). Optimal sizing of grid-connected rooftop photovoltaic and battery energy storage for houses with electric vehicle. *IET Smart Grid*, 6(3), 297–311.

[14] Liu G., Zeng J., Wu Y., & Liao S. (2022). Multi-criteria techno-economic analysis of solar photovoltaic/wind hybrid power systems under temperate continental climate. *IET Renewable Power Generation*, 16(14), 3058–3070.

[15] Nirbheram J.S., Mahesh A., & Bhimaraju A. (2023). Techno-economic analysis of grid-connected hybrid renewable energy system adapting hybrid demand response program and novel energy management strategy. *Renewable Energy*, 212, 1–16.

[16] Vakili S., & Ölçer A.I. (2023). Techno-economic-environmental feasibility of photovoltaic, wind and hybrid electrification systems for stand-alone and grid-connected port electrification in the Philippines. *Sustainable Cities and Society*, 96, 104618.

[17] Shezan S.A., Ishraque M.F., Muyeen S.M., Abu-Siada A., Saidur R., Ali M.M., & Rashid M.M. (2022). Selection of the best dispatch strategy considering techno-economic and system stability analysis with optimal sizing. *Energy Strategy Reviews*, 43, 100923.

[18] Malik P., Awasthi M., Upadhyay S., Agrawal P., Raina G., Sharma S., Kumar M., & Sinha S. (2023). Planning and optimization of sustainable grid integrated hybrid energy system in India. *Sustainable Energy Technologies and Assessments*, 56, 103115.

[19] Chaurasia R., Gairola S., & Pal Y. (2022). Technical, economic, and environmental performance comparison analysis of a hybrid renewable energy system based on power dispatch strategies. *Sustainable Energy Technologies and Assessments*, 53(PD), 102787.

[20] El-Sattar H.A., Kamel S., Hassan M.H., & Jurado F. (2022). An effective optimization strategy for design of standalone hybrid renewable energy systems. *Energy*, 260, 124901.

[21] Seedahmed M.M.A., Ramli M.A.M., Bouchekara H.R.E.H., Milyani A.H., Rawa M., Nur Budiman F., Firmansyah Muktiadji R., & Mahboob Ul Hassan S. (2022). Optimal sizing of grid-connected photovoltaic system for a large commercial load in Saudi Arabia. *Alexandria Engineering Journal*, 61(8), 6523–6540.

[22] Kharrich M., Selim A., Kamel S., & Kim J. (2023). An effective design of hybrid renewable energy system using an improved Archimedes Optimization Algorithm: A case study of Farafra, Egypt. *Energy Conversion and Management*, 283, 116907.

[23] Samy M.M., Mosaad M.I., & Barakat S. (2021). Optimal economic study of hybrid PV-wind-fuel cell system integrated to unreliable electric utility using hybrid search optimization technique. *International Journal of Hydrogen Energy*, 46(20), 11217–11231.

[24] Güven A.F., Yörükeren, N., & Samy M.M. (2022). Design optimization of a stand-alone green energy system of university campus based on Jaya-Harmony Search and Ant Colony Optimization algorithms approaches. *Energy*, 253.

[25] Güven A.F., & Samy M.M. (2022). Performance analysis of autonomous green energy system based on multi and hybrid metaheuristic optimization approaches. *Energy Conversion and Management*, 269, 116058, 1–23.

[26] Rodriguez M., Arcos-Aviles D., & Martinez W. (2023). Fuzzy logic-based energy management for isolated microgrid using meta-heuristic optimization algorithms. *Applied Energy*, 335, 1–19.

[27] Tukkee A.S., Wahab N.I.B.A., & Mailah N.F.B. (2023). Optimal sizing of autonomous hybrid microgrids with economic analysis using grey wolf optimizer technique. *E-Prime – Advances in Electrical Engineering, Electronics and Energy*, 3, 100123.

[28] Pan T., Wang Z., Tao J., & Zhang H. (2023). Operating strategy for grid-connected solar-wind-battery hybrid systems using improved grey wolf optimization. *Electric Power Systems Research*, 220, 109346.

[29] Emrani A., Berrada A., Arechkik A., & Bakhouya M. (2022). Improved techno-economic optimization of an off-grid hybrid solar/wind/gravity energy storage system based on performance indicators. *Journal of Energy Storage*, 49, 104163.

[30] Perna A., Jannelli E., Di Micco S., Romano F., & Minutillo M (2023). Designing, sizing and economic feasibility of a green hydrogen supply chain for maritime transportation. *Energy Conversion and Management*, 278, 116702.

[31] Mulenga E., Kabansh A., Mupeta H., Ndiaye M., Nyirenda E., & Mulenga K. (2023). Techno-economic analysis of off-grid PV-Diesel power generation system for rural electrification: A case study of Chilubi district in Zambia. *Renewable Energy*, 203, 601–611.

[32] Sari A., Majdi A., Opulencia M.J.C., Timoshin A., Huy D.T.N., Trung N.D., Alsaikhan F., Hammid A.T., & Akhmedov A. (2022). New optimized configuration for a hybrid PV/diesel/battery system based on coyote optimization algorithm: A case study for Hotan county. *Energy Reports*, 8, 15480–15492.

[33] Allouhi A., & Rehman S. (2023). Grid-connected hybrid renewable energy systems for supermarkets with electric vehicle charging platforms: Optimization and sensitivity analyses. *Energy Reports*, 9, 3305–3318.

[34] Singh P., Pandit M., & Srivastava L. (2023). Multi-objective optimal sizing of hybrid micro-grid system using an integrated intelligent technique. *Energy*, 269, 126756.

[35] Amole A.O., Oladipo S., Olabode O.E., Makinde K.A., & Gbadega P. (2023). Analysis of grid/solar photovoltaic power generation for improved village energy supply: A case of Ikose in Oyo State Nigeria. *Renewable Energy Focus*, 44, 186–211.

[36] Ma J., & Yuan X. (2023). Techno-economic optimization of hybrid solar system with energy storage for increasing the energy independence in green buildings. *Journal of Energy Storage*, 61, 106642.

[37] Falama R.Z., Saidi A.S., Soulouknga M.H., & Salah C.B. (2023). A techno-economic comparative study of renewable energy systems based different storage devices. *Energy*, 266, 126411.

[38] Kharrich M., Kamel S., Abdeen M., Mohammed O.H., Akherra M., Khurshaid T., & Rhee S. B. (2021). Developed approach based on equilibrium optimizer for optimal design of hybrid PV/Wind/Diesel/Battery Microgrid in Dakhla, Morocco. *IEEE Access*, 9, 13655–13670.

[39] Hashim F.A., & Hussien A.G. (2022). Snake optimizer: A novel meta-heuristic optimization algorithm. *Knowledge-Based Systems*, 242, 108320.

[40] Khan K., Rashid S., Mansoor M., Khan A., Raza H., Zafar M.H., & Akhtar N. (2023). Data-driven green energy extraction: Machine learning-based MPPT control with efficient fault detection method for the hybrid PV-TEG system. *Energy Reports*, 9, 3604–3623.

[41] Yousri R., Elbayoumi M., Soltan A., & Darweesh M.S. (2023). A power-aware task scheduler for energy harvesting-based wearable biomedical systems using snake optimizer. *Analog Integrated Circuits and Signal Processing*, 115(2), 183–194.

[42] Vivekraj A., & Sumathi S. (2023). Resnet-Unet-FSOA based cranial nerve segmentation and medial axis extraction using MRI images. *Imaging Science Journal*, 71, 1–17.

[43] Rawa M. (2022). Towards avoiding cascading failures in transmission expansion planning of modern active power systems using hybrid snake-sine cosine optimization algorithm. *Mathematics*, 10(8), 1–21.

[44] Hu G., Yang R., Abbas M., & Wei G. (2023). BEESO: Multi-strategy boosted snake-inspired optimizer for engineering applications. *Journal of Bionic Engineering*, 20, 1791–1827.

[45] Rameshar V., Sharma G.N., Bokoro P., & Çelik E. (2023). Frequency support studies of a diesel–wind generation system using snake optimizer-oriented PID with UC and RFB. *Energies*, 16(8), 3417.

[46] Li Y., Xiao L., Tang B., Liang L., Lou Y., Guo X., & Xue X. (2022). A denoising method for ship-radiated noise based on optimized variational mode decomposition with snake optimization and dual-threshold criteria of correlation coefficient. *Mathematical Problems in Engineering*, 2022, 1–21, 8024753.

[47] Dai Y., Pang J., Li Z., Li W., Wang Q., & Li S (2022). Modeling of thermal error electric spindle based on KELM ameliorated by snake optimization. *Case Studies in Thermal Engineering*, 40, 102504.

[48] Huang X., Huang Q., Cao H., Wang Q., Yan W., & Cao L. (2023). Battery capacity selection for electric construction machinery considering variable operating conditions and multiple interest claims. *Energy*, 275, 127454.

[49] Abu Khurma R., Albashish D., Braik M., Alzaqebah A., Qasem A., & Adwan O. (2023). An augmented snake optimizer for diseases and COVID-19 diagnosis. *Biomedical Signal Processing and Control*, 84, 104718.

Index

For Product Safety Concerns and Information please contact our EU
representative GPSR@taylorandfrancis.com
Taylor & Francis Verlag GmbH, Kaufingerstraße 24, 80331 München, Germany

www.ingramcontent.com/pod-product-compliance
Lightning Source LLC
Chambersburg PA
CBHW060333220326
41598CB00023B/2697